T0321973

Multimedia Services and Applications in Mission Critical Communication Systems

Khalid Al-Begain
University of South Wales, UK

Ashraf Ali
The Hashemite University, Jordan & University of South Wales, UK

A volume in the Advances in Wireless Technologies and Telecommunication (AWTT) Book Series

www.igi-global.com

Published in the United States of America by
IGI Global
Information Science Reference (an imprint of IGI Global)
701 E. Chocolate Avenue
Hershey PA, USA 17033
Tel: 717-533-8845
Fax: 717-533-8661
E-mail: cust@igi-global.com
Web site: http://www.igi-global.com

Library of Congress Cataloging-in-Publication Data

Names: Al-Begain, Khalid, editor. | Ali, Ashraf, 1983- editor.
Title: Multimedia services and applications in mission critical communication systems / Khalid Al-Begain and Ashraf Ali, editors.
Description: Hershey, PA : Information Science Reference, [2017] | Includes bibliographical references.
Identifiers: LCCN 2016054994| ISBN 9781522521136 (hardcover) | ISBN 9781522521143 (eISBN)
Subjects: LCSH: Multimedia communications.
Classification: LCC TK5105.15 .M847 2017 | DDC 621.382/1--dc23 LC record available at https://lccn.loc.gov/2016054994

This book is published in the IGI Global book series Advances in Wireless Technologies and Telecommunication (AWTT) (ISSN: 2327-3305; eISSN: 2327-3313)

British Cataloguing in Publication Data
A Cataloguing in Publication record for this book is available from the British Library.

All work contributed to this book is new, previously-unpublished material. The views expressed in this book are those of the authors, but not necessarily of the publisher.

For electronic access to this publication, please contact: eresources@igi-global.com.

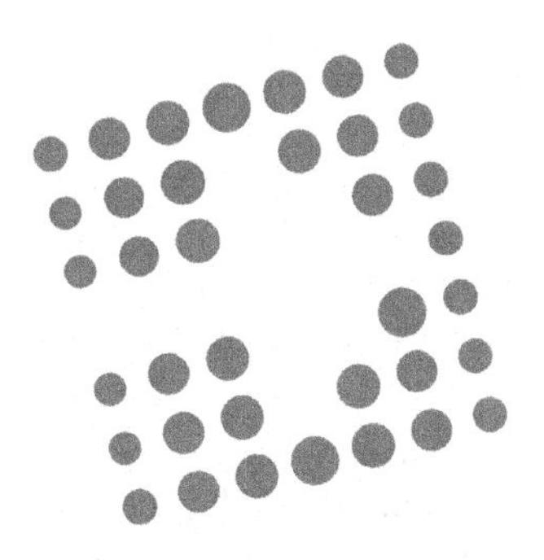

Advances in Wireless Technologies and Telecommunication (AWTT) Book Series

Xiaoge Xu
The University of Nottingham Ningbo China, China

ISSN:2327-3305
EISSN:2327-3313

Mission

The wireless computing industry is constantly evolving, redesigning the ways in which individuals share information. Wireless technology and telecommunication remain one of the most important technologies in business organizations. The utilization of these technologies has enhanced business efficiency by enabling dynamic resources in all aspects of society.

The **Advances in Wireless Technologies and Telecommunication Book Series** aims to provide researchers and academic communities with quality research on the concepts and developments in the wireless technology fields. Developers, engineers, students, research strategists, and IT managers will find this series useful to gain insight into next generation wireless technologies and telecommunication.

Coverage

- Broadcasting
- Wireless sensor networks
- Mobile Technology
- Wireless Broadband
- Mobile Communications
- Virtual Network Operations
- Telecommunications
- Network Management
- Grid Communications
- Global Telecommunications

IGI Global is currently accepting manuscripts for publication within this series. To submit a proposal for a volume in this series, please contact our Acquisition Editors at Acquisitions@igi-global.com or visit: http://www.igi-global.com/publish/.

The Advances in Wireless Technologies and Telecommunication (AWTT) Book Series (ISSN 2327-3305) is published by IGI Global, 701 E. Chocolate Avenue, Hershey, PA 17033-1240, USA, www.igi-global.com. This series is composed of titles available for purchase individually; each title is edited to be contextually exclusive from any other title within the series. For pricing and ordering information please visit http://www.igi-global.com/book-series/advances-wireless-technologies-telecommunication-awtt/73684. Postmaster: Send all address changes to above address.

Titles in this Series

For a list of additional titles in this series, please visit: www.igi-global.com

Big Data Applications in the Telecommunications Industry
Ye Ouyang (Verizon Wirless, USA) and Mantian Hu (Chinese University of Hong Kong, China)
Information Science Reference • copyright 2017 • 216pp • H/C (ISBN: 9781522517504) • US $145.00 (our price)

Handbook of Research on Recent Developments in Intelligent Communication Application
Siddhartha Bhattacharyya (RCC Institute of Information Technology, India) Nibaran Das (Jadavpur University, India) Debotosh Bhattacharjee (Jadavpur University, India) and Anirban Mukherjee (RCC Institute of Information Technology, India)
Information Science Reference • copyright 2017 • 671pp • H/C (ISBN: 9781522517856) • US $360.00 (our price)

Interference Mitigation and Energy Management in 5G Heterogeneous Cellular Networks
Chungang Yang (Xidian University, China) and Jiandong Li (Xidian University, China)
Information Science Reference • copyright 2017 • 362pp • H/C (ISBN: 9781522517122) • US $195.00 (our price)

Handbook of Research on Advanced Trends in Microwave and Communication Engineering
Ahmed El Oualkadi (Abdelmalek Essaadi University, Morocco) and Jamal Zbitou (Hassan 1st University, Morocco)
Information Science Reference • copyright 2017 • 716pp • H/C (ISBN: 9781522507734) • US $315.00 (our price)

Handbook of Research on Wireless Sensor Network Trends, Technologies, and Applications
Narendra Kumar Kamila (C. V. Raman College of Engineering, India)
Information Science Reference • copyright 2017 • 589pp • H/C (ISBN: 9781522505013) • US $310.00 (our price)

Handbook of Research on Advanced Wireless Sensor Network Applications, Protocols, and Architectures
Niranjan K. Ray (Silicon Institute of Technology, India) and Ashok Kumar Turuk (National Institute of Technology Rourkela, India)
Information Science Reference • copyright 2017 • 502pp • H/C (ISBN: 9781522504863) • US $285.00 (our price)

Self-Organized Mobile Communication Technologies and Techniques for Network Optimization
Ali Diab (Al-Baath University, Syria)
Information Science Reference • copyright 2016 • 416pp • H/C (ISBN: 9781522502395) • US $200.00 (our price)

Advanced Methods for Complex Network Analysis
Natarajan Meghanathan (Jackson State University, USA)
Information Science Reference • copyright 2016 • 461pp • H/C (ISBN: 9781466699649) • US $215.00 (our price)

Table of Contents

Detailed Table of Contents

In this chapter, the definition of Mission Critical Systems, users, and services will be introduced, then main applications of such system will be listed, the general requirements will be investigated, and finally a distinction between different types of Mission Critical Systems will be briefly discussed. Nowadays, many governmental and non-governmental organizations (NGOs) play critical roles in its operations on everyday basis. The services provided by such organization is of special nature that makes it less tolerant to executional or operational errors; such services are referred as Mission Critical Services. The criticality of the service implies that it has a set of special requirements that distinguish it from other services, and it should be available anytime, anywhere within the service operational scope, and also it is supposed to be functioning within expectation limits regardless of the operational circumstances or running conditions.

In this chapter, the most important protocols that controls the operation of the end-to-end system and has implications on the overall performance aspects will be presented. Then, performance issues will be demonstrated. the performance issues and the challenge of enhancing SIP services performance will be highlighted and briefly discussed. The protocol related performance metrics will be identified to determine the way SIP is utilizing the system resources and how to maximize it. Moreover, the architectural design challenges will be targeted to enhance the SIP performance.

In this chapter analyses the overall system capacity and scalability which is affected by added traffic introduced by more users who are trying to access the system provided services. This could happen in a mission critical communication system during natural disaster or large scale attack, where the system accessibility could be affected due to the sudden increase of number of users. The need for a more detailed study of other SIP and IMS KPIs is vital to have a better understanding of the overall system performance which will enable us to take it a step further toward system performance enhancement and optimization to avoid single point of failure of the system.

The evaluation studies need to investigate a determined performance metrics to understand and evaluate the examined scenarios. SIP-based Voice over IP (VoIP) applications over MANET, which behaves in a way similar to Direct Mode of Operation (DMO) in mission Critical Communication Systems, have two main performance categories related to the Quality of Service (QoS). The main performance metrics that are considered for the evaluation processes in this research are the SIP end-to-end Performance metrics as defined by the RFC 6076. The main performance metrics are related to the registration, the call setup, and the call termination processes. In this research study, the SIP performance metrics are based on a single SIP proxy server. For voice data, the QoS evaluation is based on two methods: The Objective method and the Subjective method. The Objective method considers the traffic throughput, end-to-end delays, packet loss, and jitter, while the subjective method considers the Mean Opinion Score (MOS), which is mostly related to the end users' experiences during voice calls.

In this chapter, analyses for the performance metrics that define the quality of service (QoS) of SIP-based VoIP will be introduced. SIP-based VoIP applications over Direct Mode of Operation (DMO), which behaves in a way similar to Mobile Ad-hoc Network (MANET) systems, have three main performance categories related to the QoS. These categories are the SIP signaling, voice data transmission, and MANET routing. The SIP signaling controls the VoIP calls initiation, termination, and modifications. The major QoS parameters of VoIP that are managed by SIP signaling are the registration intervals, call setup time, and call termination time. These QoS parameters are increased in MANET due to the nodes' mobility that affects the routing calculations and the connectivity status. These necessitate mechanisms to reduce the delays in the MANET environment. The voice packets are transferred over the Real Time Protocol (RTP) which is encapsulated in the unreliable transport protocol using the User Datagram Protocol (UDP).

With the standardization of ultra-high-definition formats and their increasing adoption within the multimedia industry, it has become vital to investigate how such a resolution could impact the quality of experience with respect to mission-critical communication systems. While this standardization enables improved perceptual quality of video content, how it can be used in mission-critical communications remains a challenge, with the main challenge being processing. This chapter discusses the challenges and potential solutions for the deployment of ultra-high-definition video transmission for mission-critical applications. In addition, it examines the state-of-the-art solutions for video processing and explores potential solutions. Finally, the authors predict future research directions in this area.

Video streaming is expected to acquire a massive share of the global internet traffic in the near future. Meanwhile, it is expected that most of the global traffic will be carried over wireless networks. This trend translates into considerable challenges for Service Providers (SP) in terms of maintaining consumers' Quality of Experience (QoE), energy consumption, utilisation of wireless resources, and profitability. However, the majority of Radio Resource Allocation (RRA) algorithms only consider enhancing Quality of Service (QoS) and network parameters. Since this approach may end up with unsatisfied customers in the future, it is essential to develop innovative RRA algorithms that adopt a user-centric approach based on users' QoE. This chapter focus on wireless video over Critical communication systems that are inspired by QoE perceived by end users. This chapter presents a background to introduce the reader to this area, followed by a review of the related up-to-date literature.

Research in network resource utilisation introduced several techniques for more efficient power and bandwidth consumption. The majority of these techniques, however, were based on Quality of Service (QoS) and network parameters. Therefore, in this study a different approach is taken to investigate the possibility of a more efficient resource utilisation if resources are distributed based on users' Quality of Experience (QoE), in the context of 3D video transmission over WiMAX access networks. In particular, this study suggests a QoE-driven technique to identify the operational regions (bounds) for Modulation and Coding Schemes (MCS). A mobile 3D video transmission is simulated, through which the correlation between receiver's Signal-to-Noise Ratio (SNR) and perceived video quality is identified. The main conclusions drawn from the study demonstrate that a considerable saving in signal power and bandwidth can be achieved in comparison to QoS-based techniques.

This chapter discusses the state of the art in dealing with the resource optimization problem for smooth delivery of video across a peer to peer (P2P) network. It further discusses the properties of using different video coding techniques such as Scalable Video Coding (SVC) and Multiple Descriptive Coding (MDC) to overcome the playback latency in multimedia streaming and maintains an adequate quality of service (QoS) among the users. The problem can be summarized as follows; Given that a video is requested by a peer in the network, what properties of SVC and MDC can be exploited to deliver the video with the highest quality, least upload bandwidth and least delay from all participating peers. However, the solution to these problems is known to be NP hard. Hence, this chapter presents the state of the art in approximation algorithms or techniques that have been proposed to overcome these issues.

The heterogeneous-based 4G wireless networks will offer noticeable advantages for both users and network operators. The users will benefit from the vibrant coverage and capacity. A vast number of available resources will allow them to connect seamlessly to the best available access technology. The network operators, on the other hand, will be benefited from the lower cost and the efficient usage of the network resources. However, managing QoS for video or voice applications over these networks is still a challenging task. In this chapter, a generalized metric-based approach is described for QoS quantification in Heterogeneous networks. To investigate the efficiency of the designed approach, a range of simulation studies based on different models of service over the heterogeneous networks are carried out. The simulation results indicate that the proposed approach facilitates better management and monitoring of heterogeneous network configurations and applications utilizing them.

This chapter provides background about Hybrid Automatic Repeat reQuest (HARQ) protocols. First, the critical situations that may be faced by wireless communication systems especially cellular mobile technologies in case of very noisy radio channels are introduced. Particularly, the chapter introduces the HARQ protocols, their main constituent components as well as some related application areas. Then, the state-of-the-art of HARQ protocols is presented. The next section explains the three basic ARQ protocols. Then, the different HARQ types are detailed. Then, a mathematical model of type II HARQ based on Rayleigh fading channel is provided. This analytical analysis is followed by a discussion of the throughput

Preface

Mission Critical Systems (MCSs) are needed mainly as All-Time-Available backbone system for emergencies, crises, and disaster scenarios. It is used in Public Protection and Disaster Relief (PPDR) operations, Utility Networks, and Intelligent Transportation Systems. Due to the critical nature of MCSs, it has a set of strict requirements to ensure accomplishing tasks and duties that are strongly associated in most times to human lives and national security. Delay, Interoperability, Availability, Reliability, Security, and resilience are some of the requirements that are needed in MCS to ensure optimum functionality as required by users. There are two main deployment options of MCS; the first one is a dedicated system that is designed to meet all the MCS requirements and operating only for mission critical operations and tasks, and the second option is a commercial general purpose mobile communication system that is used for both conventional mobile communication as well as MCS.

The dedicated mission critical communications have already been designed to meet the strict requirements of the mission critical services and communications. TErristrial Trunked Radio (TETRA) in Europe and Project25 (P25) in USA are the most common standards being used nowadays for the mission critical communications. TETRA is widely deployed in many countries due to its ability of meeting all the requirements for MCS. TETRA has short call-setup time (less than 500 ms), bidirectional authentication, it supports Direct Mode of Operation (DMO) and Trunked Mode of Operation (TMO), it is also reliable and resilient MCS, TETRA has a wide coverage due to the low frequency of the carriers, it supports interoperability with other communication networks, and it has a pre-emptive call scheduling based on the priority classes.

Despite the several features of TETRA as MCS, it can only support up to 36Kbps in Release 1 and up to 691.2 Kbps in Release 2. The afforded bandwidth by TETRA for PPDR users is considered enough for voice and basic data services. However, PPDR users nowadays need services that require more bandwidth for their routine operations, and to be able to support the newly introduced Public Safety applications such as verification of biometric data, wireless video surveillance and remote monitoring, documents scan and database check, access to buildings blueprints, and remote emergency medical services. This lead to the need for more MCS support for broadband bandwidth and accessibility for end users. One potential solution is the 4G Mobile communication technologies, which introduced to the users broadband services. They are considered the optimum migration from nowadays dedicated mission critical systems toward a more generic communication system that is used as both a MCS and conventional mobile communications system.

CHALLENGES

The requirements of public safety communication systems are considered stricter than the commercial communication networks. The requirements are governed by the following factors: the type of services provided to the end users, the technology being used, the working environment scenarios and finally the specific role of the Public Safety staff member. Different real-time and non-real-time multimedia services such as voice, data connectivity, messaging, push to talk, location and security services are the most important offered services provided by Mission Critical Communication systems. There are set of challenges and requirements associated with each service. For example, Voice services need an acceptable packet loss ratio and can tolerate up to only 5% loss ratio for acceptable voice quality. However, the end to end delay of voice streaming packets must be very small to make sure the tow-way conversation is understandable. On the other hand, Texting services do not need the aforementioned delay concerns and the band width requirements are much less than other services such as voice and video.

Another challenge is the use of commercial mobile communication networks as a public safety communication system (Balachandran, et al., 2006). It is considered a big challenge to have general purpose public commercial communication system to work as a replica of a dedicated public safety communication system with all mission critical services strict requirements maintained and satisfied. Clearly, using the dedicated system will have all the communication system resources available for only the PPDR users. On the other hand, for the commercial public safety communication system, the resources will be shared among the PPDR and public commercial mobile communication users. Therefore, for the public communication system, the need to have a distinction between the mission critical multimedia services and other non-mission critical services in terms of service priority handling and resource allocation is considered a must to make sure that the system will have enough resources and thus available for the PPDR user whenever and wherever needed.

In addition to the requirements of the multimedia services, the general framework and nowadays available architectures of different MC system will be presented and discussed. Finally, Having different abstracts in hierarchal manner for the architecture of the system as described previously, will introduce not only integration challenges but also cross layer service performance optimization challenges to improve the quality of service for the whole end-to-end mission critical communication system and to satisfy the mission critical requirements mentioned previously. The mission Critical communications and services are of special nature that needs special requirements. Therefore, there are many challenges that face implementing mission critical communication system over commercial mobile communication networks.

Technically, the Signalling using the Session Initiation Protocol (SIP) messages in the access technology domain and IP Multimedia System (MS) domain need enhancement to ensure meeting the mission critical service requirements. Moreover, test-bed experiments show that the IMS system has many limitations and stability issues that affect the availability and reliability of the mission critical communication system. Similarly, simulation results show that there are also limitations for using the SIP signalling over LTE communication system due to added overhead introduced by the increasing number of users or calls. Therefore, deploying IMS operating on top of LTE is of great challenge and needs more efforts for performance enhancement.

Part of the challenge is the need for having a cross layer communications between different system abstracts to ensure that calls, such as mission critical emergency calls, with higher priority class, will pass through the different entities of the system toward the end user and at the same time satisfying all the mission critical service requirements. In addition to that, the need to have a pre-emptive scheduling algorithm to decide which type of calls need to be dropped to flush out the allocated resources in cases where system capacity may reach to its maximum tolerable limits. Although the deployment of pre-emptive scheduling algorithms is covered of this book, a thorough evaluation of overall system capacity with different scenarios that combine the number of users along with the priority class of the service is analysed and investigated.

The Mission critical communication system is composed of different entities and interfaces to ensure the end to end connectivity between end users and overall system integrity. The end users may be connected to different network domains and access technologies, hence it needs interfaces and gateways that provides transparent connectivity between the different domains. Therefore, there is a need for standardized interfaces that connect devices regardless of the manufacturing vendors.

Finally, the need for a data-plane and control-plane protocols that ensures the connectivity and data delivery between end users. The book presents the services, protocols, framework and architecture of the Mission Critical Communication Systems and its associated multimedia services.

In conclusion, the challenges can be listed in the following broad categories:

- Access technology scalability and reliability challenges.
- System resources sharing challenge.
- System architecture design challenge.
- End-to-End QoS Challenge.
- System performance modelling and validation challenge

SEARCHING FOR A SOLUTION

For many governmental and non-governmental organizations (NGOs), Mission critical Systems play critical roles in their operations on everyday basis. The services provided by such organization are of special nature that makes it less tolerant to executional or operational errors. The criticality of the services implies that they must have a set of special requirements that distinguish them from other services, and they should be available anytime, anywhere within the service operational scope, and also they are supposed to be functioning within expectation limits regardless of the operational circumstances or running conditions. Public safety (PS) communications system (Balachandran, et al., 2006), as an example for MCSs, and its related Multimedia Services is needed in incidents where too many emergency forces, such as police, civil defence, medical staffs, and firemen, need to liaise and collaborate together to ensure dealing with critical incidents and events within strict time limits. Due to the nature of such incidents or accidents, every second counts in saving a precious life. Hence, a set of strict requirements need to be met to guarantee a reliable structure and framework that offer reliable and resilient services. The critical nature for such systems distinguishes it from other commercial mobile communications, and they are referred to as Professional Mobile Radio (PMR) or Public safety Land Mobile Radio (PLMR) or Public Protection and Disaster Relief (PPDR) radio communication or more generally as a Mission Critical Communication Systems (Blom, et al., 2008).

There are different deployment options for the mission critical systems, either to have a dedicated system or a commercial systems. Unfortunately, there are drawbacks for both options and many trade-offs that control the design and deployment process. What is important to know is that at the early deployments of dedicated mission critical systems, the voice and simple text messages was the only services that were used by PPDR for their critical missions, simply because mobile communication networks was at its early steps of evolutionary progress where the operators only thought of providing voice communication between users. Hence, when the dedicated mission critical communications was deployed, they also designed the whole system for voice and low data rate services which was fair enough at that time for providing the basic communication platform for the clients. However, due to the continuing incremental increase of mission critical data and multimedia services use to support different tasks and operations, the need for broadband data along with the traditional basic voice service has become crucial for the mission critical communication systems.

Due to the increasing demands for more bandwidth for the mission critical services, the need for access technology that is capable of providing broadband connectivity and scalable to high number of users along with broadband support in the core network is considered crucial requirement for any mission critical communication system. There is an increasing need and demand for a more bandwidth for the mission critical services. This implies that the system not only need to satisfy the general requirements for the mission critical services, but also to be a broadband capacity system with a special requirement to be scalable for the maximum number of users with their broadband services running as it should be during the crises and worst case scenarios. It is important to keep in mind that the mission critical communication system is an all-time-available system that should not be only designed for the normal daily routine PPDR tasks, but also for the worst ever case scenarios (Doumi, et al. 2013).

Using already existing 4G mobile broadband communication systems for providing the professional mission critical mobile services, has many challenges that need to be addressed and solved, especially that LTE is not optimized for voice-centric services which are mostly needed by PPDR members along with other data centric services (Budka, et al., 2011). Significant research and development are required to de-risk the compliance of the newly proposed mission critical systems to be able to meet the MCS requirements. Delays of Session Initiation Protocol (SIP) signalling between different entities must be investigated as it is heavily used by IP Multimedia Subsystem (IMS) which is supposed to be the SIP-based service manager in next generation MCS (Ferrús, Sallent, Baldini, & Goratti, 2013).

ORGANIZATION OF THE BOOK

The book presents a collection of most recent contributions to the topics related to MCS and consists of twelve chapters. A brief description of each of the chapters follows:

Chapter 1 presents the definition of Mission Critical Systems, Mission Critical Users, and Mission Critical Services. Main applications of such system are listed, are the general requirements are investigated, and finally a distinction between different types of Mission Critical Systems are briefly discussed.

Chapter 2 establishes the need for multimedia protocols that control the operation of the end-to-end system and have implications on the overall performance aspects. IP Multimedia Subsystem (IMS) is presented and Session Initiation Protocol (SIP) is investigated along with IMS performance issues, and performance issues are demonstrated.

Chapter 3 presents an IMS test-bed based performance investigation scenarios and results are presented. This is followed by an OPNET simulation based investigation where scenarios and results are demonstrated. The limitations and challenges for both experiments are also discussed.

Chapter 4 investigates a determined performance metrics to understand and evaluate different scenarios for the SIP-based Voice over IP (VoIP) applications over MANET, which behaves in a way similar to

Direct Mode of Operation (DMO) in mission Critical Communication Systems. Quality of Service (QoS) of SIP end-to-end performance metrics as defined by RFC 6076 are reviewed. The main performance metrics related to the registration, the call setup, and the call termination processes are investigated, as well. In this chapter, the main performance metrics for SIP signalling and voice data are presented to identify the evaluation methods for SIP-based VoIP applications over MANET platforms.

Chapter 5 provides analysis for the performance metrics that was defined in Chapter 4, the quality of service (QoS) of SIP-based VoIP is introduced. SIP-based VoIP applications over Direct Mode of Operation (DMO) and its major QoS parameters of VoIP such as the registration intervals, call setup time, and call termination time will be analysed. In addition, the bandwidth consumption, delays, jitter and packet loss are QoS parameters that quantify VoIP performance is reviewed. Finally, this chapter presents simulation results for evaluating SIP call processes and QoS parameters together.

Chapter 6 addresses the standardization of ultra-high-definition formats and their increasing adoption within the multimedia industry. An investigation, on how such a resolution could impact the quality of experience with respect to mission-critical communication systems, is presented. While this standardization enables improved perceptual quality of video content, how it can be used in mission-critical communications remains a challenge. This chapter discusses the challenges and potential solutions for the deployment of ultra-high-definition video transmission for mission-critical applications. In addition, it examines the state-of-the-art solutions for video processing and explores potential solutions.

Chapter 7 focus on wireless video over Critical communication systems that are inspired by the Quality of Experience (QoE) perceived by end users. This chapter presents a background to introduce the reader to this area, followed by a review of the related up-to-date literature. In the background, the elements of the triple-Q framework are first introduced. Then communication systems are demonstrated in the context of wireless video communications.

Chapter 8 suggests a QoE-driven technique to identify the operational regions (bounds) for Modulation and Coding Schemes (MCS). A mobile 3D video transmission is simulated, through which the correlation between receiver's Signal-to-Noise Ratio (SNR) and perceived video quality is identified. The main conclusions drawn from the chapter demonstrate that a considerable saving in signal power and bandwidth can be achieved in comparison to QoS-based techniques.

Chapter 9 deals with the resource optimization problem for smooth delivery of video across a peer to peer (P2P) network. It further discusses the properties of using different video coding techniques such as Scalable Video Coding (SVC) and Multiple Descriptive Coding (MDC) to overcome the playback latency in multimedia streaming and maintains an adequate quality of service (QoS) among the users.

Chapter 10 investigates the heterogeneous-based 4G wireless networks which offer noticeable advantages for both users and network operators. A vast number of available resources will allow them to connect seamlessly to the best available access technology. The network operators, on the other hand, will benefit from the lower cost and the efficient usage of the network resources. However, managing QoS for video or voice applications over these networks is still a challenging task. In this chapter, a generalized metric-based approach is described for QoS quantification in Heterogeneous networks.

Chapter 11 provides a background introduction into Hybrid Automatic Repeat reQuest (HARQ) protocols. First, the critical situations that may be faced by wireless communication systems especially cellular mobile technologies in case of very noisy radio channels are introduced. Particularly, the chapter introduces the HARQ protocols, their main constituent components as well as some related application areas. Then, the state-of-the-art of HARQ protocols is presented. The chapter explains the three basic ARQ protocols. Then, the different HARQ types are detailed. Then, a mathematical model of type II

HARQ based on Rayleigh fading channel is provided. This analytical analysis is followed by a discussion throughput which is one of the most interesting metrics used to measure the performance of HARQ systems.

Chapter 12 introduces two different algorithms to detect intrusions in mission critical communication systems to guarantee their security. The first algorithm is a classification algorithm which applies the concept of supervised learning. The second algorithm is a clustering algorithm which applies the concept of unsupervised learning. The algorithms detect intrusions using a set of detection rules that are structured in the form of decision trees. The algorithms are described in details and their results on well-known dataset are introduced.

Khalid Al-Begain
University of South Wales, UK & Kuwait College of Science and Technology, Kuwait

Ashraf A. S. Ali
Hashemite University, Jordan & University of South Wales, UK

REFERENCES

Balachandran, K., Budka, K. C., Chu, T. P., Doumi, T. L., & Kang, J. H. (2006, January). Mobile Responder Communication Networks for Public Safety. *IEEE Communication Magazine.*

Blom, R., de Bruin, P., Eman, J., Folke, M., Hannu, H., Naslund, M., & Synnergren, P. (2008). Public Safety Communication using Commercial Cellular Technology. In *International Conference and Exhibition on Next Generation Mobile Applications, Services, and Technologies*. IEEE Computer Science Press. doi:10.1109/NGMAST.2008.78

Budka, K. C., Chu, T., Doumi, T. L., Brouwer, W., Lamoureux, P., & Palamara, M. E. (2011). Public safety mission critical voice services over LTE. *Bell Labs Technical Journal*, *16*(3), 133–149. doi:10.1002/bltj.20526

Doumi, T., Dolan, M. F., Tatesh, S., Casati, A., Tsirtsis, G., Anchan, K., & Flore, D. (2013). LTE for public safety networks. *IEEE Communications Magazine*, *51*(2), 106–112. doi:10.1109/MCOM.2013.6461193

Ferrús, R., Sallent, O., Baldini, G., & Goratti, L. (2013). LTE: The technology driver for future public safety communications. *IEEE Communications Magazine*, *51*(10), 154–161. doi:10.1109/MCOM.2013.6619579

Acknowledgment

The editors would like to acknowledge the help of all the people involved in this project and, more specifically, to the authors and reviewers that took part in the review process. Without their support, this book would not have become a reality.

First, the editors would like to thank each one of the authors for their contributions. Our sincere gratitude goes to the chapter's authors who contributed their time and expertise to this book.

Second, the editors wish to acknowledge the valuable contributions of the reviewers regarding the improvement of quality, coherence, and content presentation of chapters. Most of the authors also served as referees; we highly appreciate their double task.

Khalid Al-Begain
University of South Wales, UK

Ashraf A. Ali
University of South Wales, UK & Hashemite University, Jordan

Chapter 1
Introduction to Mission Critical Systems and Its Requirements

Ashraf A. Ali
The Hashemite University, Jordan & University of South Wales, UK

Khalid Al-Begain
University of South Wales, UK & Kuwait College of Science and Technology, Kuwait

ABSTRACT

In this chapter, the definition of Mission Critical Systems, users, and services will be introduced, then main applications of such system will be listed, the general requirements will be investigated, and finally a distinction between different types of Mission Critical Systems will be briefly discussed. Nowadays, many governmental and non-governmental organizations (NGOs) play critical roles in its operations on everyday basis. The services provided by such organization is of special nature that makes it less tolerant to executional or operational errors; such services are referred as Mission Critical Services. The criticality of the service implies that it has a set of special requirements that distinguish it from other services, and it should be available anytime, anywhere within the service operational scope, and also it is supposed to be functioning within expectation limits regardless of the operational circumstances or running conditions.

INTRODUCTION

Nowadays, many governmental and non-governmental organizations (NGOs) play critical roles in its operations on everyday basis. The services provided by such organization is of special nature that makes it less tolerant to executional or operational errors; such services are referred as Mission Critical Services. The criticality of the service implies that it has a set of special requirements that distinguish it from other services, and it should be available anytime, anywhere within the service operational scope, and also it is supposed to be functioning within expectation limits regardless of the operational circumstances or running conditions.

DOI: 10.4018/978-1-5225-2113-6.ch001

In this Chapter, the definition of Mission Critical Systems, users, and services will be introduced, then main applications of such system will be listed, the general requirements will be investigated, and finally a distinction between different types of Mission Critical Systems will be briefly discussed.

"Mission Critical" is defined by TETRA (Terrestrial Trunked Radio, 2003) and Critical Communication Association (TCCA) as a function that its failure leads to a catastrophic results that places the public order or the public security at risk. A system that is providing such critical functionality must have the suitable inbuilt functionality, interoperability, security and availability. The "Mission Critical Users", are those who are responsible for welfare, health, security and safety of the public people. Law enforcement forces, fire fighters, Emergency and Medical Services (EMS), rescue services, military forces, utility staff members, transport services members are all example of Mission Critical users. The concept of "Mission Critical Communication Systems" is referred to the hardware, software and communication facilities that allow Mission Critical Users to communicate with each other and liaise with command centres securely and dependably for the sake of providing Mission Critical Services wherever the services require special communication solutions (TETRA & Critical Communication Association, 2010).

APPLICATIONS OF MISSION CRITICAL SYSTEMS

Based on the definition mentioned above, The Mission Critical Systems (MCSs) are considered as systems that provide critical services for a certain target group. MCS is needed mainly as All-Time-Available backbone system to be accessible by system users and ensuring connectivity between clients and satisfying the service requirements with acceptable service quality based on the task and service type demanded by the client. The American Public Safety Advisory Committee (PSAC) in the United States describes every system that is capable of providing: an immediate communication with instantaneous connectivity, reliability to minimize the short term disruptions, and in most cases secure in order not to be accessed by unauthorized users. Such systems are required by many applications that will be briefly introduced in the following subsections.

Public Protection and Disaster Relief

Public Protection and Disaster Relief (PPDR) is one of the MCSs that is needed by law enforcement, emergency, and medical services teams during emergencies, crises, and disaster scenarios (Baldini, Karanasios, Allen, & Vergari, 2014), the system is used mainly for liaison between different PPDR members to make sure that all needed resources are available during and after the crises that may occur anytime and anywhere. Mission Critical Communication System is needed for PPDR system since it provide services of critical nature that may make the difference in saving precious lives during harsh operating conditions. One of the main challenges is making the required resources available and dedicated for the mission critical communication system to provide an enhanced communication system that is scalable to high number of users along with diverse concurrent running applications. In later sections,

PPDR members use the talk-group function to communicate with each other or with the control room, setting the talk-group instantaneously on site is crucial for optimum teamwork functionality. Therefore, supporting group communication is one of the most important requirements in the MCS that serve PPDR members. The group communication is done via Push-To-Talk function to address another user or set of users within a group in a half-duplex communication. This implies that a special hardware terminals must be designed and manufactured to support the PTT functions.

Utilities

Utilities such as Gas, Water, and Electricity is considered of great importance nowadays for country's economy and development plans. The Electricity outage, for example, has tremendous impact on all other sectors. Therefore, the Electricity Transmission Grid is considered one of the Mission Critical Systems that is supposed to provide a service or a functionality which, in case of failure, may lead to catastrophic results. What applies to Electricity also applies on the Gas services, especially that the Gas Transmission Networks are operated across multiple countries. Similarly, Water Network is also critical by nature and need to be operated by MCS to ensure the reliability of the provided service.

All the three utilities sectors need special requirements to support the provided services by each sector. Ensuring the connectivity using a reliable system, group talk between staff members on ground, and data monitoring and collection within specific time limits to be followed by automated response, are all considered examples of services that need special requirements for a system that has a critical nature.

Intelligent Transport Systems

The wide application of technological operational methods in the transportation sector, leaded to what is called as Intelligent Transport Services (ITS). Normally, the number of applications in IT'S is more than that in Utilities and PPDR sectors. Most of the ITS applications require narrow band data, with other applications, such as video monitoring of traffic, require broadband connectivity to control centres. Controlling the transportation sector is considered a mission critical task that need to be operated via a MCS. However, in the ITS sector there are other entertainment services that are considered of non-critical nature such as providing broadband connectivity for passengers, which implies that there is a need to define the requirements of the system based on different kinds of services provided.

MCS FUNCTIONAL AND OPERATIONAL REQUIREMENTS

Public Safety (PS) communications are very important in incidents where too many emergency forces, such as police, civil defence, medical staffs, and firemen, need to liaise and collaborate to ensure dealing with critical incidents and events within strict time limits. Due to the nature of such incidents or accidents, every second counts in saving a precious life. Hence, a set of strict requirements need to be met to guarantee a reliable structure and framework that offer reliable and resilient services. Due to critical nature for such systems which distinguishes it from other commercial mobile communications, we refer to it as Professional Mobile Radio (PMR).

The requirements of public safety communication systems is considered more strict than the commercial communication networks, the requirements are governed by the following factors: the type of services provided to the end users, the technology being used, the working environment scenarios, and the specific role or the functionality of the MC users (Baldini, Karanasios, Allen, & Vergari, 2014) .

As mentioned in the previous section, there are three main sectors that applies Mission Critical functions; PPDR, Utility, and ITS. PPDR has a special operational and functional requirements (Baldini, et al., 2012) such as the need for broadband services, voice communication, back to back communication for first responders, secure communication, etc. the aforementioned functional requirements determine the technical requirements of the MCS needed for PPDR operations as will be presented in next sections.

The Utilities also has some functional and operational requirements to ensure hassle free operation such as continuous data monitoring, automated control, system restore ability, service coverage, low metering running cost, etc. Similarly, ITS has diverse applications such as traffic signals management, vehicle detection and tracking, Closed-Circuit Television (CCTV), etc. interoperability and security over narrowband communication medium are some of MCS requirements in ITS sector.

A full description of MCS requirements will be presented in the following sections.

TYPES OF MISSION CRITICAL SYSTEMS

There are three deployment options for Mission Critical Communication Systems; the first one is the dedicated mission critical systems, which, as the name implies, is only dedicated for the mission critical communications tasks and operations. The second one is the commercial Mission Critical System that is used for mobile communications in addition to the Mission Critical operations. Finally, there is a hybrid approach that combines both the commercial and dedicated MCS. (Ferrús, Sallent, Baldini, & Goratti, 2013) .

Dedicated MC Communication Systems

The dedicated mission critical communication is already designed to meet the strict requirements of the mission critical services and communications. TETRA in Europe and P25 in USA are the most common standards being used nowadays for the mission critical communications. We will investigate in depth the dedicated MC systems in the next sections.

Commercial MC Communication Systems

The commercial communication systems are considered as a general purpose system that is used for public mobile communication and also whenever needed it should be ready for being used by the PPDR clients as a mission critical communication systems. The optimum goal is to have a system that works as a replica of a dedicated mission critical communication system using commercial mobile system equipment. (Blom, et al., 2008) shows the feasibility of having Public Safety System over commercial cellular Technology. Clearly, this introduces a lot of design and enhancement challenges and constraints that will be investigate in depth in the next sections.

Hybrid MC Communication Systems

Having a dedicated MCS is of high cost and expenses. Governments are thinking about having a set of technologies to be running altogether for supporting the mission critical operations. This will provide a smooth transition towards the fully costly commercial MCS. in the hybrid approach, all the technologies such as TETRA, Commercial LTE, Wi-Fi, WiMAX, Satellite communications, legacy 2G and 3G. All can work to support the mission critical operations in the three sectors.

In Baldini et al. (2012) a framework that combines multiple communication technologies was presented to be considered for operation in Europe. The selection of the technology depends on the needed bandwidth and coverage along with the requirements needed by the MC sector. However, the management between different technologies is complex and require a clear division of responsibilities.

Migration from Current Dedicated MCSs

As described before, dedicated mission critical systems are reliable and designed for the Mission Critical operations and able to provide users with low and mid-range data services. However, due to its high operational and capital expenditures without the ability to provide broadband services with only a limited functionality to voice oriented and data limited services makes it not the best option for the future mission critical communications. In this section, the emerging of mission critical data needs will be presented followed by a case study of proposals for providing mission critical services over non-dedicated MCS structures, then the feasibility of commercial mobile communications will be discussed, and finally the need for hardening the generic commercial mobile communication to be eligible for being used as a MCS will be discussed.

Mission Critical Users Demands

The user demands for more data due to the changes in applications and users' needs has also affected the Mission Critical Systems way of operation. In (TETRA Association, 2010) the trends in the use of mobile applications in the public sector shows that the public safety sector community is following the wider society for the need to access a wider set of applications due to the change in the way of working. For example, the enhancement in the effectiveness and efficiency of the incident response requires that PPDR members access a simultaneous set of applications. A more need for data services along with voice services is also essential managing planned and unplanned major events. Nowadays PPDR operations are increasingly need more bandwidth to support the mixture of different multimedia services that support their daily operations, transmitting videos, images; high definition audio are becoming as much important as conventional voice communications.

Video conference call between the ambulance and the hospital, sending a crime scene photos and diagrams, accessing to images that are recorded as evidence by investigators, police enquiries for a suspects photos from a central information data base, video streaming of actions at incidents being transmitted to law enforcement forces for taking action, forwarding suspects biometric data to a command centre, building blueprints accessed by fire fighters, firefighters equipped by personal monitors and location tracking devices, are all applications in the PPDR sector that need more bandwidth than that provided by current dedicated MCSs.

The increase in the data demands, as described above, has motivated the governmental and non-governmental organizations to look for alternatives to current solutions and options for delivering mission critical services over systems that are capable of providing the needed operational, functional, and technical requirements in the near future.

UK Emergency Services Mobile Communications

TETRA is used in UK for emergency services as part of Public-Private Partnership contract signed between The Home Office and Airwave Ltd. Based on this contract, the UK government pays around 400 million pounds every year for Airwave Ltd to use TETRA as a nationwide dedicated PPDR network providing PPDR personnel with a narrow band reliable service.

The UK government decided already to pursue a commercial replacement option for the current too expensive and limited emergency service communication system provided by Airwave. The Emergency Services Mobile Communications Program (ESMCP) aims to replace Airwave provided solution with a commercial broadband support service for PPDR based on commercial LTE. This migration from a reliable narrowband system to a developing broadband system introduces a lot of questions and debates in terms of feasibility of commercial systems and their ability of providing reliable solution working as a replica of current dedicated MCSs. Also, there are concerns regarding the Mobile Network Operators (MNOs) commitment to maintain the agreed Quality of Service for the PPDR use that should be mentioned in the Service Level Agreements (SLAs).

The decision was taken in UK to have transition in two years to a purely commercial LTE network that support both Mission critical and non-mission critical services over the Emergency Services Network. The project aims to have the new system compatible with the current terminal equipment and applications, provide all the features that are supported by TETRA (DMO, PTT, group call, etc.), and at the same time provide a broadband and voice services at much lower cost. Hardening issues will be discussed in later sections of this report.

The Feasibility of Mobile Communications as MCS

The mobile communication technologies have evolved significantly since early 90's till nowadays. 3GPP LTE release 11is capable of providing broadband services to its users due to the advances in physical layer technologies such as Multiple Input Multiple Output (MIMO) Antennas, and Orthogonal Frequency Division Multiple Access (OFDMA). Moreover, LTE is optimized for minimum delay in both the Radio Access part and the Core Network part, which minimize the end-to-end delay of the service. However, although LTE can provide broadband services with low latency, it can only partially support the mission critical services requirements in the different sector applications. This is due to the many limitations and shortcomings of current mobile communication networks especially in the process of integrating end-to-end solution to support the mission critical operations will all needed requirements. The shortcomings and the need for hardening the overall system will be briefly introduced in the next subsection.

Commercial Mobile Technologies Concerns and Hardening Need

The main shortcomings of current mobile communication technologies will be discussed more in later sections, for now it can be summarized as follows (Forge, Horvitz, & Blackman, 2014):

- **Voice Services Support:** The lack of talk group communication, Push to Talk (PTT) and Direct Mode of Operation (DMO) support in nowadays mobile communications is considered a major issue for supporting Mission Critical Services over current mobile networks. LTE as broadband technology is considered data centric technology that is not optimized for voice services and need more effort toward supporting voice centric services, Voice over LTE (VoLTE) is contributing in this direction but needs a lot of efforts to integrate it as a solution in generic communication systems.

- **Resilience:** The limited resilience of the commercial mobile networks is unacceptable for mission critical applications in the three sectors. A lot of efforts need to be done to support the resilience in face of expectable events such power outage, natural disasters, or terrorist attacks. The resilience can be enriched by deploying backup and redundancy solutions in the access and core network parts.
- **Pre-Emptive Capability:** The PPDR applications need full access to the available resources whenever needed. Having a commercial system that is deployed for both conventional and professional use rises the resource allocation and scheduling challenge to give access to call of higher priority on expense of lower priority ones that need to be pre-empted based one certain rules.
- **Coverage:** The coverage of commercial mobile service is always enhanced for marketing purposes. The operators are interested in getting the maximum profit of their investment by building the network infrastructure and facilities near the densely populated areas with less coverage in rural areas. This would be a concern for mission critical users who may need the service anywhere and anytime. Therefore, having a 100% coverage is of great importance for having nationwide MCS. Emergency Medical Services, for example, need ubiquitous coverage in their operations.
- **Different Models:** The need for different communication models is a need for having a reliable and resilient service. Air-Ground-Air (AGA) communication, for example, is of great importance for search and rescue operations in PPDR sector. Therefore, commercial mobile communication systems need to support AGA in addition to the Ground-Ground supported communication.

Based on the previously mentioned concerns, the need for having a hardened commercial system to work as a replica of current dedicated mission critical systems is of great importance for a reliable system with broadband services support.

The emerging of the LTE capabilities along with IP Multimedia Subsystems (IMS) based services such as Push over Cellular (PoC) requires further hardening and optimization efforts in order to meet the strict requirement of MCS. Hardening efforts for mobile technologies was investigated by the research community, In (TETRA MoU Association, 2004) the potential of GPRS and CDMA for mission critical use was analyzed by TETRA Association, the study shows that the latency of call set-up time is not acceptable for MCS use. In (Kanti, 2006) IMS was added on top of GPRS network to minimize the call-setup latency, which added advantages for group calling but with added call setup overhead. In (Balachandran, Budka, Chu, Doumi, & Kang, 2006) and (Balachandran, et al. 2005), IMS with PoC on top of UMTS and CDMA2000 feasibility was investigated and shows that the overall system latency was reduced but still not enough for critical communications. In (Blom, et al., 2008) and (IPWireless, 2012) the needs of the public safety community in HSPA and LTE was addressed.

OBJECTIVES

The purpose of this project is to investigate the performance optimization issues for the mission critical communication systems that are deployed over the current LTE (4G), LTE-advanced, and 5G broadband communication Systems.

The emerging of the LTE capabilities along with IMS-based services such as Push over Cellular (PoC) based on Voice over LTE (VoLTE) system requires further integration and optimization efforts in order to meet the strict requirement of communication needed for emergency services.

In this project, the objective is to investigate more the SIP signalling over LTE to de-risk the compliance of any Mission Critical System for the End-to-End system performance targets. In addition to identifying the potential risks within the End-to-End system performance at the component performance level.

The project is contributing to the development of mission critical services over 4G and 5G broadband wireless networks with the deployment of a Large Scale Test program to de-risk the compliance of the Open Mission Critical System prototype product against the System and Software Design Description for the End-to-End system performance targets. In addition to identifying the potential risks within the End-to-End system performance at the component performance level using related test-beds and simulation tools such as OPNET.

We can summarize the objective of the project as follows:

1. To analyse and understand current and proposed MCSs by literature overview study.
2. To have a clear understanding of MCS Technical functional and user requirements.
3. To create a set of performance metrics that need to be measured for better understanding of system performance measuring criteria.
4. To test the system over a test-bed and collect statistics based on pre-defined performance measures.
5. To simulate the system over simulators and generate behavioural charts and statistics that reflect the performance of the default system.
6. To identify the areas of weaknesses in the system and suggest ways for performance improvement.
7. Validate the suggested and proposed improvement method over the test-bed and simulation tool to get the new enhanced behavioural model of the system.
8. To compare between the two collected statistics before and after applying the proposed enhancement approaches for both the test-bed and the simulation tool.

MCS REQUIREMENTS AND IMPLEMENTATIONS

The MCSs are designed to tolerate the consequences of natural disasters and abnormalities, and it is expected to be working to support the users on the ground, Otherwise it would be useless to use the whole communication system which is by nature design to operate in harsh environment. In this section, the general requirements of MCS will be discussed, then dedicated MCS implementations will be presented.

Mission Critical Services Requirements

Services such as voice, data connectivity, messaging, push to talk, location and security services are the most important offered services provided by Public Safety (PS) Communication systems. There are set of challenges and requirements associated with each service. Voice services, for example, needs an acceptable packet loss ratio and can tolerate up to only 5% loss ratio for acceptable voice quality, moreover the end to end delay of voice streaming packets must be very small to make sure the tow-way conversation is understandable. On the other hand, messaging services do not have the aforementioned delay concerns and the band width requirements are much less than other services such as voice and video. In this section, the requirements of different services over MCS will be discussed and explained.

Delay

There are many requirements for mission critical services over PSNs; one of the most important requirements is end-to-end time delay for sending the voice or data between system users. Delay is the most important requirement in PSN especially for real time services such as voice and video and non-real time services such as sending urgent push messages to report the occurrence of incident and to notify others within hundreds of mille-second time orders. There are different types of delay as shown in Figure 1. For a voice call over the network, there are mainly two types of delays: the signalling delay such as the registration, call setup, and call termination delay. The data streaming delay is mainly the end to end delay and inter-packet delay or Jitter delay. It is crucial to minimize all the delays above to guarantee the best Quality of Service (QoS) and Quality of Experience (QoE) for the end users. In later chapters, we will investigate in depth the sources of delays for any service in both the data plane and control plane.

Interoperability

Interoperability is needed for mission critical systems in places where different PPDR members from different organizations are all located in the same incident site, each has its own equipment's and they need not only to communicate within the same team only but also with other teams. For example, if there is a building of multiple stories set on fire, then it is expected to have on the ground of the site fire fighters to extinguish the fires, law enforcement members to investigate the trigger of the fire, and Emergency and Medical services members to help those who are injured on site. Clearly liaising between

Figure 1. Multimedia service sources of delay

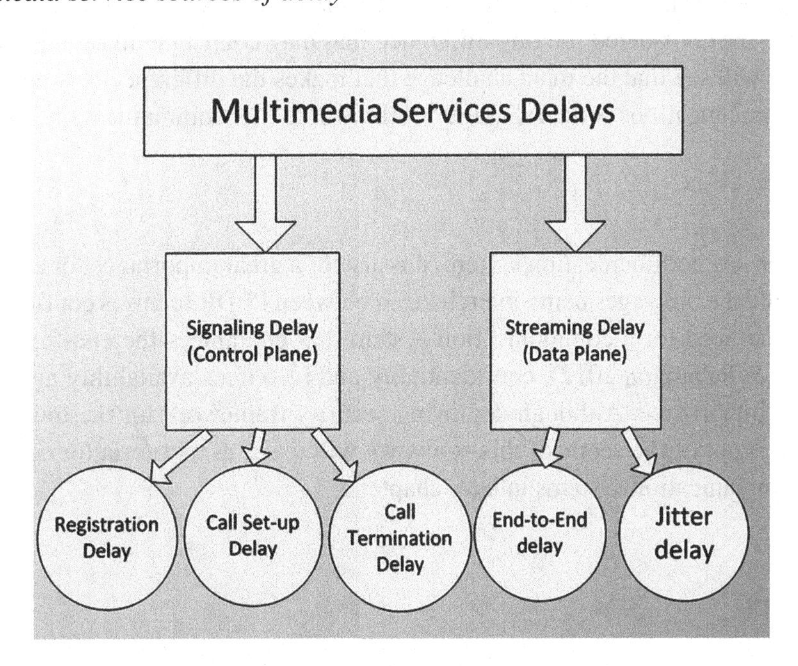

the different teams is needed to ensure interoperability for best utilization of available resources. Such interoperability can be achieved in general by following one of two approaches; the first approach is to use the same mission critical communication system by everyone. The second approach is to have different MC communication systems that are working under the umbrella of one common core standard that ensures providing the needed interfaces for seamless connectivity between the different systems. In (Balachandran, et al., 2006) a framework was proposed for a system that allows interoperability between multiple MCSs through a shared IP-core network, and interoperability with legacy mobile communication networks is presented in (Budka, et al., 2011) . In addition to what is described before, there is also interoperability issues across borders between neighbouring countries that may use different frequency bands and MC communication systems.

Priority Handling

Moreover, priority handling is important for mission critical services especially if the infrastructure being used is shared with public users over commercial communication network where the need for employing pre-emption scheduling algorithms is crucial to guarantee the availability of resources for the PSN users at the appropriate time, clearly this means that less priority calls will be dropped from the queues if the system is not able to adapt all calls at the same time.

Resilience

Another important requirement is to insure the maximum resilience of the overall system; this implies that there should be backup systems in distributed environment to ensure reliability and to avoid the single point of failure. PSN should also be able to support Direct Mode of Operation (DMO) in which users can communicate with each other using their terminals in ad-hoc manner, this is important in scenarios where the access network may fail due to a catastrophic events that may be caused by terrorist attack or natural disaster. It is sometimes referred as all-time-available communication system which is, for Public Safety services, may be considered the tiny difference that may contribute in saving a precious life. In another chapter, we will see that the main challenge that makes the difference between the major types of public safety communications is the reliability of the whole MC communication system.

Security

Security is needed in any communication system; it is also of a great importance for any mission critical service where usually the messages being interchanged between PPDR teams is confidential and critical in nature. Hence, the need for a communication system that guarantees the end-to-end data integrity (McGee, Coutière, & Palamara, 2012), confidentiality and resources availability against any possible attack is of crucial importance. Although deploying security framework for the mission critical communication system is out of the scope of this study, we will use it as a criteria for comparing different mission critical communication systems in later chapters.

TETRA

To ensure the strict requirements of the public safety communications, all governmental and non-governmental organizations nowadays rely on dedicated Public Safety (PS) communication systems due to their robustness, resilience, and reliability. Such systems ensure that, in most extreme and harsh scenarios, the network will be able to offer the service within the minimum acceptable service requirements.

As mentioned before MCSs can be mainly classified as dedicated or commercial systems. Having a dedicated PS communication system means that the system was built specifically to meet certain needs, and it also means that the operational and capital expenditures of deploying the PS system would be much higher than any other general purpose commercial communication system even though the license fees of the spectrum is usually waived out for the sake of public benefit. In this section, two dedicated MCSs and one of the expected commercial MCSs will be presented.

The Terrestrial Trunked Radio (TETRA) (Terrestrial Trunked Radio, 2003), or what is used to be named as Trans European Trunked RAdio, is a set of standards developed by the European Telecommunications Standardization Institute (ETSI). TETRA is a standard for a private Mobile Digital Radio System to meet the needs of Public Mobile Radio (PMR) organizations. TETRA is a communication system used by professional governmental and non-governmental agencies and operate independently from other commercial networks. It mainly uses Time Division Multiple Access (TDMA) with four channels each of 25 KHz bandwidth to support both voice and data communication in point-to-point or point-to-multipoint manners. Figure 2 shows the general architecture of TETRA communications system.

Figure 2. TETRA system architecture (Europian Telecommunication Standardization Institution, 2013)

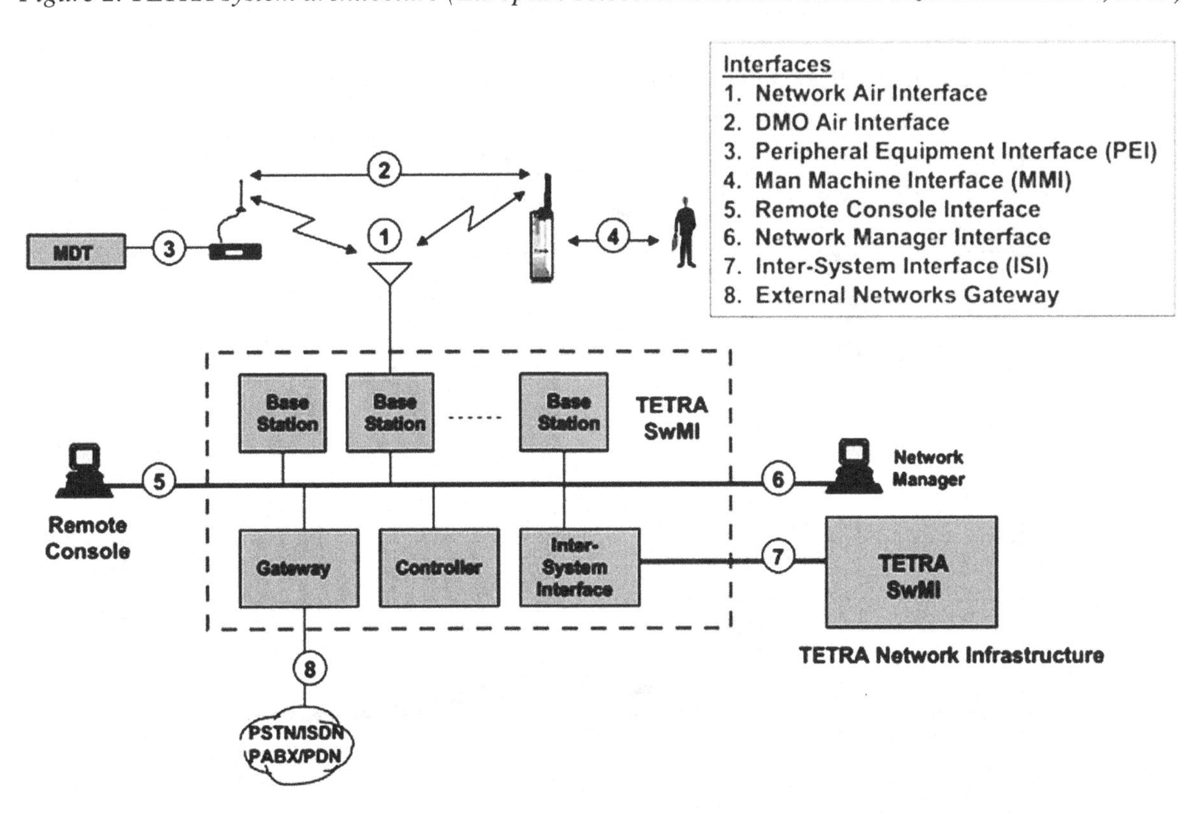

In addition to the device to base station radio interface, the handheld devices are able to communicate directly with each other without the need of the base station or core infrastructure. This device to device communication is known as Direct Mode Operation (DMO) and it is mostly needed in scenarios where public safety members are located in the same site, and there are no available nearby base station or the base station is unavailable due to an attack or natural disaster. DMO enhances the reliability of TETRA systems and widely used in nowadays deployments. On the other side, devices may access the core network using the base station in what is called Trunked Mode Operation (TMO) which uses the switching and management infrastructure (SwMI) that is made of multiple base stations.

MCS Requirements Compliance Over TETRA

As mentioned before, TETRA (Terrestrial Trunked Radio, 2003) was designed and built to be a critical communication system for public safety duties and nowadays it is widely deployed in many countries. Hence, it meets all the requirements for MCS, TETRA has short call-setup time, as shown in Figure 3, where 85% of the calls need less than 500 ms (Terrestrial Trunked Radio, 2003) that is acceptable for real time services. It also provides bidirectional authentication between terminals and the core infrastructure in addition to air interface as well as end-to-end encryption to meet the security requirements. It also supports DMO and TMO operations that enable point to point and point to multi point communication in both modes. This supports the reliability and resilience for MCS. TETRA has a wide coverage due to the low frequency of the carriers, and this will increase the range of maximum distance for communication either in DMO or TMO. Devices can work as a relay or repeaters to provide access for other devices to access the network in ad-hoc manner in scenarios where some terminals may fall outside the

Figure 3. Call set-up delay PDF and CDF (Terrestrial Trunked Radio, 2003)

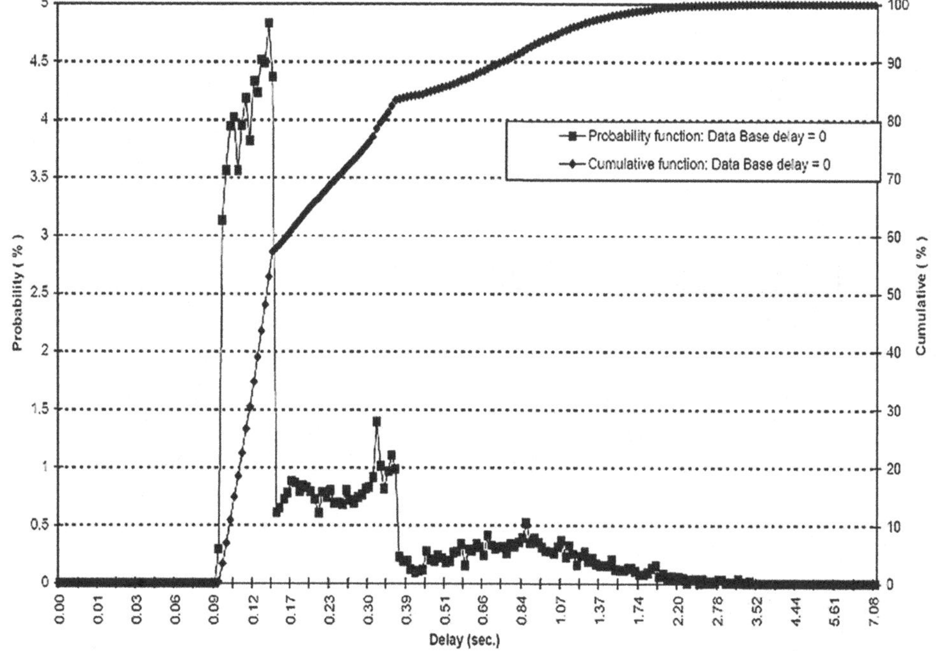

communication range, this feature enhances the both the availability and accessibility requirements of the MCS. TETRA provide transportable network solutions that are carried over vehicle that supports both reliability and capacity requirements. It also supports interoperability with other communication networks through gateways for wider accessibility domains and better interoperability enhancement between different domains and working groups. Finally, there are different levels of priorities that are defined in TETRA, pre-emptive call scheduling based on the priority class will guarantee that the more critical calls pass through the network with less priority calls schedules automatically at the bottom of the queue to be served afterwards.

TETRA Release 1

TETRA Release 1 or TETRA Voice plus data (TETRA V+D) is the original standard developed by ETSI; it forms the basis for all other TETRA systems. Release 1 of the standard introduces the voice plus data (V+D) service for TETRA systems, the Direct Mode of Operation, and the Packet Data Optimized (PDO).

TETRA uses Time Division Multiple Access (TDMA) for user multiplexing. Figure 4 shows the TDMA structure and the types of frames starting from the hyperframe level down till the subslot unit. It is important to note that the bandwidth utilization in both time and frequency domains is necessary to maximize the symbol rate or bit rate available for each user. In Release 1, π/4-shifted Differential Quaternary Phase Shift Keying (π/4-DQPSK) is the only modulation scheme used. Hence, the maximum bitrate available for users can be calculated knowing that the symbol word depth for (π/4-DQPSK) will be 2-bits per symbol and the user slot time is 14.167 ms over the 25 KHz bandwidth channel with a data rate of 36 Kbps.

Figure 4. TETRA frame structure (Terrestrial Trunked Radio, 2003)

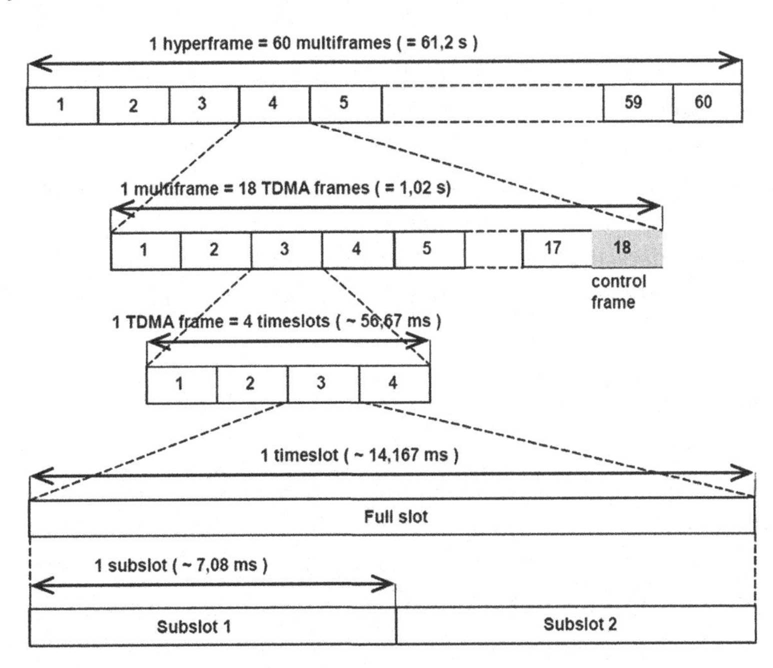

Table 1. Symbols capacity in TETRA

	Number of Symbols	
	Phase Modulation	**QAM**
Slot	255	34
Subslot	127,5	17

TETRA Release 2

I this release, the access using TDMA is done via four physical channels per carrier and there are multiple modulation schemes that is pre-allocated based on the channel width. For Phase Modulation (PM) the carrier bandwidth is 25 KHz, whereas for Quadrature Amplitude Modulations (QAM) the carrier bandwidths are 25 KHz, 50 KHz, 100 KHz or 150 KHz depending on the scheme of QAM modulation. Table 1 shows the symbols per slot or sub-slot that are available per user or physical channel at certain instant of time. The symbol rate using PM is $(18 * 10^3)$ Symbols/s and for QAM it would be $(1.2 * 10^3)$ Symbols/s.

Now we can calculate the bit rate based on the symbol rate knowing the modulation scheme used in both the PM and QAM. In PM there are two modulation schemes; π/4-shifted Differential Quaternary Phase Shift Keying (π/4-DQPSK) or π/8-shifted Differential 8 PSK (π/8-D8PSK). Hence, the symbol word depth for (π/4-DQPSK) will be 2-bits per symbol and the overall data rate is 36 Kbps, similarly the symbol word depth for (π/8-D8PSK) is 3-bits per symbol and hence the overall data rate is 54 Kbps.

For QAM modulations, there are three modulation schemes 4-QAM, 16-QAM, or 64-QAM, and there are 8 sub-carriers per 25 KHz being used. Each subcarrier has a symbol rate of 2400 Symbols/s. Hence, the maximum available bandwidth for users is shown in Table 2.

APCO 25

Similar to the role of TETRA, the Association of Public-Safety Communications Officials-international (APCO) started what is called Project 25 (P25) (Project 25, 1995) which is a standard for public safety agencies in North America. P25 was designed to address the needs for a professional Digital Mobile Radio system used by public safety agencies to meet all the mission critical service requirements. It supports DMO and the conventional, trunked mode of operations.

APCO evolved in two phases, in phase 1, the FDMA was used over channels of 12.5 KHz bandwidth that provides a maximum of 9.6 Kbps per user of which 4.4 Kbps for voice data, 2.8 Kbps for Forward

Table 2. Gross bit rates for QAM Carriers (Kbit/s)

Modulation Type	Carrier Bandwidth			
	25 kHz	**50 kHz**	**100 kHz**	**150 kHz**
4-QAM	38,4	76,8	153,6	230,4
16-QAM	76,8	153,6	307,2	460,8
64-QAM	115,2	230,4	460,8	691,2

Error Correction (FEC) and 2. Kbps is for signalling and control functions. On the other Hand, Phase 2 was developed for better spectrum utilization. It uses TDMA with two slots for medium access. A more efficient voice codec that require 6 Kbps for voice data, error correction and signalling.

In this study, P25 will not be investigated due to the similarity with TETRA and also because it has the same performance aspects of TETRA that need to be compared with the more generic proposed solution that will be presented in the next subsections.

LONG TERM EVOLUTION (LTE) AS A MCS

Introduction to LTE

The Long Term Evolution (LTE) standard was developed by the Third Generation Partnership Project (3GPP) (3rd Generation Partnership Project, 2008), it started in Release 8 and then Release 9 till Release 10 for LTE-Advanced. It is considered a normal evolutional step of the mobile technologies but using revolutionary communication techniques at the physical layer to allow higher bandwidth and less latency for the end users. It emerged due to the fact that users nowadays are more mobile, need more bandwidth, and their running services need less latency for better interactive and hassle free performance. Moreover, LTE was designed for end-to-end all-IP-Connectivity between end users for simpler design architecture and to avoid the multi-domain conversion added overhead.

There are two parts of LTE Communication Network as shown in Figure 5; the first one is the Radio Access Network (RAN) part which is known as Evolved UMTS Terrestrial Radio Access Network (E-

Figure 5. General LTE architecture (Ferrús, Sallent, Baldini, and Goratti, 2013)

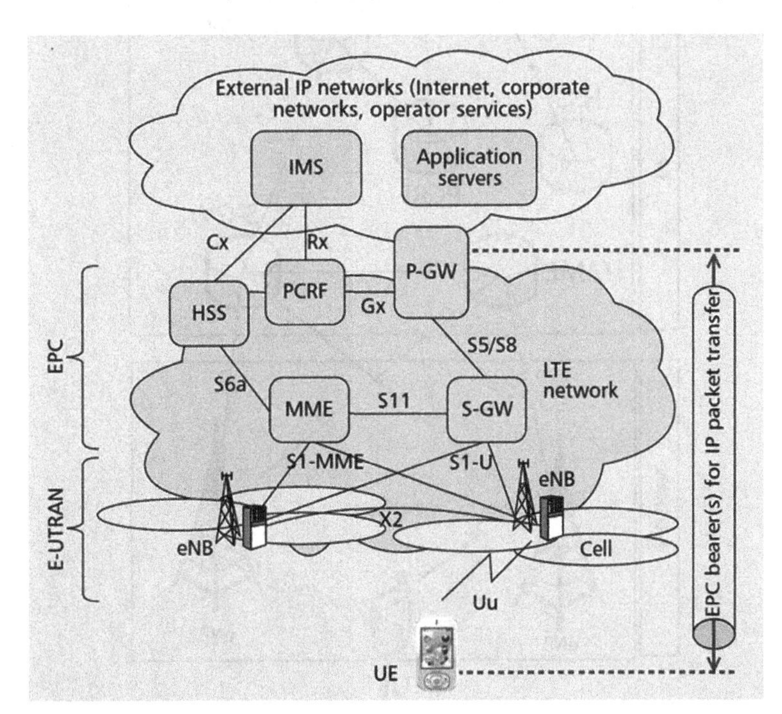

UTRAN) that has enhanced performance compared with Universal Mobile Telecommunications System (UMTS) that was used in the 3G communication networks. Evolved UMTS (E-UMTS) is mainly responsible for managing the whole radio stack signalling between the access point that is called evolved Node B (eNB) and User Equipment (UE). The other part is the core All-IP-Network part which is called the Evolved Packet Core (EPC) that is mainly responsible for managing the bearer services, mobility management, and interconnecting the interfaces and gateways with other domains and entities. Figure 5 shows the general architecture for LTE communication network

LTE as a Mission Critical System

LTE is supposed to be the first interaction point with the users as shown in Figure 6, it is a strong candidate to be considered for the Mission Critical and Public Safety Communication Systems due to its ability to provide the end users with the broadband services capacity and other MCS requirements (Doumi, et al., 2013) . Delay and bandwidth are very important because nowadays services types need more bandwidth and less latency for sufficient QoS and reliable service. There are many challenges that face the LTE as MCS deployment such as supporting DMO and providing the needed interfaces to insure interoperability with already existing MCSs, in addition to supporting group voice calls over

Figure 6. LTE as MCS

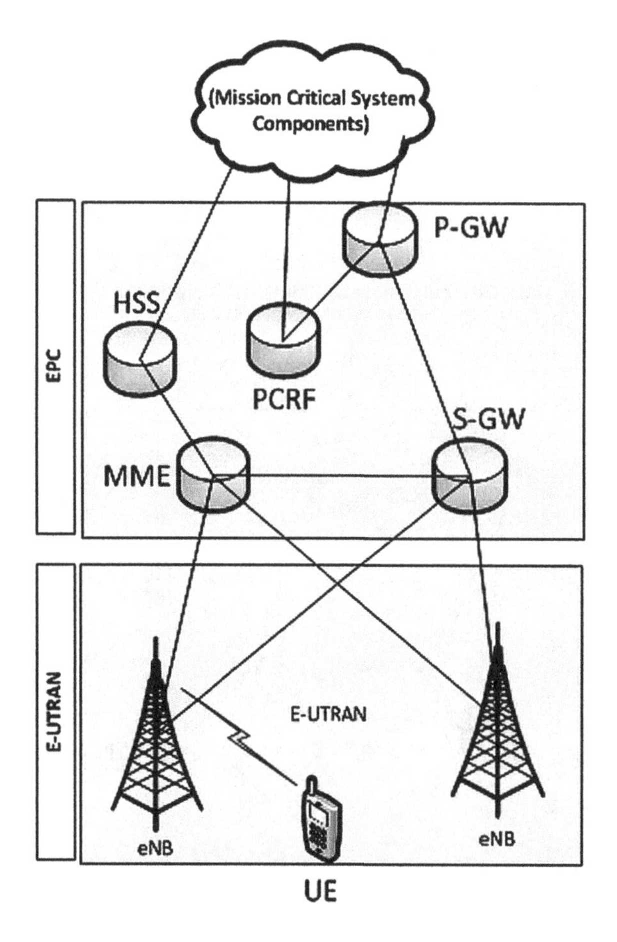

IP packet switched structure. But at the same time the All-IP-Network flat architecture is considered a big advantage for LTE to overcome the interoperability and complexity issues that may emerge in the proposed project. Moreover, the spectral flexibility and the higher spectral efficiency allows for more bandwidth utilization and better resource allocation for the end users which enhances the QoS provided, (Simic, 2012) highlights LTE capabilities in terms of providing different QoS classes at different levels.

LTE provides a set of service preferences for the user to meet a certain level of service requirements, and the requirements affect the scheduling and queueing user data, priority and preemption capabilities, and the access control treatment. All the requirements above are out of the scope of the bearer domain and managed by the access technology. At the access level, there are 16 priority classes that may be dedicated to public safety users to overpass the network overload access issues and at the network level, the Evolved Packet System (EPS) bearer, which is a logical channel between the UE and the P-GW at the far edge of the EPC, has two types; the first one is the Guaranteed Bit Rate (GBR) bearer where the user has a reserved resource during admission, and the Non-Guaranteed Bit Rate (None-GBR) bearer that is based on best effort service without guarantees. There is also the Allocation ad Retention Priority (ARP) which determines the priority class of the connected bearer in addition to two other flags; the preemption capability flag, that determine if it is allowed to preempt another lower priority bearer. The pre-emption vulnerability flag, which determines if the bearer may be preempted by another higher priority bearer. The APR will facilitate the decision of managing the bearer connection in overload conditions. In order to run MCS, the LTE needs to be coupled with the IP Multimedia Subsystem (IMS) to create an environment capable of supporting voice and video traffic in a shared packet data network. The next section speaks about IMS and its coupling with LTE.

REFERENCES

Balachandran, K., Budka, K.C., Chu, T.P., Doumi, T.L., & Kang, J.H. (2006, January). Mobile Responder Communication Networks for Public Safety. *IEEE Communication Magazine*.

Balachandran, K., Budka, K.C., Chu, T.P., Doumi, T.L., Kang, J.H., & Whinnery, R. (2005). *Converged Wireless Network Architecture for Homeland Security*. Military Communications Conference, IEEE MILCOM2005, Atlantic City, NJ.

Balachandran, K., Budka, K. C., Chu, T. P., Doumi, T. L., & Kang, J. H. (2006). Mobile responder communication networks for public safety. *IEEE Communications Magazine*, *44*(1), 56–64. doi:10.1109/MCOM.2006.1580933

Baldini, G. (2012). The Evolution of Public Safety Communications in Europe: the results from the FP7 HELP project. *ETSI Reconfigurable Radio Systems Workshop*.

Baldini, G., Karanasios, S., Allen, D., & Vergari, F. (2014). Survey of Wireless Communication Technologies for Public Safety. *IEEE Communications Surveys and Tutorials*, *16*(2), 619–641. doi:10.1109/SURV.2013.082713.00034

Blom, R., de Bruin, P., Eman, J., & Folke, M. (2008). Public Safety Communication Using Commercial Cellular Technology. *The Second International Conference on Next Generation Mobile Applications, Services and Technologies*. doi:10.1109/NGMAST.2008.78

Blom, R., de Bruin, P., Eman, J., Folke, M., Hannu, H., Naslund, M., & Synnergren, P. (2008). Public Safety Communication using Commercial Cellular Technology. In *International Conference and Exhibition on Next Generation Mobile Applications, Services, and Technologies*. doi:10.1109/NGMAST.2008.78

Budka, K. C., Chu, T., Doumi, T. L., Brouwer, W., Lamoureux, P., & Palamara, M. E. (2011). Public safety mission critical voice services over LTE. *Bell Labs Technical Journal*, *16*(3), 133–149. doi:10.1002/bltj.20526

Das, S. K. (2006). *Feasibility study of IP Multimedia Subsystems (IMS) based Push To Talk over Cellular for Public Safety and Security Communications* (Master's Thesis). Department of Electrical and Communication Engineering. Helsinki University of Technology (HUT).

Doumi, T., Dolan, M. F., Tatesh, S., Casati, A., Tsirtsis, G., Anchan, K., & Flore, D. (2013). LTE for public safety networks. *IEEE Communications Magazine*, *51*(2), 106–112. doi:10.1109/MCOM.2013.6461193

Europian Telecommunication Standardization Institution. (n.d.). Retrieved from http://www.tandcca.com/about/page/12024

Ferrús, R., Sallent, O., Baldini, G., & Goratti, L. (2013). LTE: The technology driver for future public safety communications. *IEEE Communications Magazine*, *51*(10), 154–161. doi:10.1109/MCOM.2013.6619579

Forge, S., Horvitz, R., & Blackman, C. (2014). *Is Commercial Cellular Suitable for Mission Critical Broadband? Academic Press.*

IPWireless. (2012). *LTE addressing the needs of the Public Safety Community*. 3GPP RAN Workshop on Rel-12 and Onward RWS-120030.

McGee, A. R., Coutière, M., & Palamara, M. E. (2012). Public safety network security considerations. *Bell Labs Technical Journal*, *17*(3), 79–86. doi:10.1002/bltj.21559

Project 25. (1995). *Project 25 System and Standard Definition, TIA, TSB102-A*. Retrieved from http://www.project25.org

Public Safety mobile broadband and spectrum needs. (2010). *Report for the TETRA Association.*

Simic, M. B. (2012). Feasibility of long term evolution (LTE) as technology for public safety. In *2012 20th Telecommunications Forum (TELFOR)*, 158–161. http://doi.org/ doi:<ALIGNMENT.qj></ALIGNMENT>10.1109/TELFOR.2012.6419172

Terrestrial Trunked Radio (TETRA). (2003). *Voice plus Data (V+D); Part 2: Air Interface (AI), ETSI, EN 300 392-2 v2.3.10*. Author.

TETRA and Critical Communication Association. (2010). Retrieved from http://www.tandcca.com/

TETRA MoU Association. (2004). *Push To Talk over Cellular (PoC) and Professional Mobile Radio (PMR)*. TETRA.

Third Generation Partnership Project, Organization, 3GPP. (2008). *3rd Generation Partnership Project;Technical Specification Group Radio Access Network*. Evolved Universal Terrestrial Radio Access(EUTRA).

Chapter 2
IP Multimedia Subsystem and SIP Signaling Performance Metrics

Ashraf A. Ali
The Hashemite University, Jordan & University of South Wales, UK

Khalid Al-Begain
University of South Wales, UK & Kuwait College of Science and Technology, Kuwait

ABSTRACT

In this chapter, the most important protocols that controls the operation of the end-to-end system and has implications on the overall performance aspects will be presented. Then, performance issues will be demonstrated. the performance issues and the challenge of enhancing SIP services performance will be highlighted and briefly discussed. The protocol related performance metrics will be identified to determine the way SIP is utilizing the system resources and how to maximize it. Moreover, the architectural design challenges will be targeted to enhance the SIP performance.

INTRODUCTION

In this section, the most important protocols that controls the operation of the end-to-end system and has implications on the overall performance aspects will be presented. Then, performance issues will be demonstrated.

There are mainly two types of protocols; the control plane set of protocols and the data plane protocols. The Session Initiation Protocol (SIP) is one of the control domain protocols and was fully standardized and specified by Internet Engineering Task Force (IETF) in RFC 2543 for the first version SIP 1.0 and in RFC 3261 for the second version SIP 2.0 (Rosenberg, et al., 2002). SIP operates over IP protocol and considered as a communication protocol for signalling of real-time multimedia services such as voice and video and non-real-time services such as text messages and presence notifications. The protocol, which is text-based, mainly defines the signalling order between end users for call initiation, termina-

DOI: 10.4018/978-1-5225-2113-6.ch002

tion, in addition to modifying the call setup instantly during the call. It is also used for registering the users before the call being initiated.

In this report, the SIP message headers and signalling details will not be presented. But the performance issues and the challenge of enhancing SIP services performance will be highlighted and briefly discussed. Figure 1 shows the signalling diagram for registering users, initiation and termination of a call between caller, callee, and back to back SIP server.

SIP PERFORMANCE ISSUES

The protocol related performance metrics need to be identified to determine the way SIP is utilizing the system resources and how to maximize it. Moreover, the architectural design challenges need to be targeted to enhance the SIP performance. Some of the protocol-related metrics in addition to implementation related metrics is discussed in (On SIP Performance, 2004), it shows different set of tests that measures the processing time for SIP messages, memory allocation, thread performance, and call-setup time. The results show that the performance of the proxy server changes by varying SIP related parameters and thus affecting the number of calls by seconds that the proxy server can handle at a time. It also shows that the performance of SIP related architectures, such as the IP Multimedia Subsystems (IMS) that will

Figure 1. SIP signalling diagram

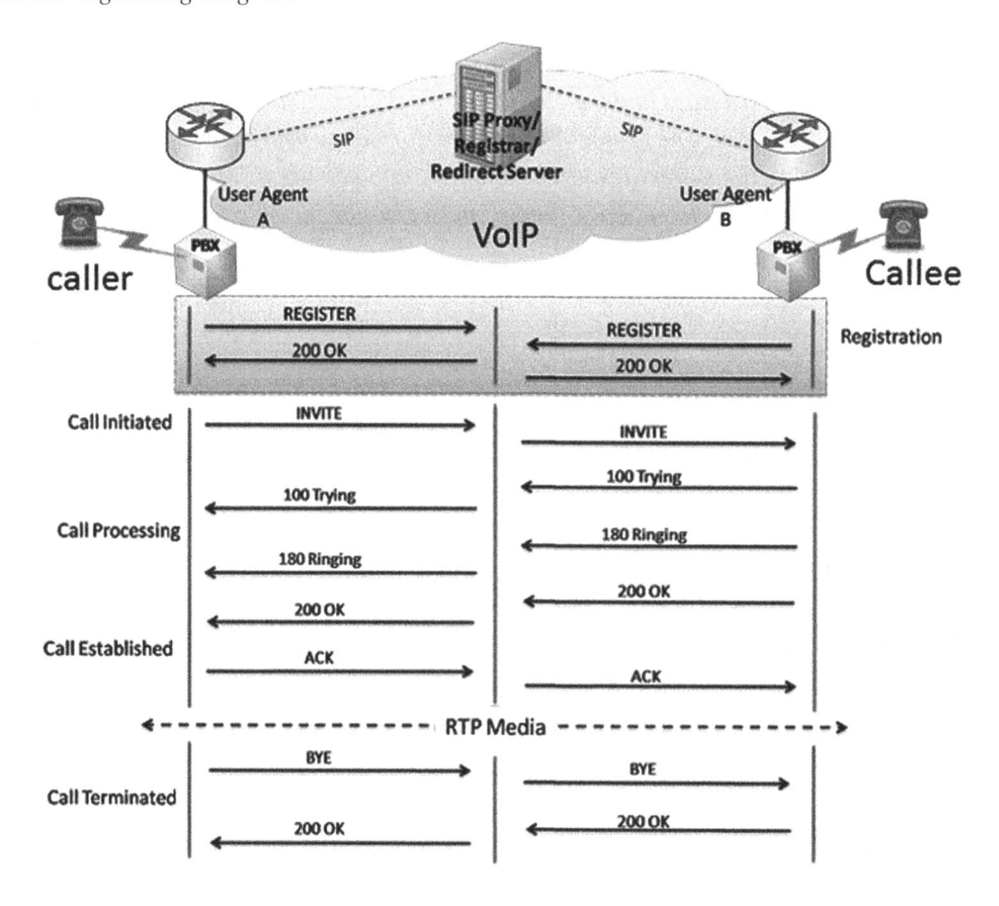

be presented later, is more affected due to the heavy dependence of SIP signaling and SIP messages structure compared with a simple Proxy/Registrar Server. Furthermore, it is important to note that the performance of SIP signalling is significantly affected by the delay at different stages of registration, call initiation, and call termination processes. The performance of SIP signalling will also affect the QoS of the offered service. Hence, the need to define the metrics that identify the performance measure for SIP is crucial for evaluation and performance comparison purposes.

IETF proposed the criteria for the end-to-end SIP performance measures in RFC 6076 (Malas & Morton, 2011).

Due to the lack of a SIP benchmarking numbers to define the baseline performance of SIP signaling, RFC 6076 defines the performance metrics for SIP in VoIP applications to provide Key Performance Indicators and Service Level Agreement (SLA) indicators for best network resources utilization and best end user Quality of Experience (QoE). As shown in Figure 2, the main metrics defined in RFC 6076 are the Registration Request Delay (RRD), Session Request Delay (SRD), and Session Disconnect Delay (SDD). RRD is the time needed for the user to finish the registration process successfully. SRD is the time needed to get a reply from the server side regarding the requested call setup from the user side; it

Figure 2. SIP signalling flow and performance metrics

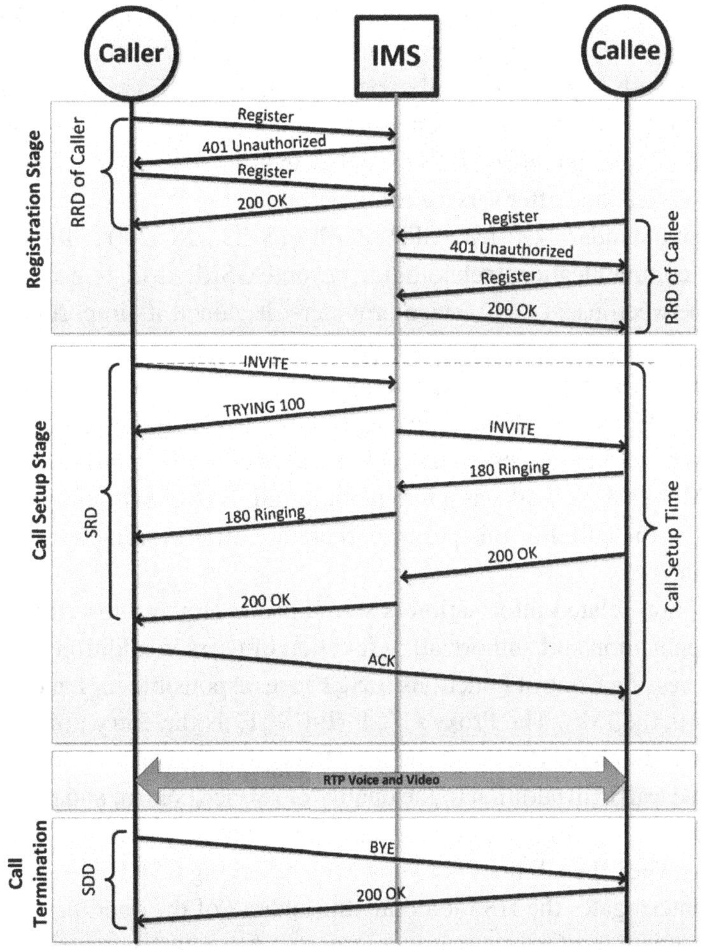

is counted for both successful and unsuccessful call requests. If the call requested was successfully set then the call setup time will be simply SSD in addition to acknowledgment sending time. The SDD is the time difference between sending BYE message from the user side and the time of receiving 200OK confirmation from the server.

In this report, the call setup delay will be used to measure the system performance due to its importance in real-time multimedia services in general and in mission-critical communications specifically. The QoE for SIP-based systems is significantly affected by the call setup value. Based on (ITU-T, 2004) the call setup time means value can be up to 800 ms. However for LTE-based Mission Critical Systems which are supposed to work as a replica for traditional dedicated Mission Critical Systems such as TETRA, the call setup time need to be within 500 ms delay.

IP MULTIMEDIA SUBSYSTEM (IMS)

Due to the evolution of legacy mobile communication networks and the next generation systems which are able to provide end users with new set of applications and services, the need of having a system working on top of the access technology domain for providing the needed signaling for multimedia services is of great importance, IP Multimedia Subsystem (IMS) (Technical Specification Group Services and System Aspects, 2006) is considered the core networks for Next Generation Networks (NGNs) plays important role in helping the service operators in merging the multimedia services in cellular networks providing the end users with key features and services within certain Quality of Service (QoS) levels set by operators. The multimedia services such as messaging, instant voice, video conferencing, group management, and push services all rely on IMS to control the signaling and the state of the service before initiation, during the service, and after service termination.

IMS is designed and standardized by 3GPP (3GPP TS 23.228, 2005) for providing multimedia services over mobile communication technologies beyond GSM. IMS is used for delivering the IP multimedia services between users and service providers. It gained its importance as an architectural framework for multimedia communications. SIP is the primarily signalling protocol used within IMS to create, modify and destroy multimedia sessions. Figure 3 shows the IMS, architecture model. As defined by the standard, the IMS is operating as an interface between the service/application layer and the transport layer which enables the service providers and operator to control the user QoS based on its subscription profile. Moreover, it works as a hub point for all the SIP signalling that need to take place before, during, and after the call. For this purpose, there are different functions that are connected by interfaces to ensure integrity.

The user's subscription-related information is stored in the Home Subscriber Server (HSS) which also perform the authentication and authorization function of users in addition to registration status record update. The Call Session Control Functions (CSCF) are responsible for handling the SIP signalling messages and packets in the IMS. The Proxy CSCF (P-CSCF) is the entry point into the IMS system. All SIP messages flow through the P-CSCF. The P-CSCF may also apply security or compression algorithms over the received traffic in addition to the quality of service control and bandwidth management. The interrogating CSCF (I-CSCF) is one of the main elements of the IMS systems. It is used during the registration process when the UE does not know which Serving CSCF (S-CSCF) should receive the request. The I-CSCF interrogates the HSS to obtain the address of the appropriate S-CSCF that should process the request. S-CSCF performs session control that has interfaces with the HSS to check and

Figure 3. IMS architecture

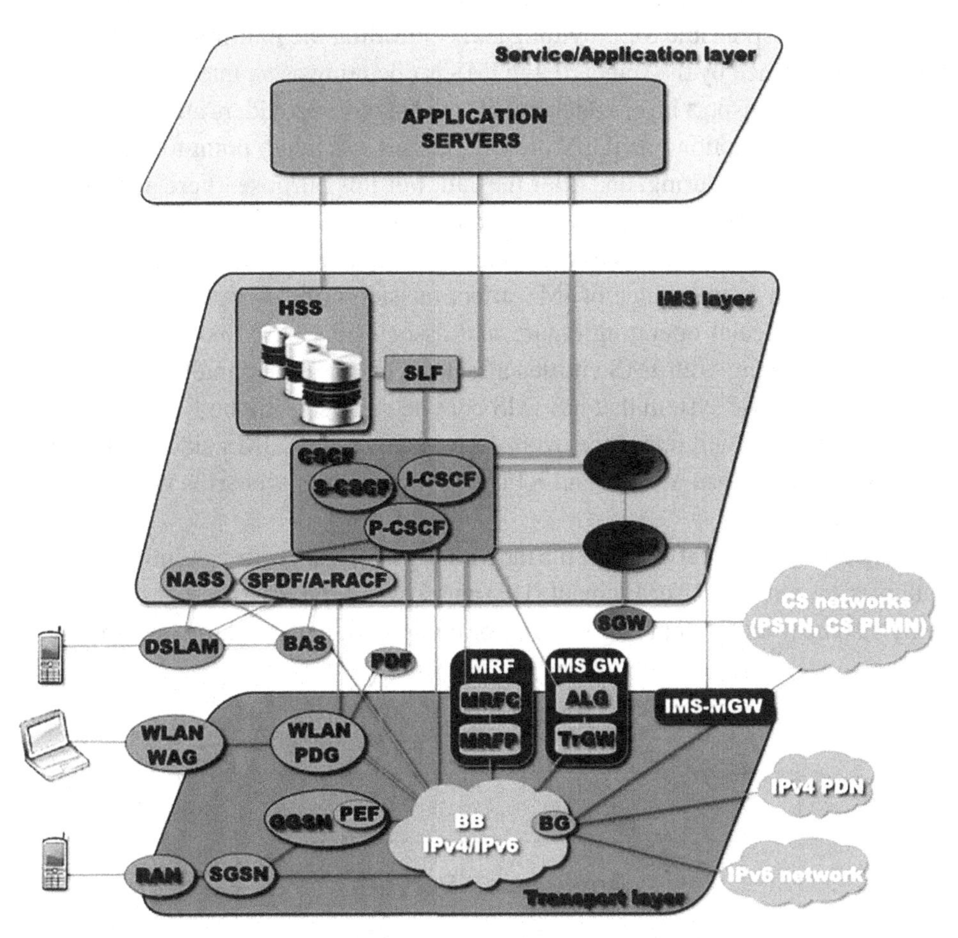

download the user profile information and also assign the Application Server (AS) for the user for further services in addition to enforcing the operator policy control. The SIP AS has an SIP interface with S-CSCF, and it is used to host specific IMS services. After the registration process is completed and the S-CSCF is allocated to the UE, the I-CSCF is no longer used for any further communication. All future communication happens between the UE, P-CSCF, and the S-CSCF.

The performance of the IMS is very critical in affecting the overall system performance due to its hierarchal position in the stack and core functional role. In the next sections, the performance of IMS will be investigated.

IMS PERFORMANCE ISSUES

After migrating from the circuit switched Second Generation Mobile Networks (2G) toward the packet switched domain in fourth generation (4G) communication networks and beyond, the need for supporting the multimedia services in all IP network infrastructure is essential to ensure the convergence of data, multimedia services, and mobile networks technologies. IP Multimedia Subsystem (IMS) is being developed by Third Generation Partnership Project to be considered as a key part for Next Generation Networks (NGN) that is responsible for providing and controlling the multimedia services in the packet switched domains. As defined by the standard, the IMS is operating as an interface between the service/ application layer and the transport layer which enables the service providers and operator to control the user QoS based on its subscription profile. Moreover, it works as a hub point for all the SIP signalling that need to take place before, during, and after the call. For this purpose, there are different functions that are connected by interfaces to insure the operation integrity.

In the hierarchal stack, IMS resides between the application layer and the transport layer as shown in Figure 4. Therefore, the performance of IMS affect mission critical applications which has a strict requirements due to its special operating nature and associated tasks. End-end Delay of the access technology components along with IMS entities affect the end-to-end signalling and QoE offered for the end user. Having a generic system that has IMS core network serving both legacy mobile users and mission critical system users will introduce overload especially at the IMS side. In this report, evaluation of IMS and SIP performance metrics and KPIs was investigated scenarios with increasing number of clients was investigated

The performance of the IMS is affected by the individual performance of each entity inside it; (P-CSCF) is the first that interacts with the User Element (UE) and forward the SIP messages to other functions in the IMS, it may also be used for applying security or compression algorithms over the received traffic

Figure 4. IMS charging functions

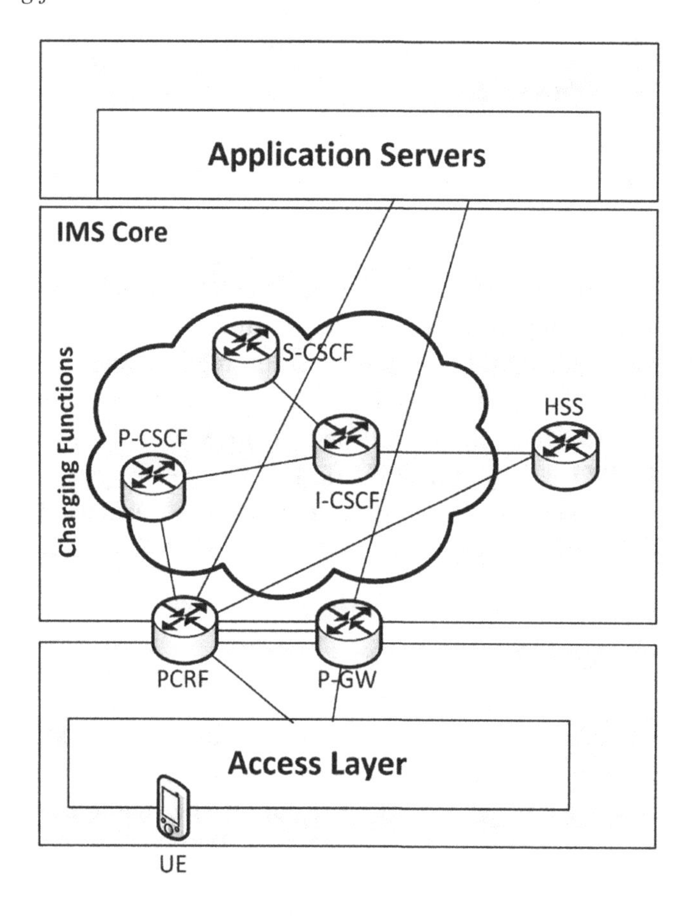

in addition to quality of service control and bandwidth management. The I-CSCF is a SIP server that is assigned by the HSS to the user when it is enquired by the I-CSCF, S-CSCF performs session control that has interfaces with the HSS to check and download the user profile information and also assign the Application Server (AS) for the user for further services in addition to enforcing the operator policy control. The SIP AS has a SIP interface with S-CSCF, and it is used to host specific IMS services. There are also gateways that are working as interfaces with other domains.

IMS Registration Performance

In the IMS registration, the user request authorization to gain access to the IMS services. Figure 5 shows the registration steps in IMS. First, the UE sends SIP Register request to the P-CSCF, which forwards the SIP Register request to the I-CSCF. Then the I-CSCF send User Authentication Request (UAR) over Diameter protocol to the Home Subscriber Station (HSS) to check the user profile for authorization process and to determine the S-CSCF address that is allocated to the user. Then, the HSS reply with User Authentication Answer (UAA) to the I-CSCF and authorize the user. After getting the S-CSCF form the HSS, The I-CSCF then forwards the SIP Registration request to the S-CSCF, which in turn sends Multimedia Authentication Request (MAR) to the HSS to download user authentication data. Then the HSS reply with Multimedia Authentication Answer (MAA) to the S-CSCF, which reply to the user a SIP 401 Unauthorized response that embed a challenge for the UE. UE generate another SIP Register request following the same steps described before but the only difference is when the authentication this time is successful, the S-CSCF sends Server Assignment Request (SAR) over Diameter to the HSS which replies back with Server Assignment Answer (SAA) over Diameter. Finally, the S-CSCF sends SIP 200OK message to the UE to complete the registration process.

As mentioned before, the Registration Request Delay (RRD) is the time needed for the registration process to complete, which is in this case the time difference between sending the first SIP Register message by the UE until the reception of SIP 200 OK message.

During the registration process, the end user need to be subscribed to the IMS to be authorized for making calls and being able to initiate other services instantly during the call. The Registration process is needed before initiating calls as shown in Figure 6, and re-registration is needed for the SIP device to re-authenticate itself after receiving a notification from the SIP server, this is needed in order to avoid compromising the SIP device.

The delay in the registration process is one of the SIP performance metrics and criteria that was defined by IETF in RFC 6076 (Malas and Morton, 2011), Due to the lack of a SIP benchmarking numbers to define the baseline performance of SIP signalling, RFC 6076 defines the performance metrics for SIP in VoIP applications to provide a key performance indicators and Service Level Agreement (SLA) indicators for best network resources utilization and best end user Quality of Experience (QoE). The main metrics defined in RFC 6076 are the Registration Request Delay (RRD), Ineffective Registration Attempts (IRA), Session Request Delay (SRD), and Session Disconnect Delay (SDD). RRD is the time needed for the user to finish the registration process successfully and plays a major role in the overall QoS.

In addition to the delay of Registration process, the rate at which registration messages are being sent to the server is of great importance especially in Mission Critical operations where the number of users or devices that are expected to request for access to the SIP/IMS server may be of a large number that affects or even deteriorates the performance of the provided service. In (Rankin, Costaiche, & Zeto, 2013) different real-world scenarios for registering users/devices over LTE are presented; the first sce-

Figure 5. IMS registration process

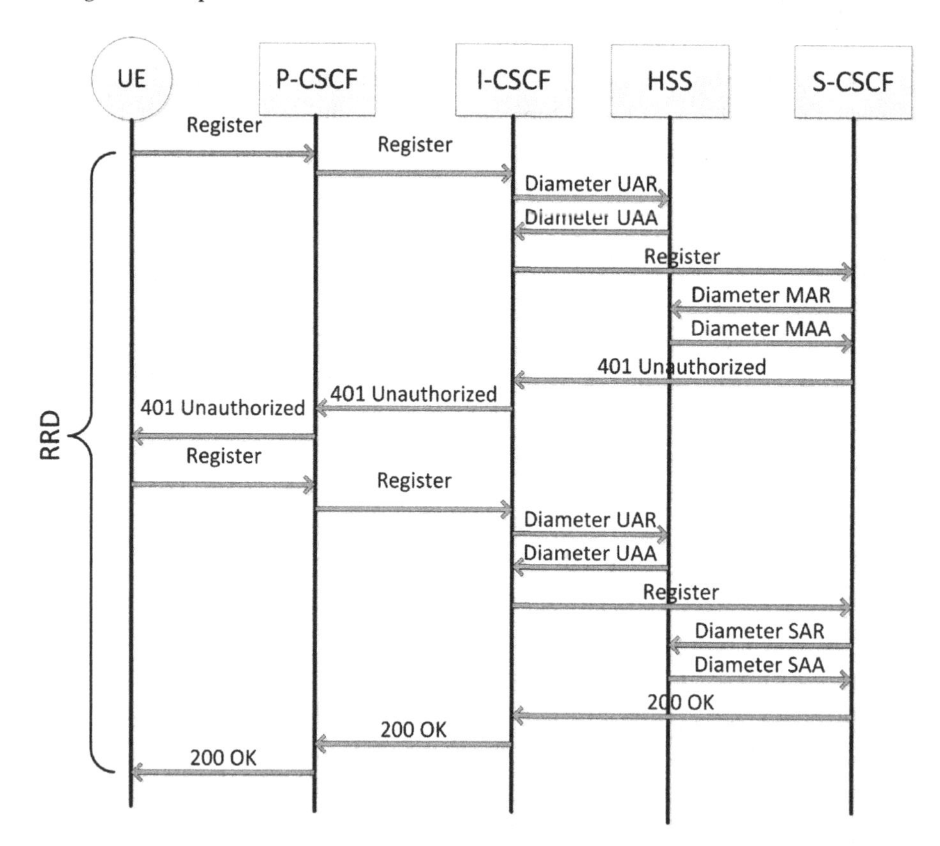

Figure 6. User registration with IMS

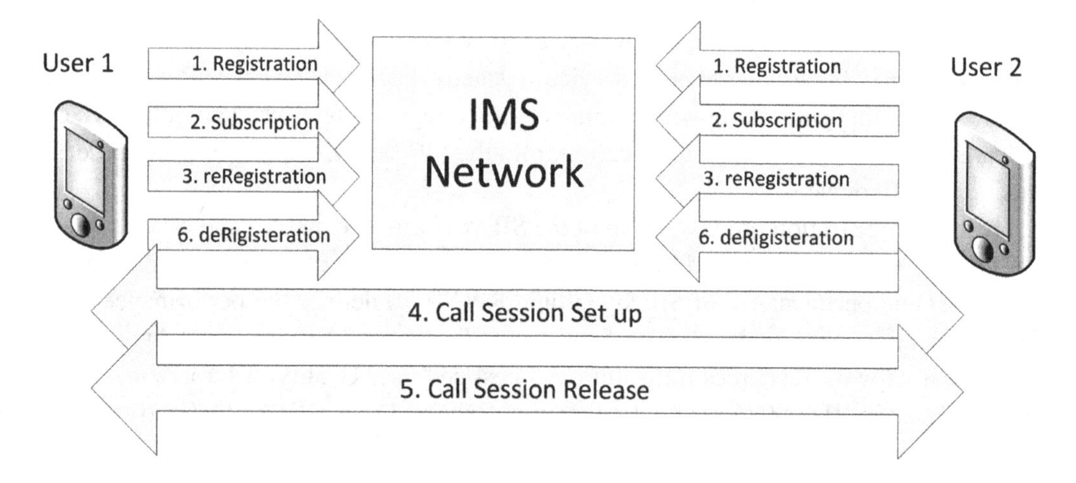

nario is for Airbus A380 landing at Heathrow airport with 1000 passengers each is commencing IMS Registration over LTE at a rate of 10 Registrations per second. In the second scenario, a regional power outage is restored at a big city which causes a storm of IMS registration requests to the network, 100 registrations per second is expected in this scenario. In both scenarios, it is important to handle all the requests to successfully finish the registration process within acceptable time limits regardless of the received traffic especially over a Mission Critical system that is by nature should be highly available, resilient and reliable.

PoC is one of the mission critical services that is supported by IMS, it is supposed to work as a replica of Push To Talk (PTT) service in the conventional dedicated mission critical communication systems such as TErristrial Trunked Radio (TETRA) networks in and P25. Therefore, the investigation of the IMS added delay in the registration process is of great importance for mission critical services which is less delay tolerant especially if operating over generic communication system that is not only dedicated for mission critical communications.

In (Arsalan, & Ann, 2010) analysis of IMS registration over WiMax and 3G interworking architectures shows that the access technology along with IMS design affect the registration signalling delay, in (Nadir, Sarhan, Heng, & Chaitra, 2012) a performance analysis of IMS signalling, including registration signalling, in multimedia networks, the study shows that the maximum signalling delay is affected by both the number of User Elements (UEs) and the available bandwidth available in the core network, however it was conducted for up to 400 users only. In (Daniel, David, Andres, Florina, Peter, 2010) a general IMS registration protocol for wireless network was proposed to reduce the registration delay via combining both the registration processes with the access technology and IMS network. In (Farahbaksh, Varposhti, & Movahhedinia, 2007) the re-registration procedure modification was proposed to reduce the transmission delay in IMS, re-registration is needed to refresh the registration state of the user or to update the user registration profile due to change in user capabilities. In (Jiri, Lubos, Vit, Paval, & Domonik, 2006) the impact of SIP signalling load over IMS was investigated, and it was found that the SIP signalling load, including registration, has a tremendous impact over the Key Performance Indicators (KPI) for IMS.

The target of this report is to investigate the delay in the registration process introduced by IMS for increasing number of UEs, the effect of increasing the load over the KPI and related QoE will be also presented.

SIP Key Performance Indicators

A set of SIP KPIs was defined in (3GPP TS 24.228, 2006) such as the Registration Request Delay (RRD) which is the difference between time of final successful registration arrival and the time of sending the registration request. Ineffective Registration Attempts (IRA) is a metric that indicate the ratio of unsuccessful registration attempts to the total number of registration requests sent. Session request Delay (SRD) is the time difference between the reception of status indicative response and the time of sending the SIP invite message. Session Disconnect Delay (SDD) is the time difference between sending the SIP Bye message and the reception of confirmation or timeout message. The Session Duration Time (SDT) is the time the time interval between the reception of 200OK SIP message of the Invite Request and at the originating UE till the reception of BYE timeout message. Session Establishment Ratio (SER) which defines the ability of terminating the User Agent or Proxy Server during the session establishment. Session Establishment Effectiveness Ratio (SEER) is the ration between the number of SIP Invite Requests that result 200OK response and number of SIP Invite Requests that result other responses than 200OK. The Ineffective Session attempts (ISAs) is a metric that is calculated as a ratio of failed session setup requests to the total number of session requests. Finally, the Session Completion Ratio (SCR) is the ratio between successful completed sessions over the total number of sessions. In this report, we are interested in both the RRD and IRA as a KPI metrics that reflect the performance of the SIP Registration process.

IMS Key Performance Indicators

The IMS performance via test-beds is defined via European Telecommunications Standards Institute (ETSI) in (ETSI TS 186 008, 2012) and by 3GPP in the technical standard (3GPP TS 32.409, 2012) which defines a set of KPIs for IMS that are classified as follows:

- **Accessibility KPIs**: Metrics that reflect the accessibility of IMS for the users
  - Initial Registration Success Rate (IRSR) of S-CSCF reflect the accessibility performance provided by IMS and it describes the percentage of the number of User successful initial registration (UR.SuccInitReg) of S-CSCF to the number of attempted User initial registrations (UR.AttInitReg) based on the following equation:

$$\text{IRSR} = \frac{\sum_{s-cscf} \text{UR.SuccInitReg}}{\sum_{s-cscf} \text{UR.AttInitReg}} * 100\% \quad (1)$$

- **Mean Session Setup Time**: This KPI is for the mean time of session setup starting from sending the SIP Invite message to the P-CSCF till the reception of 200OK.
- **Session Establishment Success Rate**: Calculated by I-CSCF as the ratio of the number of successful session initiation requests to the total number of session initiation attempts. The ration of the number of successful session termination attempts to the total session termination attempts.

$$\text{SESR\_Orig} = \frac{\sum_{Type} \text{SC.SuccSessionOrig}.type}{\text{SC.AttSessionOrig}} \quad (2)$$

$$\text{SESR\_Term} = \frac{\sum_{Type} \text{SC.SuccSessionTerm}.type}{\text{SC.AttSessionTerm}} \quad (3)$$

  - Type is for SIP 180 and SIP 200OK message types
- **Third Party Registration Success Rate**: This KPI is obtained by successful third party registration procedures divided by attempted third party registration procedures.

$$\text{TPRSR} = \frac{\sum_{s-cscf} \text{UR.Succ3rdPartyReg}}{\sum_{s-cscf} \text{UR.Att3rdPartyReg}} * 100\% \quad (4)$$

- **Re-Registration Success Rate of S-CSCF**: The ratio of successful re-registrations of S-CSCF to the total number of attempted re-registrations.

$$\text{RRSR} = \frac{\sum_{s-cscf} \text{UR.SuccReReg}}{\sum_{s-cscf} \text{UR.AttReReg}} * 100\% \quad (5)$$

- **Mean Session Setup Time Originated from IMS (MSSTOI):** The mean setup time of successful originated call from IMS.
- **Retainability KPI**: It has one value which is Call Drop Rate of IMS Sessions which is calculated as ratio of dropped sessions to the number of successful session establishments.

$$\mathrm{SEDR} = \frac{\mathrm{SC.DroppedSession}}{\sum_{type} \mathrm{SC.SuccSession}.type} \tag{6}$$

- **Utilization KPI**: Only one value for Mean Session Utilization (MSU) that is calculated as the percentage of mean number of simultaneous online and answered sessions to the maximum number of sessions provided by IMS network.

$$\mathrm{MSU} = \frac{\mathrm{SC.NbrSimulAnsSessionMean}}{\mathrm{Capacity}} \times 100\% \tag{7}$$

The QoE, which is the quality of the service as perceived by the end user, is affected by the KPI values of both the SIP and IMS. In order to improve the QoE, the related KPI values need to be continuously monitored and improved. The KPI measure, if combined together, will help in determining the failure points in the IMS network and how to improve it to avoid the bottleneck and single point failure in case of traffic overload scenarios.

The KPI may also be used Load Detection Function (LDF) which is described in (3GPP TR23.812, 2013). The LDF may be used to support load balancing for the S-CSCF and, therefore, reduce the service request latency.

VOICE APPLICATIONS

There are different types of applications and applications data types that are used in the mission critical communication system. Voice, video, and push messages are some of many applications that are considered killer applications due to the bandwidth requirements especially at the core and access point side which are considered as a bottle neck for the overall system. Push To Talk (PTT) which, as the name implies, enable the caller to push a button before sending the voice burst to the callee side uses half duplex channels to save the bandwidth and minimize the bandwidth capacity and hence users scalability concerns that may emerge due to tow way or full duplex communication methods. Therefore, for the broadband mobile communication systems the concept of having PTT over Cellular (PoC) was adopted for the same reason mentioned before. However the adoption of PoC introduced additional delays especially if it was applied over LTE, (Over & Etworks, 2012) shows the delay analysis of PoC over LTE.

Voice over IP (VoIP) is used for the delivery of voice bursts over IP networks. Due to the emerging of All-IP-Networks, VoIP has gained a great importance to be considered on top of real-time streaming protocols such as RTP as the best voice streaming solution for voice calls.

VoIP services are considered Bandwidth killing applications for mobile broadband Access Technologies which provides the end users with broadband services similar to that provided by DSL services in the traditional wired Networks. However, the wireless access introduces additional challenges and constraints on the data rate and delay requirements. Therefore, the need for bandwidth requirements study for VoIP is needed to make sure that the mission critical service that uses VoIP will operate as expected.

It is important to note that since VoIP is the killer application of Broadband access technologies, knowing the bandwidth to support the largest number of user within acceptable QoS guarantees is of a great importance.

PROBLEM DEFINITION

Based on what we discussed before, there are different deployment options for the mission critical systems, either to have a dedicated system or a commercial systems. Unfortunately, there are drawbacks for both options and many trade-offs that control the design and deployment process. What is important to know is that at the early deployments of dedicated mission critical systems, the voice and simple text messages were the only services that were used by PPDR for their critical missions, simply because mobile communication networks was at its early steps of evolutionary progress where the operators only thought of providing voice communication between users. Hence, when the dedicated mission critical communications was deployed, they also designed the whole system for voice and low data rate services which was fair enough at that time for providing the basic communication platform for the clients. However, due to the exponential increase of mission critical data use to support different services and tasks, the need for broadband data along with the traditional basic voice service has become crucial for the mission critical communication systems.

Due to the increased demands for more bandwidth for the mission critical services, the need for access technology that is capable of providing broadband connectivity and scalable for high number of users along with broadband network support in the core is considered crucial requirement for any mission critical communication system.

SCOPE OF THE STUDY

As shown in Figure 7, the study will focus on the cross-layer optimization of SIP signalling between the LTE-IMS interface in addition to IMS-Application Server interface. Optimization for less delay is needed to enhance the QoE for the mission critical services.

CHALLENGES AND PROPOSED SOLUTIONS

Access Technology Challenges

As discussed before, there is an increasing need and demand for a more bandwidth for the mission critical services. This implies that the system not only need to satisfy the general requirements for the mission critical services that we have discussed before, but also to be a broadband capacity system with a special requirement to be scalable for the maximum number of users with their broadband services running as it should be during the crises and worst case scenarios. It is important to keep in mind that the mission

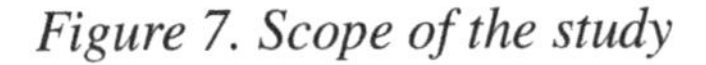

Figure 7. Scope of the study

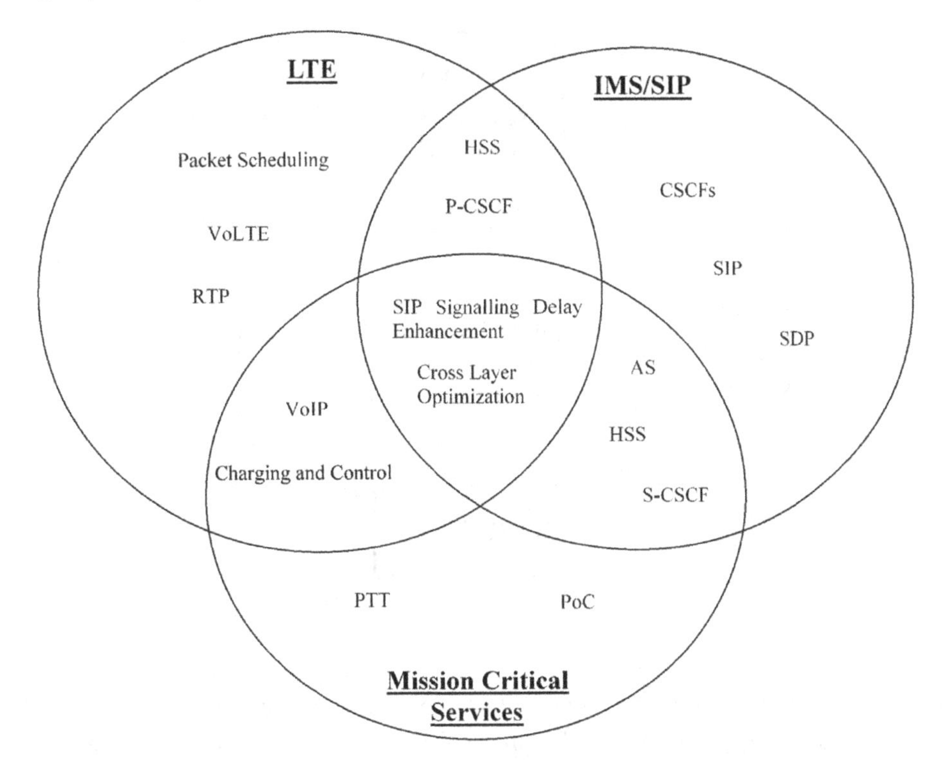

critical communication system is an all-time-available system that should be design for the worst case ever scenarios not for the normal daily routine PPDR tasks.

Increasing the bandwidth of the system is not a simple task in the physical layer of any communication system. As shown in Figure 8 there are multiple factors and trade-offs that complicates the evolution process of any communications system.

From business perspectives, the mobile network operators need to utilize every Hertz of the bought licensed frequency band to ensure sustaining the maximum profit and getting the most out of the bought frequency. Technically speaking, the frequency spectrum should be best utilized to have the highest possible bandwidth to support the broadband services. Hence, the need to deploy methods for the best frequency exploitations such as: adaptive modulations, Multiple Input Multiple Output (MIMO) Antennas, Orthogonal Frequency Division Multiple Access (OFDMA), Code Division Multiple Access (CDMA), Frequency Division Duplex (FDD), Time Division Duplex (TDD) and many other technologies in both physical and MAC layers.

As shown in Figure 8 there are trade-offs between the frequency, transmission power and coverage area or cell size. Also, there is another trade-off between the frequency and maximum system throughput based on the Bit Error Rate value. Finally, the system throughput controls the downlink and uplink capacity of the resource blocks which limits the maximum number of users that may use the system at the same time or the scalability of the whole system.

Nowadays we have an operational 4G mobile communications networks, Long Term Evolution (LTE), which is standardized by the 3GPP project, is widely used as 4G communication standard. LTE will be presented as a public safety system in the next section. But it is important to know for now that all the

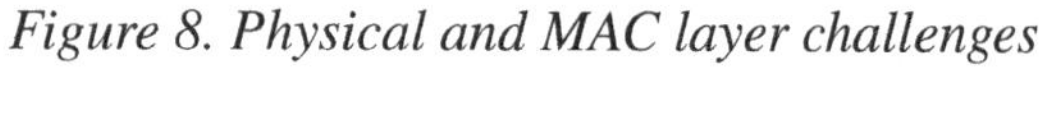

Figure 8. Physical and MAC layer challenges

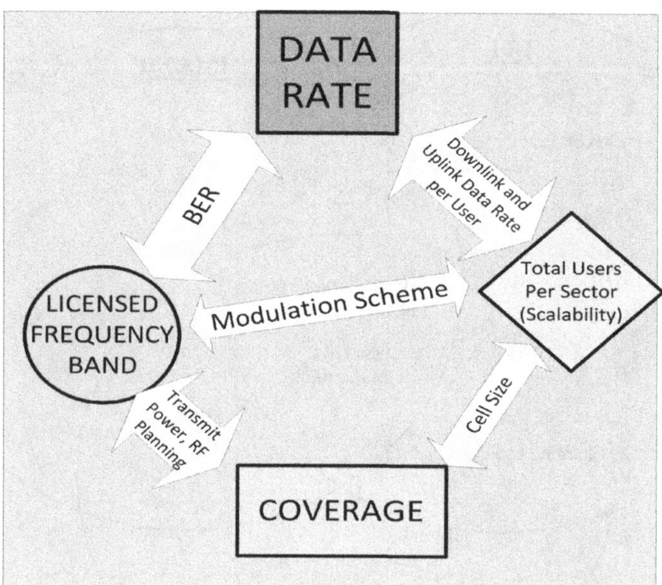

trade-offs that was introduced before are issues within the access technology abstract, and LTE or other technologies provided enough solutions in the physical and MAC layers to maximize the capacity of the system.

System Resources Sharing Challenge

Another challenge is to use commercial mobile communication networks as a public safety communication system. As mentioned before, we may either have a dedicated system or commercial system for the mission critical services. Clearly, it is considered a big challenge to have general purpose public, commercial communication system to work as a replica of a dedicated public safety communication system with all mission critical services strict requirements maintained and satisfied. Clearly, using the dedicated system will have all the communication system resources available for only the PPDR users. On the other hand, for the commercial public safety communication system, the resources will be shared between the PPDR and public, commercial mobile communication users. Therefore, for the public communication system the need to have a distinction between the mission critical services and other non-mission critical services in terms of service priority handling and resource allocation is considered a must to make sure that the system will have enough resources and thus available for the PPDR user whenever and wherever needed.

Part of this study will focus on the cross layer communications between different system abstracts to ensure that calls, such as mission critical emergency call, with higher priority class, will pass through the different entities of the system toward the end user and at the same time satisfying all the mission critical service requirements. In addition to that, the need to have a pre-emptive scheduling algorithm to decide which type of calls need to be dropped to flush out the allocated resources in cases where system capacity may reach to its maximum tolerable limits. Although the deployment of pre-emptive scheduling algorithms is out of the scope of this study, however a thorough evaluation of overall system capacity with different scenarios that combine the number of users along with the priority class of the service will be analysed and investigated.

System Architecture Design Challenge

Mission critical communication system that is composed of different entities and interfaces to ensure the end to end connectivity between end users and overall system integrity. The end users may be connected to different network domains and access technologies, hence the need for an interfaces and gateways that provides transparent connectivity between the different domains. In addition to that the need for standardized interfaces that connects devices regardless of the manufacturing vendors. Finally, the need for a data-plane and control-plane protocols that ensures the connectivity and data delivery between end users.

Figure 9 shows the general layered architecture model. Regardless of the details of any mission critical communication system, all systems share the same hierarchal architectural structure. End users are the ones who are holding the mobile device that is running the needed applications for initiating the needed services through the available access technology. The access technology that is being used to reach the core network can either be only one predetermined access technology or a more or any available access technology that is available at time for the user. Clearly, having more than one supported access technology to access the core network will support the reliability requirement of requested mission critical service. On the other hand, if the access technology being used is predetermined and fixed this will simplify the complexities and any added overhead to the overall system that may be introduced and affecting other mission critical requirements such as the end-to-end delay.

Regardless of the access technology that is being used, the need to have a common platform working between the access technology and the mission critical application servers is a must for two reasons. Firstly, the platform need to work as intermediate between both the control signalling domain and the data streaming signalling domain. Secondly, the platform is needed to provide seamless connectivity for the end-users along with the application servers regardless of the access technology being used. Not to forget mentioning that there is no clear distinction between the data domain and control domain signalling protocols. One of the main objectives for this study is to propose a cross layer optimization solutions between both the interfaces and protocols running in the mission critical system for the sake of enhancing the overall performance and its associated metrics.

In this study, a framework for the mission critical communication system will be proposed and explained. The different layers and interfaces for both data plane and control plane will be demonstrated. The layered architecture model of the system and the details for each layer will be investigated and demonstrated.

End-to-End QoS Challenge

Having different abstracts in hierarchal manner for the architecture of the system as described previously, will introduce not only integration challenges but also cross layer service performance optimization challenges to improve the quality of service for the whole end-to-end mission critical communication system and to satisfy the mission critical requirements mentioned previously. Therefore, in this study, a cross layer optimization and integration of the signalling used in both control and data domains will be proposed.

Figure 9. Layered abstract model

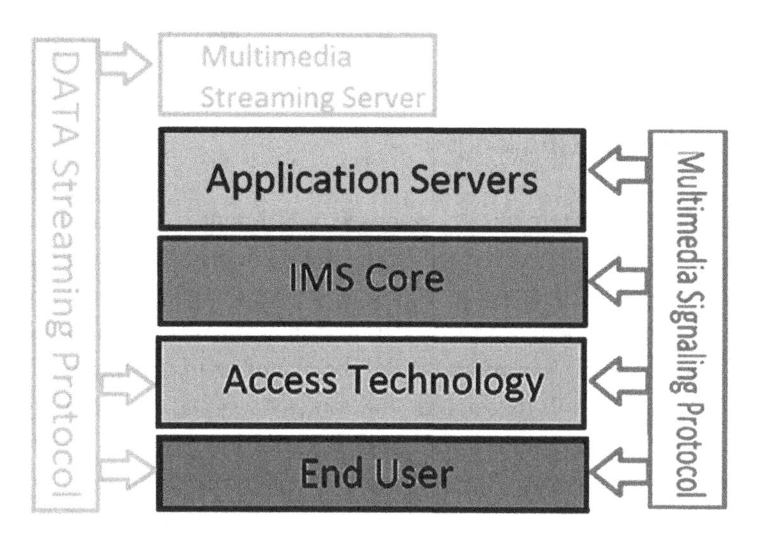

System Performance Modelling and Validation Challenge

Finally, to test the proposed framework and the overall proposed system architecture model in addition to validating the proposed cross layer optimization to enhance the end-to-end overall Quality of Service, the whole system will be modelled using simulation tools. The Quality of Service performance measures will be compared with the performance metrics of a test-bed for the mission critical system that will be used as a benchmark.

REFERENCES

3GPP TS 23.228. (2005). *Service Requirements for the Internet Protocol (IP) Multimedia Core Network Subsystem (IMS), Stage 1*. 3GPP.

3rd Generation Partnership Project. (2012). *Technical Specification Group Services and System Aspects, Performance Management (PM); Performance measurements; IP Multimedia Subsystems (IMS) (Release 11)*. 3GPP TS 32.409 (v 11.4.0). 3GPP.

3rd Generation Partnership Project. (2013). *Technical Specification Group Services and System Aspects. Feasibility study on IP Multimedia Subsystems (IMS) evolution (Release 11)*. 3GPP TR23.812 (v12.1.0). 3GPP.

Diaz-Sanchez, Proserpio, Marin-Lopez, Mendoza, & Weik. (2010). *A General IMS Registration Protocol for Wireless Network Interworking*. Academic Press.

ETSI, European Telecommunication Standard Institute, IMS Network Testing (INT). (2012). *IMS/NGN Performance Benchmark*. ETSI TS 186 008. Author.

Farahbaksh, R., Varposhti, M., & Movahhedinia, N. (2007). Transmission Delay reduction in IMS by Re-registration Procedure modification. *The Second International Conference on Next Generation Mobile Applications, Services and Technologies.*

Hosek, J. (2006). *Performance Analysis: Impact of Signalling Load over IMS Core on KPIs. Recent Advances in Circuits, Systems.* Telecommunications and Control.

ITU-T TR Q-series supplements 51 signaling requirements for IP-QoS. (2004). ITU-T.

Malas, D., & Morton, A. (2011). *Basic Telephony SIP End-to-End Performance Metrics*. Technical Report RFC 6076. Internet Engineering Task Force (IETF). Retrieved from http://tools.ietf.org/html/rfc6076

Mir, Musa, Gao, & Shivakumar. (2012). Performance Analysis of IMS Signaling Multimedia Networks. *Information Engineering, 1*(1).

Munir, A., & Gordon-Ross, A. (2010, May). SIP-Based IMS Signaling Analysis for WiMAX-3G Interworking Architectures. *IEEE Transactions on Mobile Computing*, *9*(5), 733–750. doi:10.1109/TMC.2010.16

On SIP Performance. (2004). *Bell Labs Technical Journal.*

Rankin, Costaiche, & Zeto. (2013). *Validating VoLTE – A Definitive Guide to Successful Deployments.* IXIA.

Rosenberg, J., Schulzrinne, H., Camarillo, G., Johnston, A., Peterson, J., Sparks, R., … Schooler. (2002). *SIP: Session Initiation Protocol.* RFC 3261.

Signalling flows for the IP multimedia call control based Session Initiation Protocol (SIP) and Session Description Protocol; Stage3 (Release 5). (2006). 3GPP TS 24.228 v5.15.0. 3GPP.

Technical Specification Group Services and System Aspects. (2006). *IP Multimedia Subsystem (IMS), Stage 2, TS 23.228, 3rd Generation Partnership Project TS 23.228 v8.2.0.* Author.

Chapter 3
Session Initiation and IP Multimedia Subsystem Performance Evaluation

Ashraf A. Ali
The Hashemite University, Jordan & University of South Wales, UK

Khalid Al-Begain
University of South Wales, UK & Kuwait College of Science and Technology, Kuwait

ABSTRACT

In this chapter analyses the overall system capacity and scalability which is affected by added traffic introduced by more users who are trying to access the system provided services. This could happen in a mission critical communication system during natural disaster or large scale attack, where the system accessibility could be affected due to the sudden increase of number of users. The need for a more detailed study of other SIP and IMS KPIs is vital to have a better understanding of the overall system performance which will enable us to take it a step further toward system performance enhancement and optimization to avoid single point of failure of the system.

INTRODUCTION

IP Multimedia Subsystems (IMS) (Technical Specification Group Services and System Aspects, 2006) and Session Initiation Protocol (SIP) (Rosenberg, et al., 2002) performance play major role in multimedia communication networks by altering the Key Performance Indicators (KPIs) related to the Quality of Experience (QoE) metrics of the end-to-end service. Registration Request Delay (RRD) is one of the SIP KPIs that also has impact over both IMS KPIs and end user QoE. Therefore, it is crucial to have performance evaluation of both SIP and IMS based on RRD metric to give indication of the overall system capacity and scalability potentials.

In this Chapter analyse the overall system capacity and scalability which is affected by added traffic introduced by more users who are trying to access the system provided services. This could happen in a mission critical communication system during natural disaster or large scale attack, where the system accessibility could be affected due to the sudden increase of number of users.

DOI: 10.4018/978-1-5225-2113-6.ch003

It was also found that the system ability to process the registration requests per unit time is increasing exponentially to a limit with the linear increase of the number of users. After this limit, the number of processed requests will start to decrease and will eventually degrade and lead to system failure. The simulation results shows that the system was able to handle a maximum of 7400 registrations per second which could happen during nationwide disaster with users trying to access the Mission Critical System (MCS).

The need for a more detailed study of other SIP and IMS KPIs is vital to have a better understanding of the overall system performance which will enable us to take it a step further toward system performance enhancement and optimization to avoid single point of failure of the system.

The research methodology that was presented by John W. Creswell (Creswell, 2003) was followed for both the qualitative and quantitative approaches to set the broad lines foe all measurements and simulations. The methodology to decide the qualitative values that need to be investigated can be summarized as follows:

- Determine the challenges that need to be investigated within the scope of the study. As presented in the previous section, it was found that the project embed several challenges and the focus of the project will be in the signalling domain especially between the end user and the core network in addition to the signalling interface between the core network and IMS.
- Determine the benchmark for what is considered accepted SIP performance and decide on the metrics that will be measured to judge and compare the performance of the suggested setup.
- Decide the appropriate simulation tools to get the results from multiple sources to have the appropriate comparison criteria based on the selected tool.
- Determine the key factors that affect the SIP signalling in addition to multimedia services operation in LTE and IMS that affect the overall QoS for the Mission Critical system.

The Quantitative Methodology to get the needed quantitative measures can be summarized as follows:

- Develop a test-bed for the IMS to test the performance of the system. Then decide the performance metrics that need to be collected to be compared with other implementations and scenarios. The details of the experiment will be presented in later sections.
- Develop a virtual machine to generate a virtual clients and running IMS to compare the performance of the system with the running test-bed., then again compare the results with the test-bed outcomes.
- Develop a simulation project for both LTE and IMS over OPNET to investigate the performance of the system and compare it with the test-bed implementation and decide the benchmark that need to be followed for the different performance metrics.
- Determine the variables and parameters in OPNET models and scenarios that need to be changed and manipulated to evaluate the performance of the overall system.

In summary, as shown in Figure 1, there are three simulation/testing options.

This section will present the test-beds and simulation results that are related to the project. IMS test-bed scenario and results will be presented first, and then the OPNET simulation Scenarios and results will be demonstrated at the end. Finally, the limitations and challenges for both experiments will be discussed.

Figure 1. Available simulation and testing tools

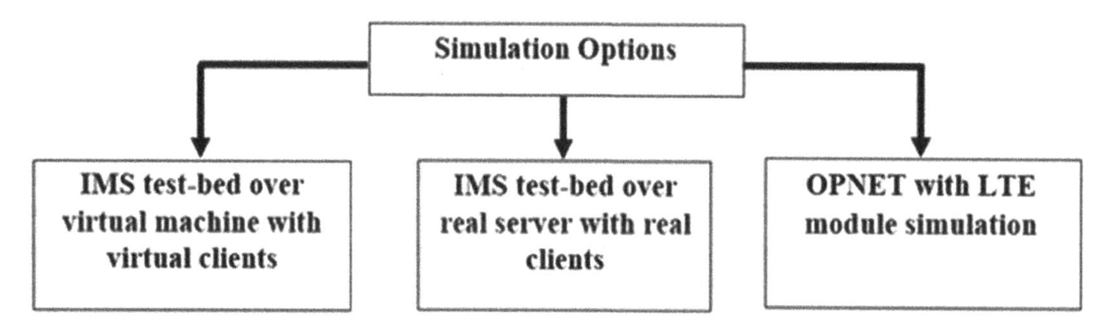

TEST-BED EXPERIMENT

As shown in Figure 2 which shows the experimental topology of the test-bed, the test-bed is composed of four main parts; the IMS core which is based on Open-IMS-Core model (Fokus, 2004), Packet Generator, Domain Name Server (DNS) and finally Wireshark which is a traffic analysing tool.

Packet Generator

The packet generator which is responsible for simulating virtual clients for the purpose of generating concurrent calls that could be transmitted in serial or parallel manner by theoretically unlimited number of users. Due to the focus on SIP and IMS performance, the Packet Generator is designed to send SIP Register Message as defined by RFC 3261 in addition to SIP invite and bye messages. All the messaged are carried over UDP transport protocol, and the sender port addresses is dynamically allocated to avoid using restricted ports at the sender or server sides. The GUI interface of the Packet Generator enables the user to select a predefined set of users and the Proxy IP address. Finally, the SIP request sending pattern is selected to be either serial or parallel.

The test-bed is constructed based on the following entities:

- **IMS Network:** The IMS core is based on OpenIMSCore (Fokus, 2004) which is open source implementation developed by FOKUS Institute for Open Communication System. The server embed the IMS Call Session Control Functions CSCFs; such as PCSCF, I-CSCF, and S-CSCF. In addition to the home Subscriber Station HSS, which are all considered the core architecture for the next generation networks that was specified by 3GPP. The purpose of the experiment is to test the capacity of the IMS in terms of the maximum number of users that can be adapted by the system without any stability issues.
- **Packet Analyser:** The sent packets by the Packet Generator was monitored using Wireshark as a packet analyser at the sender side. The trace files extracted from the packet analyzer will help in calculating the KPI values for both SIP and IMS.
- **DNS:** an external Domain Name Server (DNS) to resolve the IP addresses of all servers in the system setup.

Figure 2. Experiment test-bed

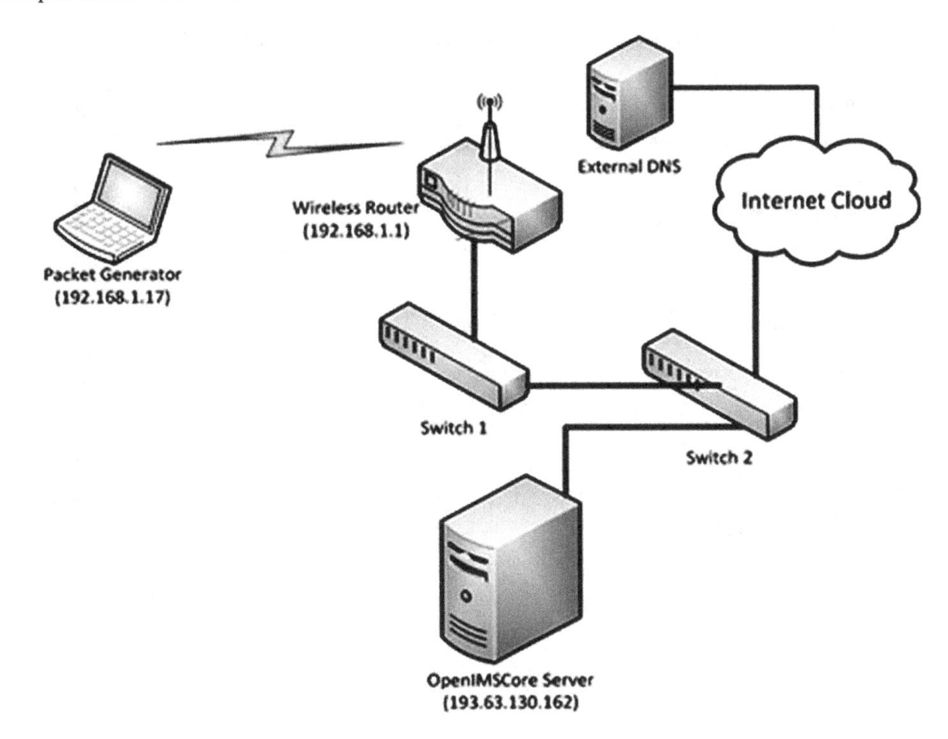

Based on the previous setup the experiment aimed to evaluate the SIP performance over the IMS using either a wired or wireless connectivity with the server. For this purpose, the register message delay was calculated by running Wireshark at the packet generator side and calculating the difference between the sent registration request time and the 200OK response reception time. Then using MATLAB the data was exported and manipulated to get the Probability Density Function (PDF) and Cumulative Density Function (CDF) curves for better understanding of the variance in Registration delay within the same scenario and among different running scenarios.

The GUI interface for the Packet Generator is shown in Figure 3.

The user select first a predefined set of users, and the Proxy IP address (which is the IMS server IP) in addition to the domain name is filled, and finally the SIP request sending pattern is selected to be either series or parallel, and finally the send button is pressed to send the registration requests (one per user) consecutively and dynamically. The Packet Generator sent packets was monitored using Wireshark at the sender side and the log screen at the packet generator will record the time stamp of the sent requests and received responses which helps in calculating the end to end application delay (the time between peer application layers).

Based on the previous setup the experiment aimed to evaluate the SIP performance over the IMS using either a wired or wireless connectivity with the server. For this purpose, the register message delay was calculated by running Wireshark at the packet generator side and calculating the difference between the sent registration request time and the 200OK response reception time. Then using MATLAB the data was exported and manipulated to generate curves for better indication of the variance in Registration delay with time. The experiment was run multiple times, each time the number of users was incremented in both the wired and wireless scenarios.

Figure 3. Packet generator GUI

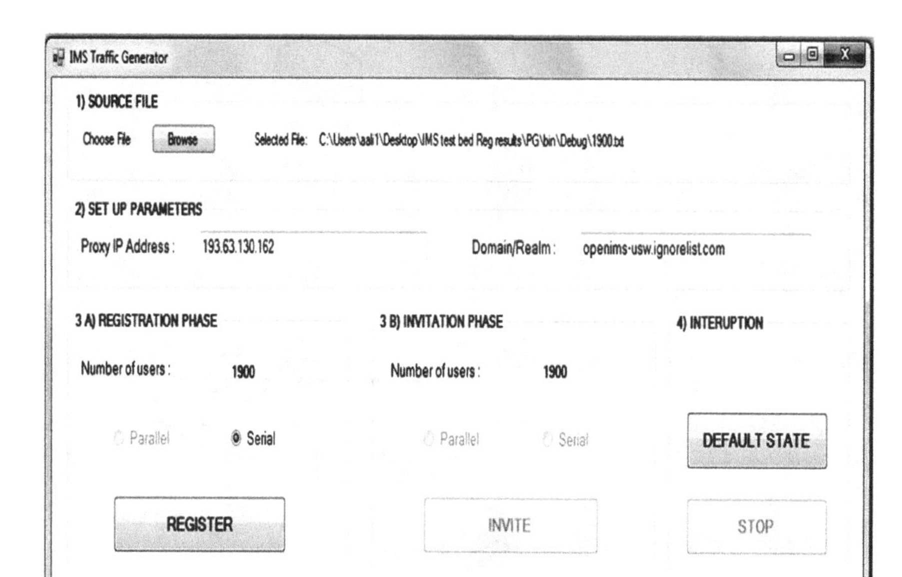

RESULTS

In this scenario, the packet generator was wired directly to router and the number of users sending the registration request were incremented in steps of from 100 users till 1300 users in a step of 200 users each time. Figure 4 shows the PDF and CDF of the registration delay for 100 and Figure 5 shows the PDF and CDF for 500 users.

It is clear from Figure 4 and Figure 5 that the registration delay for 500 users increased by more than double value of that for 100 users, which even increased more while increasing the number of users in each step. Similarly, seven scenarios were implemented over the test bed, in the first scenario 100 user registration requests were transmitted, and in the subsequent scenarios the number of users was incremented in a step of 200 until the seventh scenario with 1300 users. It was decided to increase the number of users gradually to test the scalability of the system to have a better understanding of the relation between the number of users and the KPI values for both SIP and IMS.

Figure 6 shows the Probability Distribution Function (PDF) and Figure 7 shows the Cumulative Distribution Function (CDF) for the RRD of the first scenario with only 100 users each sending one registration request at a time in sequential order. As shown in Figure 7, 90% of Registration requests need less than 40 ms to be completed which meets the requirements of mission critical applications and real-time services. as shown in Figure 6, the highest frequency of the registration trials needs in average 20 ms to be completed. This is considered the best case scenario that will be used as a benchmark for the other scenarios to compare both the RRD time and the percentage of trials that finish at certain time threshold.

Figure 4. CDF and density functions of the registration delay for 100 users

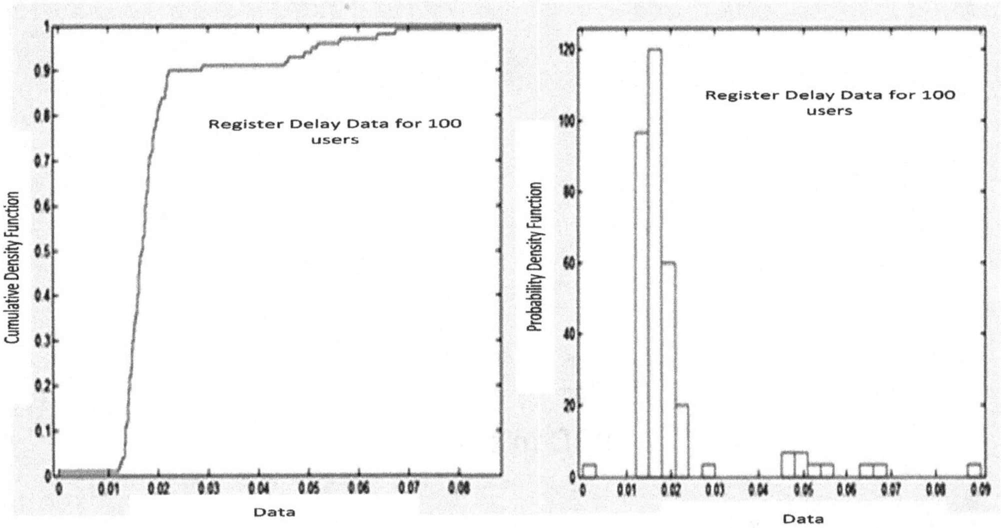

Figure 5. CDF and density functions of the registration delay for 500 users

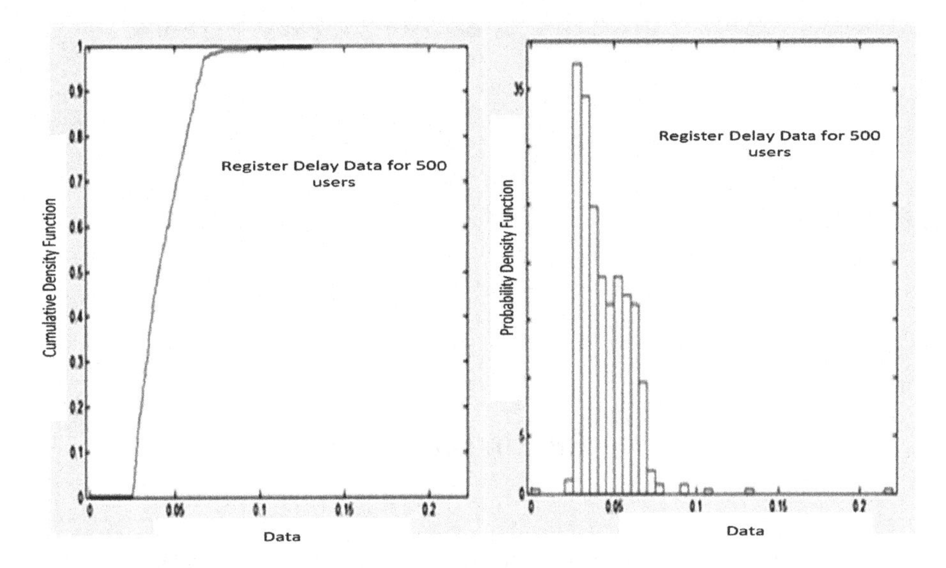

Similarly, the PDFs and CDFs for all the seven scenarios were generated as shown in Figure 8 and Figure 9. It is clear that when the number of clients increase, then the system will need more time to serve the registration requests. This happens due to accumulation of both SIP and DIAMETER signalling messages in the queues of the CSCFs interfaces (especially in S-CSCF) and the HSS interface. Both S-CSCF and HSS are considered bottleneck points of congestion that are affected significantly via linear increase of the number of registration requests. This leads to a queuing delay that emerge rapidly in the system interfaces which eventually even cause system failure.

Figure 6. PDF of RRD for 100 users

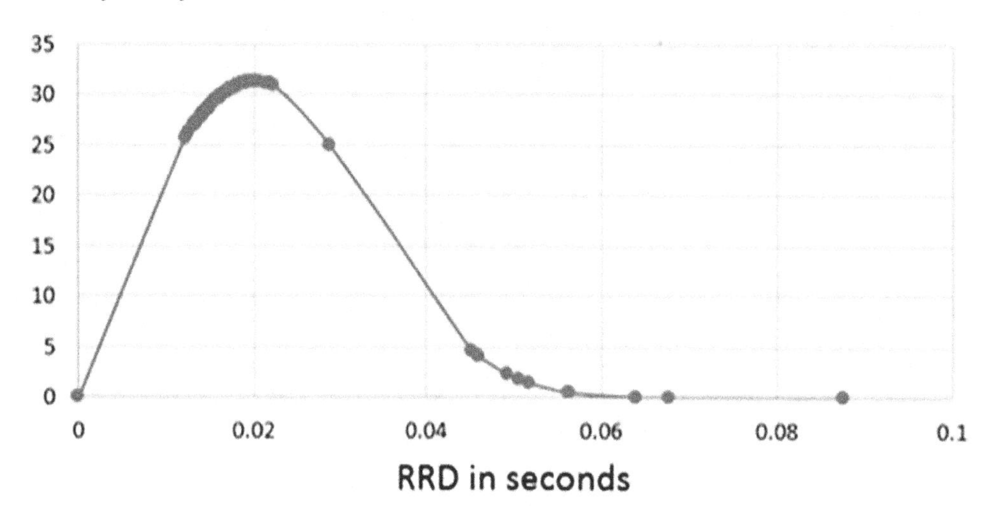

Figure 7. CDF of RRD for 100 users

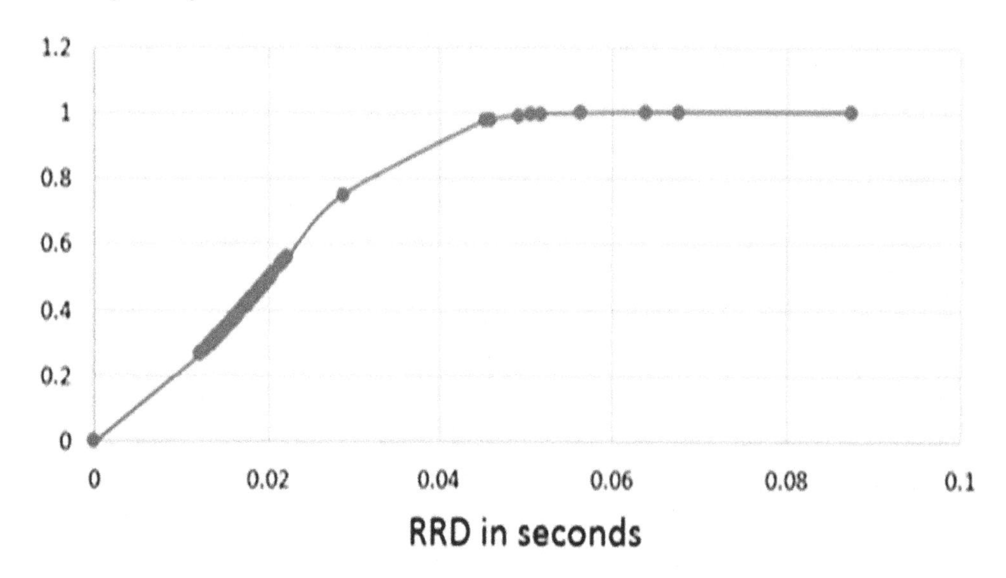

For the sake of comparing the performance of the seven scenarios, two performance metrics were defined. The first one is the time needed to process successfully 90 percent of requests, which will be referred to as 90% completion time (90CT). The second one is the percentage of successfully completed registration requests within 40 ms seconds (which is the maximum RRD time needed to process 90% of requests in the 100-users scenario) and will be referred as 40ms Completion Ratio (40msCR).

Table 1 shows the average RRD values along with 90CT and 40msCR for each Scenario in addition to the registration requests per second (RRps) processing rate which is the number of registration requests that are successfully processed by the test-bed in unit time. The RRps values replicate the real world disaster scenario where thousands of users may send registration request to gain access to the system which is supposed to be scalable and reliable at the same time.

Figure 8. PDF of RRD values for all scenarios

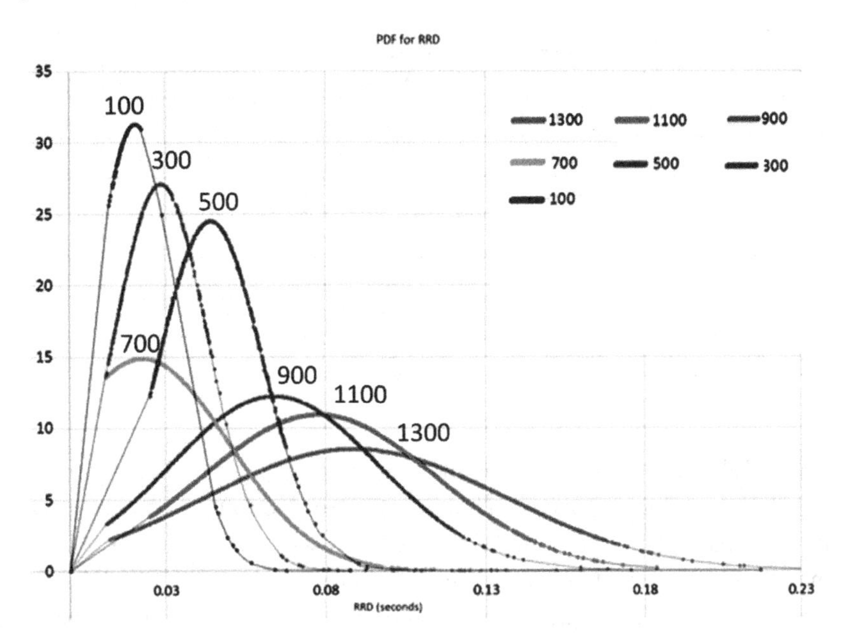

Figure 9. CDF of RRD values for all scenarios

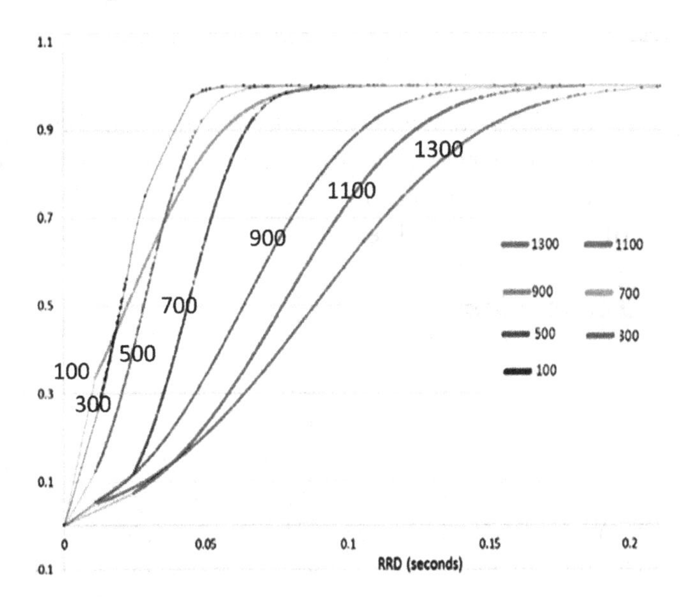

Based on the results, it was found the RRps increases exponentially with linear incremental of number of users up to a limit (which is 1100 users), then the RRps start to fall down which eventually leads to a system degradation and failure. This is clearly shown by RRps for both scenario 6 and scenario 7, where RRps of scenario 7 is much less than RRps for scenario 6 although the number of users increased by 200 users.

Table 1. Calls statistics from simulation results

Scenario no.	RRps	RRD Avg. Value	90CT (ms)	40msCR (%)
Scenario 1 (100 users)	1800	20	40	100%
Scenario 2 (300 users)	3600	28	47	80%
Scenario 3 (500 users)	5900	44	65	40%
Scenario 4 (700 users)	4900	22	55	73%
Scenario 5 (900 users)	5500	64	105	23%
Scenario 6 (1100 users)	7400	77	125	17%
Scenario 7 (1300 users)	5800	88	150	15%

As expected, the 90CT is increasing with the linear incremental of number of users, starting from 40 ms for scenario 1 till 150 ms for scenario 7. This is expected due to the more processing time needed for the increasing received registration request. Moreover, it was found that the 40msCR is decreasing with the increasing number of users. Comparing the values with scenario 1, which is used as benchmark, shows that only 15% of registration requests needed less than 40 ms RRD value to be completed, which again implies that the system is not able to process the received request within very strict time limit.

OPNET SIMULATION

For purpose of investigating the LTE system and especially the SIP signalling performance over LTE communication network, OPNET simulator was used to create a scenario with multiple users initiating calls and then investigating the SIP performance metrics to check the efficiency of the system and its capacity tolerance. Figure 10 shows the created setup.

Simulation Setup and Scenarios

In this research study, OPNET Modeller was considered as a simulation tool as it provides the required level of simulation capabilities to implement and model different multimedia applications over LTE. The system design that was implemented and investigated in this report is shown in Figure 10 based on the configuration parameters on Table 2. The implementation of the LTE network system is based on a single (EPC) that serves two eNBs where each has four clients. The clients in eNB1 are making SIP-based VoIP calls with the clients in eNB2 through the EPC in a Normal distribution call generation system using a fixed length for all VoIP calls duration. The EPC then connected to the SIP server, which is supposed to reflect the performance of the P-CSCF in the IMS, that manage the registration, call initiation and call termination processes using the SIP signaling system and only through the IP cloud. In this study the implementations was conducted without any background traffic in the LTE system and IP cloud to check and study the actual performance level for SIP-based VoIP applications within a best effort environment which helps with the results accuracy. In addition, the LTE implementations in this research study have not considered any mobility issues for its clients along with all the implemented scenarios.

The simulation implementations has considered four scenarios based on the introduced design at Figure 10 and the simulation parameters in Table 2. The first scenario represents the basic implementation for VoIP applications over LTE using a single pair of UEs between client A-1 in eNB 1 and client B-1 in eNB 2. This scenario examines the best-case implementation of the assigned network system with only one single call at the time. The second scenario has an additional connection with multiple calls with another pair of UEs (client A-2 and client B-2) added to the first scenario. The same thing with the third scenario where additional pairs of calls been added (client A-3 and client B-3). Finally, the fourth scenario has a fourth additional pair between client A-4 and B-4. This gradual increase in the pairs of SIP-based VoIP calls from the first to the fourth scenarios aimed to check the performance of the SIP signalling system over LTE based communications with additional VoIP calls between different clients. The highest load of VoIP calls is represented in the fourth scenario that consumes higher bandwidth over LTE where all clients in each eNB are calling one single client in the other eNB. Therefore, the results of these implemented scenarios will be compared and studied throughout the research study in terms of the performance for SIP signalling and efficiency for LTE system.

Simulation Results

As the main considerations in this study is for the SIP signalling and LTE performance for mission critical systems, the results representation focused on the call setup time and related LTE performance metrics. The optimum number of initiated calls for each pair of calls is falling between 150 to 180 calls for 30 minutes of simulation time with the uniform based distribution system for calls initiation. Table 3 shows the number of rejected calls in the overall system for the four scenarios with the implemented normal based system. The number of rejected calls has increased with the increased number of initiated call pairs. For calls implemented from Caller A-1, the number of failed calls *initiation* processes had

Table 2. Simulation Parameters in OPNET

A. LTE Network System			
Number of Simulations	4	**Simulation Seed Number**	128
Simulation Duration:		30 Minutes = 1800 Seconds	
Number of EPC:	1	**Background Traffic**	0%
Number of eNB:	2	**Number of nodes for each eNB:**	4
Antenna Gain for eNB:	15dBi	**eNB Maximum Transmission Power:**	0.5 W
eNB Receiver Sensitivity:	- 200 dBm	**eNB Selection Threshold:**	- 110 dBm
B. Applications: SIP Based VoIP			
VoIP Calls (Unlimited)	**Call Duration**	**Caller**	**Callee**
	10 Sec	Node A	Node B
Maximum Simultaneous Calls	**SIP Server**	**User Agent (Caller/Callee)**	**Voice Codec:**
	Unlimited Call/ Second	1 call at time between each pair	**GSM** 13 Kbps
Calls Start Time Offset:		Normal (150 sec, 100 sec)	
Calls Inter-repetition Time:		Normal (20 sec, 5 sec)	

Figure 10. System design and implementation for SIP-based VoIP applications over LTE network system in OPNET

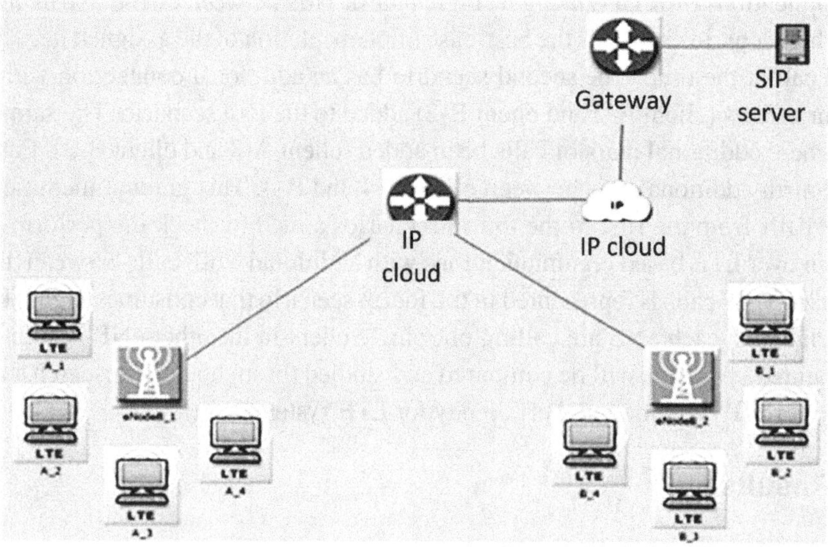

been increased with the increased number of call pairs with scenarios S2, S3, and S4, where the total initiated calls over all scenarios is 56 calls. This increase in the failed initiation processes results from different causes even related to the SIP servers' performance or LTE system performance.

Call Setup Performance

The importance of studying the call setup time is to analyse the SIP signalling performance during the main SIP signalling stage over different call sessions. As long as the call setup time for the majority of initiated SIP-based calls were in the acceptable range, the performance of

The SIP signalling system falls in its acceptable level (Malas and Morton, 2001) (Voznak and Rozhon, 2010). Figure 11 represents the average call setup time for all successful VoIP calls for the four implemented scenarios.

Table 3. Calls statistics from simulation results

SIP calls statistics for the Implemented Scenarios				
Scenario	**S1:** 1Pair	**S2:** 2Pairs	**S3:** 3Pairs	**S4:** 4Pairs
Number of Calls Rejected in the overall system	45	95	152	218
Number of Calls Initiated from Caller A-1	56	56	56	56
Number of failed calls initiation for calls from Caller A-1	27	30	38	34

Figure 11. Call setup delay

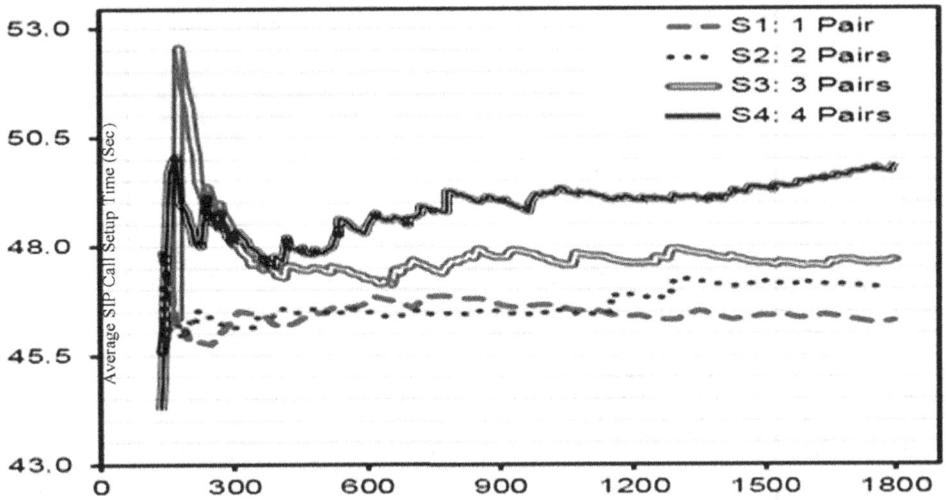

The results show that the scenario with only one pair of calls had the lowest average of call setup time from 46 to 47 ms and increased up to 48.5ms with the scenario of two pairs. With three pairs of VoIP calls, the average call setup time has increased from 47ms to 49.5ms. The longest call setup time registered with the fourth scenario in which four pairs of VoIP connections were active at the same time. These simultaneous calls affected the SIP signalling performance and increased the average delay to reach up to 50.5ms. In general, the call setup time for successfully initiated calls over all scenarios is still in its acceptable level regarding the performance of the SIP signalling system. This came as a result of the representation for the LTE network system that implemented in its best effort without any assigned delays or background applications.

LTE Downlink Packets Dropped

The LTE parameters of the implemented system have a direct effect on the performance of the running applications. Moreover, real-time applications could be enhanced in case related LTE system behaviour been considered with the required level of performance. Therefore, the LTE nature most considered during the studies of SIP-based VoIP over LTE. The average of packets dropped begun with 1 to 3 packets/sec for the single pair scenario and increased up to 6 to 18 packets/sec for the scenario with four pairs as shown in Figure 12. The downlink packets dropped of LTE system shown a relation with successful SIP sessions where as long as the number of calls increased as the percentage of packet dropped relatively increased.

The LTE system delays in the transferred data between LTE components affect the performance for real-time applications. The average LTE delays with one and two pairs of VoIP calls is between 2ms to 2.7ms as shown in Figure 13. The average LTE delays for three pairs of VoIP calls is from 2.4ms to 3.5ms, and between 2.5ms to 4.3ms with four pairs of calls. The longest delays mostly happen at the system start-up time and come to the stability level later during the simulation time.

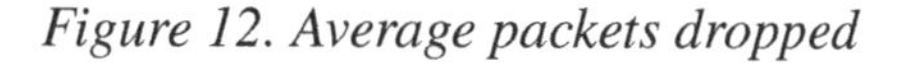

Figure 12. Average packets dropped

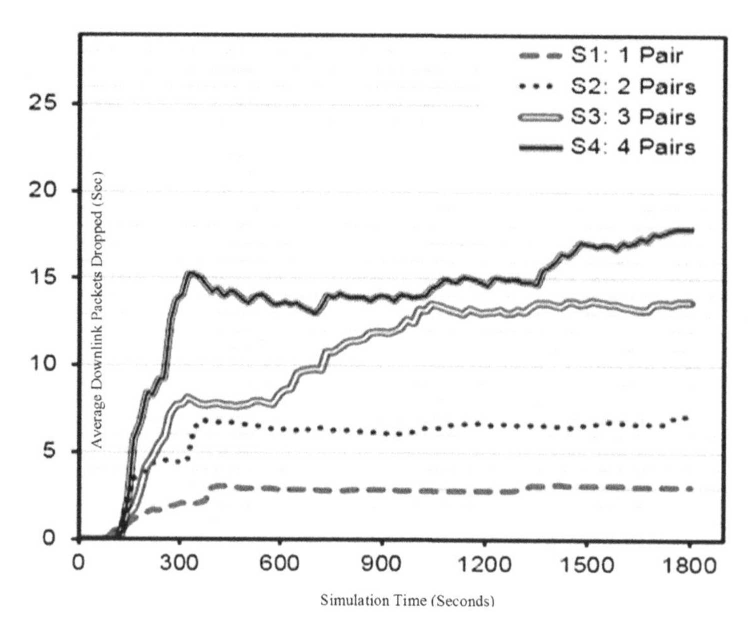

Figure 13. Average LTE delays in second for caller A-1 node

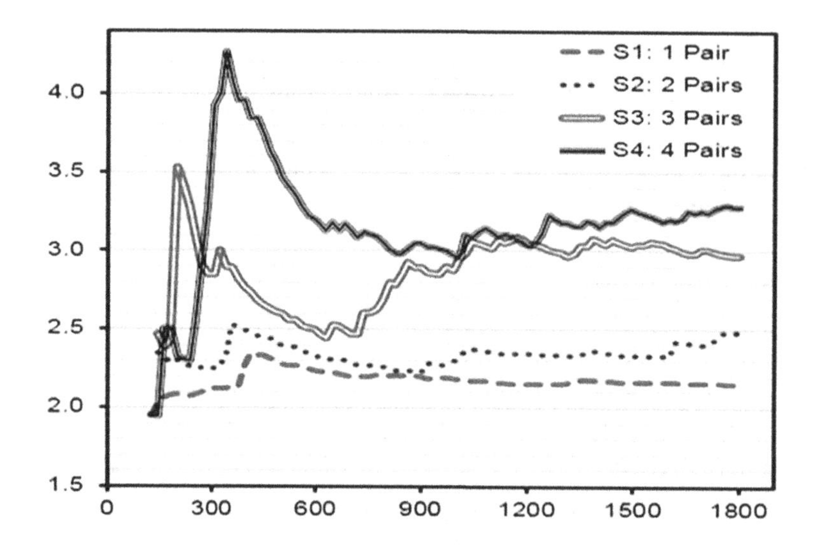

CONCLUSION AND FUTURE WORK

Based on the simulation results, it is clear that there is increasing delay in the call setup if LTE communication system was used. This delay is increasing with the number of served client, which indicate that the Delay requirement or the maximum number of users that can be served at a time may not meet the mission critical service requirements. Hence, the need for decreasing the gap of call setup delay for commercial broadband systems compared with other dedicated mission-critical communications systems is of great importance and considered one of the main challenges for mission-critical communications.

This means that there is a need for a new mechanism with minimum possible overhead to minimize access delay by exploring the LTE and IMS domains in addition to the interfaces between LTE and IMS and the interface between LTE and User Element.

So far a simulation was implemented with static mobile nodes, the positions of the nodes are fixed. Hence, there is no handoff added complexity for the nodes moving between two cell domains. By mobility, it is not meant only having a moving nodes, but also a dynamic topology that supports handoff mechanisms between the subscriber stations and different base stations. Hence, testing different communication scenarios for an end to end connectivity over LTE communication system is needed. For such dynamic topology, the need for measuring the overall performance of the system in terms of SIP signalling and data streaming delay is crucial.

REFERENCES

Creswell, J. W. (2003). *Research Design: Qualitative, Quantitative, and Mixed Methods Approaches*. Research Design Qualitative Quantitative and Mixed Methods Approaches.

Fokus, F. (2004). *Open IMS core*. Fraunhofer Institute FOKUS Available: http://www.openimscore.org

Malas, D., & Morton, A. (2011). *Basic Telephony SIP End-to-End Performance Metrics*. Technical Report RFC 6076. Internet Engineering Task Force (IETF). Retrieved from http://tools.ietf.org/html/rfc6076

Rosenberg, J., Schulzrinne, H., Camarillo, G., Johnston, A., Peterson, J., Sparks, R., Handley, M., & Schooler. (2002). *SIP: Session Initiation Protocol*. RFC 3261.

Technical Specification Group Services and System Aspects. (2006). *IP Multimedia Subsystem (IMS), Stage 2, TS 23.228, 3rd Generation Partnership Project TS 23.228 v8.2.0*. Author.

Voznak, M., & Rozhon, J. (2010). SIP back to back user benchmarking. *Wireless and Mobile Communications (ICWMC), 20106th International Conference on*, 92-96.

Chapter 4
Performance Metrics for SIP-Based VoIP Applications Over DMO

Mazin I. Alshamrani
Ministry of Haj and Umra, Saudi Arabia

Ashraf A. Ali
The Hashemite University, Jordan & University of South Wales, UK

ABSTRACT

The evaluation studies need to investigate a determined performance metrics to understand and evaluate the examined scenarios. SIP-based Voice over IP (VoIP) applications over MANET, which behaves in a way similar to Direct Mode of Operation (DMO) in mission Critical Communication Systems, have two main performance categories related to the Quality of Service (QoS). The main performance metrics that are considered for the evaluation processes in this research are the SIP end-to-end Performance metrics as defined by the RFC 6076. The main performance metrics are related to the registration, the call setup, and the call termination processes. In this research study, the SIP performance metrics are based on a single SIP proxy server. For voice data, the QoS evaluation is based on two methods: The Objective method and the Subjective method. The Objective method considers the traffic throughput, end-to-end delays, packet loss, and jitter, while the subjective method considers the Mean Opinion Score (MOS), which is mostly related to the end users' experiences during voice calls.

INTRODUCTION

The evaluation studies need to investigate a determined performance metrics to understand and evaluate the examined scenarios. SIP-based Voice over IP (VoIP) applications over MANET, which behaves in a way similar to Direct Mode of Operation (DMO) in mission Critical Communication Systems, have two main performance categories related to the Quality of Service (QoS). The main performance metrics that are considered for the evaluation processes in this research are the SIP end-to-end Performance metrics as defined by the RFC 6076 (Malas & Morton, 2011) . The main performance metrics

DOI: 10.4018/978-1-5225-2113-6.ch004

are related to the registration, the call setup, and the call termination processes. In this research study, the SIP performance metrics are based on a single SIP proxy server. For voice data, the QoS evaluation is based on two methods: the Objective method and the Subjective method. The Objective method considers the traffic throughput, end-to-end delays, packet loss, and jitter, while the subjective method considers the Mean Opinion Score (MOS), which is mostly related to the end users' experience during voice calls. The voice codec represents the compression system for voice data that is used during the calling session and affects the construction of the voice traffic volume. The other related metrics for SIP-based VoIP over MANET are the routing performance metrics. The routing mechanism of the considered MANET routing protocol is responsible for the average bandwidth consumed for routing data, the average routing traffic sent and the average routing received. In this chapter, the main performance metrics for SIP signaling and voice data are presented to identify the evaluation methods for SIP-based VoIP applications over MANET platforms

VOICE CODECS

The voice applications used to compress the analog voice signals into digital signals uses different types of voice codecs. Voice codecs are audio data compression algorithms for use for different types of voice based applications (Ganguly & Bhatnagar, 2008) . This basic stage happens on the caller's side to make the voice data transferable over the PSTN or the Internet for far distances. For wireless based VoIP applications, the voice compression is critical, as the voice signal needs to be compressed as much as possible to fit with the loose nature of wireless communications. This compression effectively reduces the bandwidth consumption and transmission power over wireless network systems. In addition, the voice compression systems create smaller packets, which reduce the packet loss ratio, and end-to-end delays that support the voice quality as the number of received voice packets relatively increase (Sinnreich, & Johnston, 2012). The present researcher studied the SIP-based VoIP applications over MANET using four common voice codecs:

- G.723.1 (ITU-T Rec. G.723.1, 2006) is one of the most common voice codecs for VoIP applications that operates at 5.3 Kbit/s or 6.3 Kbit/s and is officially known as Dual Rate Speech Codec for Multimedia Communications Transmitting at 5.3 and 6.3 Kbit/s.
- G.729 (Schulzrinne, Casner, Frederick, and Jacobson, 2003) is another common voice codec for VoIP applications because of its low bandwidth requirements. It operates at 8 Kbit/s and formally known as Coding of Speech at 8 Kbit/s Using Conjugate-Structure Algebraic Code-Excited Linear Prediction Speech Coding (CS-ACELP).
- The GSM voice codec is developed by the European Telecommunication Standards Institute (ETSI). It is widely used in mobile telecommunications as it operates at 13 Kbit/s and has good performance over CPU demands that support the nodes' mobility nature (GSM, 2000) .
- G.728 (ITU-T Rec. G.728, 2012) is a speech coding algorithm which operates at 16 Kbit/s and is described by the International Telecommunication Union – Telecommunication Standardization (ITU-T) as the Coding of Speech at 16 Kbit/s Using Low-Delay Code-Excited Linear Prediction (LD-CELP).

In this research, the GSM voice codec is considered as the main voice codec in the simulation results investigation for research study scenarios. The main outcomes of the evaluation study in this research are meeting with the implementation results for other voice codecs such as G.723.1, G.729, and G.728. The encapsulated voice data of VoIP applications are transferring over a special multimedia transport protocol known as the Real Time Protocol (RTP), which runs over the User Datagram Protocol (UDP) (Sinnreich, & Johnston, 2012). As RTP is depending on an unreliable transferring process, the voice packets have a losing ratio which depends on the network health and connectivity status between source and destination. In general, voice data has the priority over different types of network traffic to support the QoS of voice application in different network systems. For SIP-based VoIP applications, the call initiation stage activates the media data transfer process for RTP directly between the caller and callee using one of the supported voice codecs. The payload of voice packets and the headers are shown in Figure 1, where IPv4 packets have an overhead of 20 Bytes and IPv6 packets have an overhead of 40 Bytes. This difference between IPv4 and IPv6 packet size influences the traffic transferring and the QoS for voice applications.

SIP SIGNALING PERFORMANCE METRICS

The signaling performance of SIP plays an important role in affecting the overall Quality of Service (QoS) and Quality of Experience (QoE) in next generation networks. SIP signaling delays are relating to the connectivity status between the SIP call parties, which are the Caller, Callee, and SIP proxies. These delays mainly happen during the Caller's registration process, call initiation, call termination, and/or call management (Sinnreich, & Johnston, 2012) . In addition, SIP signaling is affected by the behaviour of the Transportation Protocol (TCP or UDP) that SIP relies on during the different connectivity processes of SIP calls. Many standards have been proposed for the performance evaluation of telephony signaling protocols, however none of these metrics were used to address the SIP signaling performance until the IETF proposed the RFC 6076, the SIP end-to-end Performance Metrics (Malas & Morton, 2011) . However, there are no numerical values or benchmark objectives for the RFC 6076 SIP performance metrics.

SIP Signaling Performance Metrics in the Literature

As many approaches were introduced to evaluate and standardize the telephony signaling protocols, none of these approaches studied the SIP signaling performance until the appearance of RFC 6076 in 2011 (Malas & Morton, 2011). The concerns about SIP signaling performance began with assumptions and investigations for a set of related performance metrics for SIP signaling over different platforms. The ITU-Recommendations proposed two SIP performance metrics for circuit-switched network systems, the Post Selection Delay (PSD) to measure the call setup times for SIP calls, and the Network Effectiveness

Figure 1. IPv4/IPv6 voice packet (Ganguly, Bhatnagar, 2008)

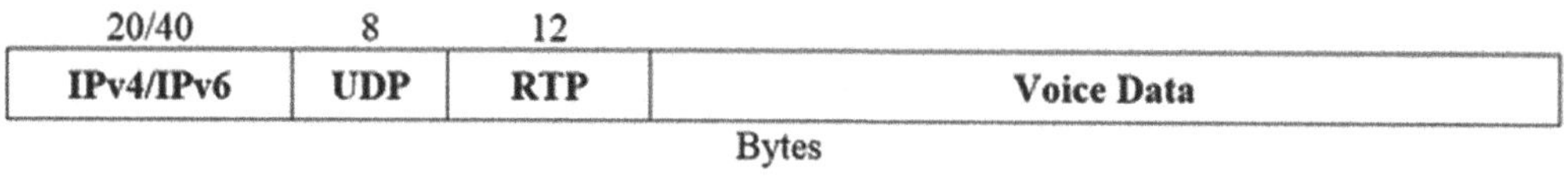

20/40	8	12	
IPv4/IPv6	**UDP**	**RTP**	**Voice Data**

Bytes

Ratio (NER) to detect the ability of the User Agent or the proxy server during the session establishment of SIP calls (Eyers & Schulzrinne, 2000) .

Beside the standardization efforts, a number of studies tried to identify different performance metrics for SIP signaling using different related network elements for SIP sessions. One of the early studies (Eyers & Schulzrinne, 2000) proposed to examine the main processing entities for SIP network elements and identify a number of SIP performance metrics in 3GPP IMS systems. The study provided some experimental results related to the processing efforts for SIP sessions regarding CPU usage and memory consumption with an analysis of SIP servers' performance during SIP sessions. In (Eyers & Schulzrinne, 2000), an alarm message was proposed for the sent INVITE requests for SIP calls known as Session Setup Delay (SSD). In the same way, the Call Setup Delay (CSD) performance metric for IMS supports end-to-end QoS over fixed multimedia applications and converged mobile services (Agbinya, 2010) . The CSD depends on three main metrics: the Post Dialing Delay (PDD), Answer Signal Delay (ASD) and Call Release Delay (CRD). The PDD is the time interval between the call initiation by the caller and the time the caller hears the callee's terminal ringing while the time interval between when the callee picks up the line and the caller receives the callee's response is known as ASD. The CRD is the time interval between call disconnect and initiating/receiving a new call by the same calling party. The values of these metrics are affected by the number of hops and the distance between the call parties. In (Happenhofer, Egger, & Reichl, 2010), a performance evaluation concept was proposed for a single SIP transaction known as Quality of Signaling (QoSg) that consists of the following metrics: User to User Delay (UUD), Processing Delay (PD) and Response Delay (RpD). The UUD is the time interval between sending an INVITE request until it is received by its destination. The PD is the time needed by the User Agent Server (UAS) to process a request and send its response message, while the RpD is the time interval between sending a request by a User Agent Client (UAC) and receiving its response message.

In (Brajdic, Suznjevic, & Matijaševic, 2009), two performance metrics were proposed, the Session Negotiation Time (SNT) and the Session Re-Negotiation Time (SRNT). The SNT is used in the negotiation process of a calling session to measure the required time interval between a sent SIP INVITE request until an applicable 200 OK response is received. The SRNT is used to describe the negotiation process initiated by sending an INVITE request or by a SIP update request that carries updated information regarding SIP sessions. A few studies tried to benchmark SIP performance metrics for IMS core components by identifying different metrics such as Registration Time, Initial Ringing Time, Initial Response Time, Disconnect Request Time, and other system estimation metrics like CPU and power consumption (Thisen, Espinosa Carl'in, & Herpertz, 2009) (Tang, Davids, & Cheng, 2008). All the reviewed studies considered non-standardized performance metrics for SIP-based applications. They used different metrics with limited levels of usability that make it difficult to be applied over different platforms and network systems. In addition, the limitations of applying these performance metrics provided non-comparable and incompatible results for the performance metrics.

RFC 6076: SIP End-to-End Performance Metrics

The Internet Engineering Task Force (IETF) adopted standardized end-to-end Performance Metrics for a basic SIP-based signaling system as defined in RFC 6076 (Malas & Morton, 2011). These metrics provide Key Performance Indicators (KPIs) and Service Level Agreement (SLA) indicators to support the SIP-based telephony systems and enhance the network utilization. The RFC 6076 defined the following SIP end-to-end Performance Metrics:

Registration Request Delay (RRD)

The RRD is used to determine the response delay time for the User Agent REGISTER request. The RDD helps to measure and analyze the successful Registration requests and at the originating User Agent as represented in Figure 2. The output values for RRD should be in milliseconds (ms). This metric is calculated using Equation 1:

$$RRD = Time\ of\ Final\ Response - Time\ of\ REGISTER\ Request\,\mathbf{RRD} \quad (1)$$

The RRD is calculated only for successful registrations. In addition, when the load of SIP calls increases in the network systems, the value of RRD also increases. When there is low load in the network system the value of RRD will be in the range of the lowest values.

Ineffective Registration Attempts (IRAs)

The IRA is a metric that detects the failures or the impairments that cause the inability of the registrar to receive the User Agent REGISTER request. The IRA is a percentage parameter of the total number of unsuccessful registrations of the REGISTER requests and is calculated using the following Equation 2:

$$IRA\left[\%\right] = \frac{Nubmer\,of\,IRAs}{Total\,number\,of\,REGISTER\,Requests} X100 \quad (2)$$

Session Request Delay (SRD)

The SRD is a metric designed to detect the faults or defects that cause delays in responding to INVITE requests. SRD considers both successful and unsuccessful session setup requests where the duration for success and failure responses are varied. A simple representation for SRD related to the SIP flows is shown in Figure 2. The SRD is calculated using the following Equation 3:

$$SRD = Time\ of\ Status\ Indicative\ Response - Time\ of\ INVITE \quad (3)$$

Session Disconnect Delay (SDD)

The SDD is designed to calculate the time interval between the time that the session completion message (BYE) is sent and the last subsequent acknowledgement of the session completion response received (2xx). The SDD is used to detect the failures or impairments that cause the delays for a session to end. The SDD measures both successful and failed session disconnects where the output values are in milliseconds (ms). The SDD is calculated using the following Equation 4:

$$SDD = Time\ of\ \left(2xx\right)\ or\ Timeout - Time\ of\ Completion\ Message\ \left(BYE\right) \quad (4)$$

Session Duration Time (SDT)

The SDT is designed to define the time interval between the receipt of the (200 OK) response to the INVITE request and the receipt of the last associated (BYE) request. The SDT is measured at the originating User Agent and the terminating User Agent. Therefore, the SDT value measured at the originating User Agent is different from the value that is measured at the terminating User Agent. The SDT is calculated using the following Equation 5:

$$SDT = Time\ of\ BYE\ or\ Timeout - Time\ of\ \left(200\ OK\right)\ response\ to\ INVITE \qquad (5)$$

Session Establishment Ratio (SER)

The SER is considered to define the ability of terminating the User Agent or proxy server during the session establishment process. This metric is used to calculate the ratio of the INVITE requests that result in the (200 OK) responses and the difference of the total number of the INVITE requests and INVITE requests results from the 3xx responses. The Equation 6 of this metric is as following:

$$A = \frac{Total\,Number\,of\,INVITE\,Requests}{Associated\left(200\,OK\right)}$$

$$B = Total\ Number\ of\ INVITE\ Requests - \frac{Number\,of\,INVITE\,Requests}{3XX\,Respones} \qquad (6)$$

$$SER\left[\%\right] = \frac{A}{B} \times 100$$

Session Establishment Effectiveness Ratio (SEER)

The SEER metric is complementary to the SER, however it is used to exclude the potential effects of an individual user of the target User Agent from the metric. The SEER is defined as the ratio of INVITE requests resulting in a (200 OK) response and the INVITE requests resulting in a 480, 486, 600 or 603 response, to the total number of the initiated INVITE requests less the INVITE requests resulting in a 3xx response. The SEER is calculated using the following Equation 7:

$$C = \frac{Total\,Number\,of\,INVITE\,Requests}{Associated\left(200, 480, 486, 600, or\,603\right)} \qquad (7)$$

$$SEER\left[\%\right] = \frac{C}{B}\ X100$$

Ineffective Session Attempts (ISAs)

The ISA is a metric that is used when SIP entities are damaged or overloaded. The ISA is calculated as a percentage of the total session setup requests using the following Equation 8:

$$ISA[\%] = \frac{Number\ of\ ISAs}{Total\ number\ of\ Session\ Requests} * 100 \quad (8)$$

Session Completion Ratio (SCR)

The SCR is used to represent the percentage of the successfully completed sessions over the total number of sessions. This metric is similar to the Call Completion Ration (CCR) in the telephony applications of SIP. The SCR is calculated using following Equation 9 where the output indicates the percentage of successfully completed sessions:

$$SCR[\%] = \frac{Number\ of\ Successfully\ completed\ Sessions}{Total\ number\ of\ Session\ Requests} X\ 100 \quad (9)$$

Figure 2. SIP performance metrics over SIP call stages (Malas and Morton, 2011)

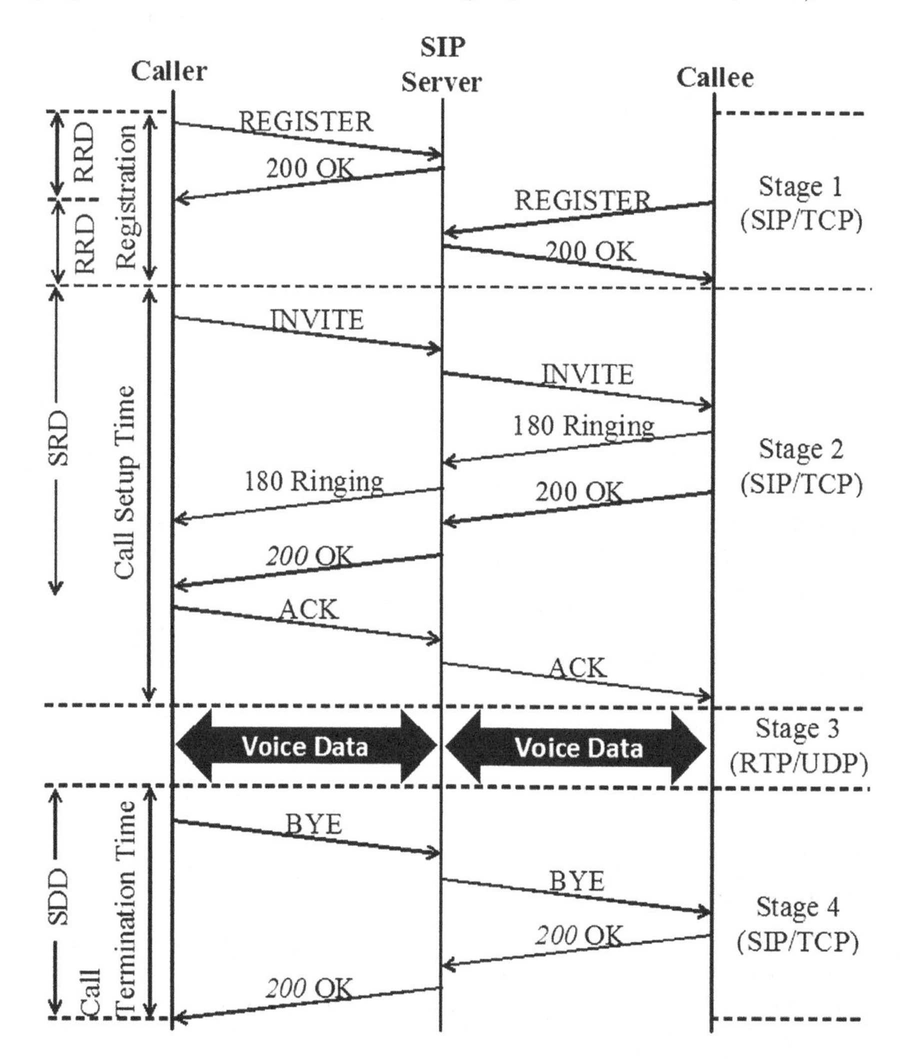

RFC 6076 Related Research

A limited number of research efforts studied the RFC 6076 SIP performance metrics for VoIP applications. The early efforts to identify a specific set of performance metrics to evaluate the SIP signaling performance was proposed in (Voznak & Rozhon, 2010). The research measured the overall performance of a SIP server by defining the parameters and methodology for benchmarking the SIP-based VoIP infrastructure. This method was designed to investigate the performance of the SIP server which is acting as a Back-to-Back User Agent (B2BUA) for VoIP applications. The study proposed and examined two parameters, Registration Request Delay (RRD) and Session Request Delay (SRD). The study was updated in (Rozhon, Jan, Voznak, Tomala, & Vychodil, 2012) and proposed new methods of stress testing for the SIP management and control approaches that generated high SIP-based traffic. The efforts had extended later to study the B2BUA-based SIP server and proposed a stress test of SIP signaling and benchmarking of the performance metrics based on the behavioural analysis of the SIP environment (Voznak & Rozhon, 2013). The SIP registration burst load for B2BUA-based SIP servers was examined in (Voznak & Rozhon, 2011). The research studied the RRD performance metric for a B2BUA-based SIP server where the effectiveness of handling the burst loads for SIP Registration requests was investigated in an Asterisk PBX system. In (Rezac, Voznak, Partila, & Tomala, 2013), the SIP performance metrics were used in the methodology of the performance tests for calls and registrations for the Interactive Video and Audio System (IVAS) which had been developed for the INDECT project which aims to develop the tools for enhancing the security of citizens and protecting the confidentiality of recorded and stored information. The study analysed the central elements of the IVAS system in VoIP PBX Asterisk by investigating the call registrations and initiations at the end users' devices.

In (Voznak & Rozhon, 2012), the RRD and SRD SIP performance metrics of the RFC 6076 were employed to measure and evaluate the SIP signaling performance on B2BUA-based SIP signalling system using an Asterisk based implementations. The research used open-source applications such as jQuery, Python, JSON and the cornerstone SIP generator SIPp. The results show that the SIP performance metrics could provide accurate values in busy network systems. The values collected for the SIP performance metrics in this research are limited to the RRD and SRD values, and could applied with simple SIP-based applications. However, no other performance metrics had investigated in this research and the results are limited to the LAN systems only. In (Husic, Hidic, Hadzialic, & Barakovic, 2014), a simulation-based optimization algorithm of SIP signaling procedures in IMS is presented to improve the SIP signaling performance by assigning a high priority value for SIP messages to reduce the network congestion and improve the overall QoS. The simulation efforts were conducted with ns-2 and the results were analysed in terms of the RRD, SRD, and SDD. The results have not compared with other related published results because the measurements used are implemented under different conditions and environments. In (Grasic & Kos, 2012), the SIP performance metrics were used to evaluate the implementation of the Rich Communication Suite (RCS) services in IMS platforms. The study employed the RRD, SRD, SDD, and SDT as timing parameters to optimise the SIP signaling services for the proposed system during its implementation. In (Kulin, Kazaz, & Mrdovic, 2012), the SIP performance metrics were used to measure and compare the overhead of using Transport Layer Security (TLS) compared with TCP and UDP for secure SIP implementations. The experimental results show a noticeable decrease in the performance of VoIP services based on the RRD and SRD when using SIP over a TLS based signaling system. The RFC 6076 performance metrics had been used in (Grgic & Matijasevic, 2013) as performance metrics for context based charging in a 3GPP network environment. The metrics were used to measure the response times of the messages related to the resource reservation and online charging procedures to support the QoS. The RRD was used in (Fabini, Hirschbichler, Kuthan, & Wiedermann, 2013) as a performance metric to evaluate the SIP Retransmission timers in HSPA 3G Networks for both the SIP server and the User Agents. The comparison with other work is still a challenging issue as no unified approach is provided to support the achieved studies.

User Agent Registration

The first stage for SIP-based VoIP calls is the Client's registration with the SIP server for non Peer-to-Peer SIP applications as shown in Figure 2. The SIP server is applying the Back-to-Back User Agent (B2BUA) for the exchanged SIP signaling system. The registration process is initiated whenever the User Agent joins the network system or updates its status. The SIP calls cannot setup if the other call parties are not registered with the SIP server. Whenever a Caller wants to call an unregistered UA, the UAS will fail the call setup process. A few number of research studies investigated the performance of SIP registration processes while other research focused on the related security issues. An optimization approach had been proposed for the SIP registration process over the mobile environment where the network access and IP addresses are changing frequently (Tzvetkov, 2008). The SIP clients need to proactively register with the SIP server and update its current parameters using identified dynamic registration intervals relating to the probability of links' disconnection to reduce the wasted resources and overloaded links. This approach is based on using an extended version of the Kalman filter to indicate the dynamic registration version. The simulation results show an enhanced performance for SIP registration processes of using the proposed registration method compared with the constant update intervals method of the registration processes. In (Voznak & Rozhon, 2012) and (Voznak & Rozhon, 2010), the registration process is measured using the RRD performance metric of RFC 6076 for a simple scenario using a SIP testing platform designed in the VSB – Technical University of Ostrava. The research is in the context of benchmarking SIP performance metrics to improve the performance evaluation methodology of a SIP signaling system by using an Asterisk testing platform. The study investigated the performance of the SIP server during the registration process by collecting the RRD performance metrics for the clients. The examination used a separate registration stress over an Asterisk PBX server during the registration process for a number of simultaneous registrations. The test results show that the SIP server has the main effectiveness of handling the burst loads for the registration requests.

The registration process investigated within different proposed approaches to enhance the SIP signaling performance over MANET platforms. In (Chang, Sung, Chiu, & Lin, 2010), a design of Ad hoc VoIP using an embedded p-SIP server provided an enhanced algorithm for the registration process over the SIP server. This approach reduced the registration delays by handling the route discovery mechanism over clients and SIP server for the registration and call setup processes. The study succeeded with reduction of the registration delays by improving a p-SIP Register algorithm to reduce the discovery time and the hops number between the call entities. In (Pack, Jeong, & Kim, 2010), the design of the Proactive Route Optimization (PRO) in SIP mobility considered the registration processes to maintain the client's location information to reduce the call setup time. In (Yahiaoui, Belhoul, Nouali-Taboudjemat, & Kheddouci, 2012) and (Mourtaji, Bouhorma, Benahmed, & Bouhdir, 2014), developed designs for the Registration process in MANET platform employed the client location and the relaying priority to enhance the SIP server performance.

Call Setup Delay

The Call Setup Delay is defined as the elapsed time between sending the initial INVITE request and receiving a 180 RINGING response (Mourtaji, Bouhorma, Benahmed, & Bouhdir, 2014) . The call setup time for each call is the difference between the absolute times of the INVITE and corresponding ACK message that belong to the same call. The call identifier headers are used in addition to the To and From headers, to categorise each SIP session as the SIP Method header can identify the type of the message, e.g., INVITE, and ACK. The Call Setup Delay is also known as the Post Dial Delay (PDD) and is considered as one of the required parameters for QoE evaluation. A number of studies have evaluated the call setup time for SIP-based network systems. The SIP call setup time was analysed for a reliability model over a SIP server in (Mourtaji, Bouhorma, Benahmed, & Bouhdir, 2014) .

In (Fathi, Chakraborty & Prasadm, 2006), the SIP session setup delay was investigated for correlated fading channels for 3G wireless networks. The SIP call setup time and the Real Time Protocol (RTP) were evaluated for one-way delay using IPv4 and IPv6 for basic network systems (Hoeher, Petraschek, Tomic, & Hirschbichler, 2007) . In (Munir, 2008), the SIP-based IMS establishment sessions for WiMax-3G networks had considered. The call setup latency for SIP-based VoIP over wireless local area networks was analysed using an ns-2 simulator in (Pack, and Hojin Lee, 2008). In (Pirhadi, Hemami, & Tabrizipoor, 2009), a call setup model is presented for SIP-based stateless calls for next generation networks. The model was designed based on the queuing models and the call setup delays for single domain and multiple domain scenarios. In (Ali, Liang, Sun, & Cruickshank, 2011), an evaluation was made of SIP-based call setup time and other QoS parameters of VoIP over IPv4 and IPv6 satellite environment based on the unreliable User Datagram Protocol (UDP) at the transport protocol layer. The call setup time is increased in the satellite networks environment due to the long propagation delay. The research recommended that an improvement for SIP signaling should be applied to reduce the number of SIP messages and call setup time over satellites. In (Babar & Malliah, 2011) the call setup delay had been surveyed for a SIP-based VoIP LANs Cisco environment. The study concluded that the average SIP call setup delay for various network loads is in the range of 200 to 300 milliseconds. On the other hand, the simulation and analytical results in (Mourtaji, Bouhorma, Benahmed, & Bouhdir, 2014) proved that the call setup latency is sensitive to the number of mobile nodes in a WLAN.

The previous approaches investigated and enhanced the SIP setup time for VoIP applications. The results could be applied or examined over different network systems and scenarios, however, these approaches could not considered as benchmark results for SIP-based call setup processes because it is not dependent on standard performance metrics for SIP signaling systems. The ITU-T recommendations defined considerable constraints for SIP Calls Setup Delay where the mean value is equal to 800 ms and the maximum value is equal to 1500 ms (ITU-T TR, 2004). In (Mourtaji, Bouhorma, Benahmed, & Bouhdir, 2014), the SIP call setup delay had been investigated based on the values in (ITU-T TR, 2004) by using different QoS and QoE factors for the SIP signaling evaluation over IPv4 and IPv6 IMS networks. In addition, it provided some theoretical analysis of SIP setup delays and provided results that show that the values of Call Setup Delays when using IPv6 are greater than those using IPv4 for radio access links with a bit rate less than 128 kbps. In (Tuong et. al, 2010), the Call Setup Delay is represented by using the RFC 6076 performance metrics. The research tried to consider both the RRD and SRD values to evaluate the Call Setup Delays for SIP-based VoIP Services in the context of the high loss, high latency, and bandwidth constrained Airborne Network environment using the OPNET®

simulation tool. In (Gutkowski & Kaczmarek, 2012), a model and algorithm for end-to-end call setup time calculations had been proposed for SIP-based VoIP applications. This model was designed to check all possible situations during the call setup processes that affect the general performance of the SIP call when dealing with highly loaded and congested networks. The study presented numerical calculations and simulation comparisons for the presented approaches using a cumulative distribution function for a trapezoid model. Furthermore, the study had shown that the probability of successful call setup in networks with a large number of nodes is very time-consuming. However, with low traffic networks, the SIP retransmission values and the call setup time depend on transmission times.

Call Termination Delay

The call termination represents the closing signal for SIP call sessions between the caller and the callee. The call termination process is initiated by one of the callers to end the call. In addition, it happens when the voice QoS begins to drop or when the connectivity is lost between both ends. The long delay for the call termination process could hold the caller's status as busy for a longer time which affects its availability in the network system as other callers may be trying to contact them through the SIP server. A limited number of research efforts studied the performance of termination processes within the general investigation of the SIP system. The impact of end users' response delays on the SIP server was examined in (Nahum, Tracey, & Wright, 2007) . The study provided an analytical performance evaluation for an open source SIP Proxy server by using the proposed SIP Performer testing tool with a central test host and multiple distributed test agents to simulate different traffic models including the users' response delays. The SIP Performer considered the call termination delays in its testing platform. The study proposed an enhanced termination mechanism for SIP calls within the overall proposed SIP signaling system. However, this approach has not been evaluated within different platforms where the proposed design is tight to a simple scenario that consists of two nodes and a SIP server.

A SIP-based QoS management system was proposed to provide a consistent QoS control system for multimedia applications over WLAN by considering the users' priorities and the providers' objectives (Tebbani, Haddadou, & Pujolle, 2009) . This approach employed both data and control planes to detect the congestion events of SIP processes and apply suitable actions to overcome the system deficiency within bandwidth and end-to-end delays for SIP-based multimedia applications. Furthermore, the approach architecture considered the session termination process to enhance the SIP applications QoS by sending BYE messages whenever delays or congestion was detected. The simulation efforts of the study show that the overall delays of the session termination process should not exceed one second (1s) to provide a good level of QoS.

VOICE QUALITY

Voice Performance evaluation for VoIP applications has two main measurement methods: Objective and Subjective methods (De Rango, Tropea, Fazio, & Marano, 2006) . The Objective methods are mathematical-based methods that are used to measure different physical quantities of voice traffic such as Packet Loss, Delays, Jitter, amount of Sent and Received Traffic (Kazemitabar, Ali, Nisar, Md Said, & Hasbullah, 2010) . In general, measuring VoIP applications is more objective and based on the performance calculations of the traffic transferred on the IP networks. On the other hand, the Subjective methods are human-based methods that depend on the average user's perception of the voice quality. It investigates the callers about the quality of the voice through simple questionnaires with limited classified choices.

Objective Methods

The objective methods depend on two mathematical testing techniques for voice quality: the intrusive or the non-intrusive techniques. The intrusive technique is used to inject a testing voice signal into the network system without the existence of the application users. This technique is mainly used during the system development stages when the users are offline or not beginning to use the provided services. On the other hand, non-intrusive techniques are used to examine live real-time traffic without the need to determine a reference signal. It depends on the network impairment parameters such as delays, packet loss ratio, and jitter. In general, the non-intrusive techniques provide larger numbers of live and real-time tests, however it has less accuracy compared with the intrusive techniques. In this research effort, the non-intrusive technique is used with the objective methods to measure the end-to-end or Mouth-To-Ear voice quality. The main performance metrics in this investigation are summarized in Figure 3 (Karapantazis, & Stylianos), where each metric has a direct influence on the voice transmission for VoIP applications (Sinnreich & Johnston, 2012) .

Throughput

The throughput is identified as the maximum number of bytes that are received in the receiver side out of the total sent voice traffic during an interval of time. For wireless networks, IEEE 802.11 standards have different data rates for different wireless applications. In this research work, the IEEE 802.11n standard has been used with a bit rate of 13 Mbps. In addition, the required bandwidth for VoIP calls ia mainly determined by the voice codec system used. Limited bandwidth could cause traffic congestions and packet delays that affect the general performance of the voice data. The complexity of the voice codec system affects the coding speed and the required bandwidth. As long as the coding complexity increased, the coding speed reduced and the required bandwidth increased. The amount of the consumed

Figure 3. Main factors affects the mouth-to-ear voice quality. (Adopted from Karapantazis and Stylianos, 2009)

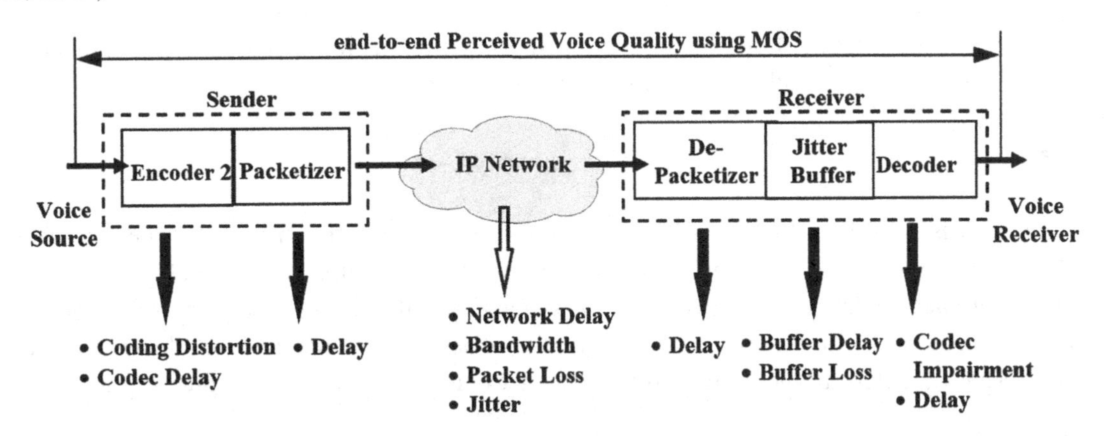

IP bandwidth for a voice call using one of the identified voice codecs in the previous section are computed by the following equations (Cisco, 2009):

$$Total\ packet\ size\ =\ IP\ /\ UDP\ /\ RTP\ header\ +\ Voice\ payload\ size \tag{10}$$

Number of Packets needed per second to deliver the codec rate:

$$PPS\ =\ Codec\ bit\ rate\ /\ Voice\ payload\ size \tag{11}$$

$$Bandwidth\ =\ Total\ packet\ sizeXPPS \tag{12}$$

where PPS is the number of packets required per second to deliver the amount of data for the voice codec.

Table 1 represents the voice codecs that were used in this research study, the payload size, and the Mean Opinion Score (MOS) for each codec (Kondoz, 2004). The MOS parameter will be explained in detail in later sections. The bandwidth calculations for a voice call are computed using Equations 10, 11, and 12. The header size for IPv4 and IPv6 voice packets are 20 Bytes and 40 Bytes, respectively, as represented in Figure 1. The table also shows the percentage of difference between IPv4 and IPv6 for the maximum required bandwidth for a single voice call. Figure 4 shows the maximum required bandwidth for a Full Rate single voice call over IPv4 and IPv6 Best Effort static MANET scenarios with different voice codecs as represented in OPNET® simulations. The Full Rate voice call is a call that uses the full voice rate of the voice codec to improve the performance of the voice codec and it is mostly used to examine the maximum possible bandwidth of the voice channel (Kondoz, 2004). The Full Rate call is used to show the difference in the bandwidth consumptions. IPv6 consumes more bandwidth compared with IPv4 over all voice codecs with different ratios, as calculated in Table 1. The transmission throughput is much below the maximum defined bit rate over different wireless standards because of the overhead caused by the network protocols. For example, the maximum throughput achieved in WLANs is between 50% and 70% of the maximum transmission rate. This is low compared with the maximum throughput achieved in Ethernet, which is in between 80% and 90% of the transmission rate (Rodellar, Gannoune & Robert, 2003). The throughput investigation is used to estimate the VoIP performance; when low throughput of voice packets is received that does not meet the expected amount of voice traffic, a concern about the voice performance must be raised (Eyers & Schulzrinne, 2000) . Low throughput indicates that the traffic has problems in the network connection such as a high percentage of traffic congestion. Another metric is the number of the reported dropped packets on the receiver side, which indicates the buffer overflow or problems with the network signaling system.

Delays

The voice delay is defined as the time taken for a person communicating with another person to speak a word and for it to be heard at the other end. As represented in Figure 3, Voice data delays in VoIP applications are accumulative delays from several parameters, algorithms, and transferring processes (Metha & Udani, 2001). The delays can be classified into: delays at source, network delays, and delays at receiver. The voice algorithmic delays result from voice compressing and decompressing at the source

Table 1. Voice codecs and bandwidth consumption; adopted from (Cisco, 2009) and (Kondoz, 2004)

Codec Information			Bandwidth Calculations				
Voice Codecs	**Codec Bit Rate**	**Mean Opinion Score (MOS)**	**Max. Voice Payload Per Packet**	**Max. IPv4 Bandwidth**	**Max. IPv6 Bandwidth**	**Difference between IPv4 and IPv6**	
							Percentage of Increase
G.723.1	5.3 Kbps	3.80	0.15625 Kbit/Packet	15.840 Kbps	21.120 Kbps	5.280 Kbps	33.33%
	678.4 Bps		20 Byte/Packet	1980 Bps	2640 Bps	660 Bps	
G.729	8 Kbps	3.92	0.15625 Kbit/Packet	24.000 Kbps	32.000 Kbps	8.000 Kbps	33.33%
	1024 Bps		20 Byte/Packet	3072 Bps	4096 Bps	1024 Bps	
GSM	13 Kbps	3.50	0.257813 Kbit/ Packet	28.616 Kbps	36.456 Kbps	7.840 Kbps	27.39%
	1664 Bps		33 Byte/Packet	3577 Bps	4557 Bps	980 Bps	
G.728	16 Kbps	3.61	0.46875 Kbit/ Packet	26.400 Kbps	31.680 Kbps	5.280 Kbps	20.00%
	2048 Bps		60 Byte/Packet	3300 Bps	3960 Bps	660 Bps	

Kilo bit per second (Kbps); Bytes per second (Bps)

Figure 4. The maximum required bandwidth for IPv4/IPv6 VoIP calls using full rate voice with different voice codecs

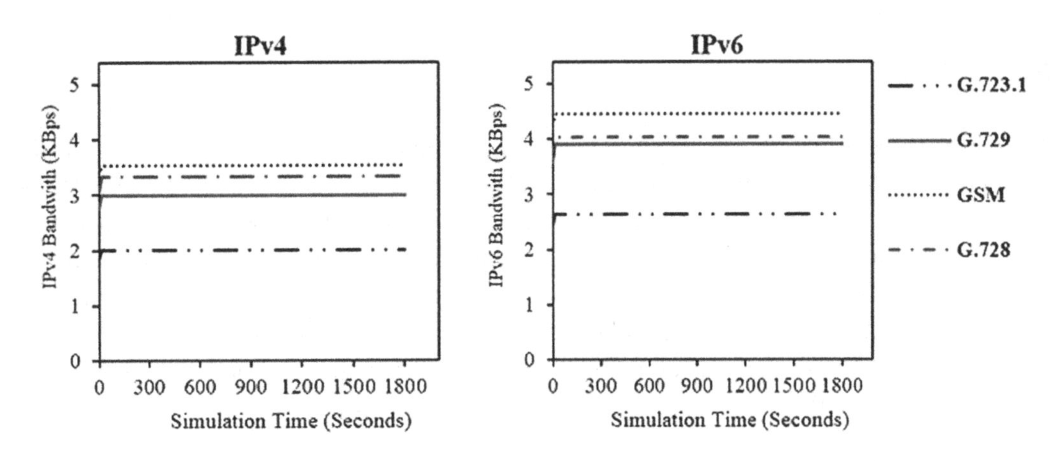

and destination from analog voice signals into digital signals and *vice versa*. The delays at the source and destination depend on the complexity and speed of the codec system used (Kondoz, 2004). In addition, the length of the voice frames and its headers count in the total delays for VoIP applications during the packetization and de-packetization processes. Furthermore, in wireless network systems, there are extra delay factors for VoIP applications compared with wired network systems because of its connectivity nature. In MANET, the delays vary as VoIP data transportation is affected by the features of Mobile

Ad hoc nodes such as the hops numbers between source and destination, Ad hoc routing protocol, node movement, mobility model, and traffic congestion. In summary, the total delays for voice data transport from source to destination are considered as a One-way delay. The One-way delay is the time difference for the same packet at source and destination (Metha & Udani, 2001). VoIP is a real-time application where RTP transmits the Voice packets. The delays of voice data packets should not exceed the acceptable level of One-way delays as identified by the ITU-T recommendations and shown in Table 2 (Metha & Udani, 2001) (ITU-T Rec, 2003).

Jitter

The successful arrival of sequential voice packets to the destination with different time delays is known as jitter. As voice packets transmit by RTP, it identifies the voice stream by using a unique Synchronization Source identifier (SSRC), port numbers, sequence numbers and timestamps. The variations in delay happen because of the different delays in the sender side or in the network (Sinnreich, & Johnston, 2012). The buffer in the receiver side used to overcome part of the jitter problem by reordering the received packets depending on the RTP timestamps and then it plays out the reordered received packets. When the buffer overflows because of the jitter, any more received packets will be dropped. Because of that, jitter could seriously affect the quality of voice as it is involved in the overall delays and packet drops of voice traffic. The jitter can be estimated using the Equation 13 below, where R represents the arrival time of a packet, S represents the RTP timestamp, and D (x,y) represents the arrival difference between two packets, packet x and packet y (Minoli, 2011) . The acceptable jitter for VoIP applications is 50 ms (Sun, 2005) .

$$D(x,\ y) = (R_y - R_x) - (S_y - S_x) = (R_y - S_y) - (R_x - S_x) \tag{13}$$

Packet Loss

The Packet loss for voice data is identified as when one or more voice packets fail to reach its destination which degrades the voice quality at the receiver side. Voice packets are carried on UDP, which is an unreliable transport layer protocol where packet loss could happen because of the traffic congestion, signal interference, or signal noise in the network systems. It also happens at the receiver side because of the buffer delay or the buffer overflow. In addition, delayed voice packets that arrive to the receiver side behind the scheduled voice playtime will be discarded (Voznak, Kovac, & Halas, 2012) . The probability of a packet being lost depends on the network and receiver status. The Packet loss percentage for mouth-to-ear voice traffic can be calculated by Equation 14 below. As the percentage of the packet

Table 2. One-way delay constraints for voice data (Metha & Udani, 2001)

ITU-T Recommendations G.114 for one-way voice data delays	
Under 150 ms	Acceptable
From 150 ms to 400 ms	Acceptable with limitations
Over 400 ms	Unacceptable

loss increases, the voice quality on the receiver side decreases. In VoIP calls, up to 10% of packet loss is acceptable (Metha & Udani, 2001). There are different ways to reduce the percentage of packet loss for voice traffic such as sending redundant phases of voice packets and using interleaving packets (Sinnreich, & Johnston, 2012).

$$Packet\ Loss\ = \left(\left(Packet\ Sent\ -\ Packets\ Received\right) / \left(Packets\ Sent\right)\right)\ X100 \quad (14)$$

Other Related Performance Metrics for Objective Methods

The packet latency that happens when it travels using a longer route until reaching its destination can cause the packet to be dropped because of its late arrival. The packet latency can be used as a performance metric for VoIP traffic. It could be avoided by using a shorter and faster route between the sender and the receiver. In addition, the out-of-order packet arrival is a performance metric which indicates that the received voice packets are using different routes. The out-of-order packet arrival causes the buffer to overflow, and packets drop. This could be resolved by using other supportive protocols for RTP that enhance the routing performance for voice packets (Kazemitabar, Ali, Nisar, Md Said, & Hasbullah, 2010) . Another QoS metric for VoIP applications is voice echo. Voice echo happens when the caller at the sender side hears his voice after he speaks where the callee does not notice this echo. Echo results from the reflection of the voice traffic that is sent back to the caller as an acoustic coupling between the phone's microphone and speaker. In addition, echo results from the voice packets' delays and becomes noticeable when its round trip delay becomes more than 50 ms (De Rango, Tropea, Fazio, & Marano, 2006) . Echo cancellers can be used when the round trip time for VoIP applications exceeds 100 ms to enhance the performance of the VoIP.

SUBJECTIVE METHODS

The Subjective method is a human-based approach to evaluate the voice performance. For SIP-based VoIP, the subjective method used to evaluate the VoIP applications performance depends on the average user's perception about the voice quality.

Mean Opinion Score (MOS)

Most of the telephony network systems have been using MOS as a subjective performance measurement tool for a long time. It is used to evaluate the voice quality by obtaining the human listener's opinion regarding the quality of the voice heard during a call session. The ITU-T Recommendation P.800 introduced an MOS subjective determination of voice quality based on a user's perception using a ranged measurement scale from 1 to 5, where 1 is the lowest perceived voice quality, and 5 is the highest perceived voice quality (ITU-T P.800, 1996). ITU-T proposed different objective testing methods using MOS. In this research study, the E-model was used to investigate the MOS values during the simulation stages as becomes the most widely used tool for objective evaluation of speech quality. The E-model had been introduced by ITU-T Recommendation G.107 and improved by the European Telecommunications

Standards Institute (ETSI), and the Telecommunications, and Industry Association (TIA). This model combines different impairment factors using an analytical equation to predict the voice QoS (ITU-T Rec. G.107, 2003). It estimates the R value which represents the voice quality in the range from 0 to 100 as shown in Equation 15:

$$R = R_0 - I_s - I_d - I_e + A \tag{15}$$

where: R_0 represents the signal to noise ratio, I_s represents the impairments such as too loud a speech level, I_d represents the mouth-to-ear delay, I_e represents impairments due to low bit rate voice codecs, and A represents the advantage of access that some systems have in comparison with PSTN which is determined by the G.107 standardization factors. R_0 and I_s values are related to the voice signal itself and not depending on the voice transmission process over IP networks. The R value of the E-model directly converted to its relative MOS value is shown in Table 3.

In this research effort, MOS is used as a subjective performance method for SIP-based VoIP applications. The default values of MOS measurements for OPNET® simulations are based on the G.107 and GSM recommendations.

Routing Protocol Performance Metrics

Real-time applications like VoIP, video and audio streaming, and collaborative group work, are attaining remarkable attention of researchers and practitioners to be implemented over Mobile ad-hoc networks for establishing communication with remote locations, and emergency areas where there is no availability of predefined infrastructure. However, the implementation of these applications over such environments is a challenging task due to the nature of MANET, like dynamic topology, resource limitations, vulnerable wireless links, and without a central controlling authority. Routing is an important strategy in MANET and various routing protocols are used for it. Each routing protocol needs to manage these MANET features, when it transfers the packets from source to destination via a multi-hop network. Moreover, these protocols need to provide and satisfy the Quality of Service (QoS) requirements of real-time applications to provide reliable communication with higher performance. At the same time each real-time application has different characteristics, so different requirements in the network are needed to provide reliable and synchronized data communication.

Table 3. ITU-T Rec. G.107 E-model, and MOS voice quality scores and descriptions (ITU-T P.800, 1996)

R: E-model	MOS	Voice Quality	Impairment Description
81 - 100	5	Excellent	Imperceptible
61 - 80	4	Good	Perceptible but Not Annoying
41 - 60	3	Fair	Slightly Annoying
21 - 40	2	Poor	Annoying
0 - 20	1	Bad	Very annoying

The performance of any routing protocol can be analysed using various metrics like qualitative metrics and quantitative metrics. Qualitative metrics are beneficial for evaluating internal efficiency of the routing protocol like bandwidth utilization, memory utilization and power consumption. Quantitative metrics are used for a comparative analysis between routing protocols on the basis of various parameters like mobility model, latency, and routing overhead. Some performance metrics are (Macker & Corson, 1998):

1. **Packet Delivery Ratio:** Packet delivery ratio is the ratio of total packet delivered to destination over total packet sent from source.

$$\text{Packet delivery ratio } \% = \frac{\text{total packets reached at destination}}{\text{total packet sent from source}} \times 100 \tag{16}$$

   It is an important metric for real-time applications as applications are using the UDP protocol like multimedia communication in Mobile ad-hoc networks; a lower packet delivery ratio may reduce the quality of communication to the end user.
2. **Traffic Packet Counts:** The total number of packets in relation to CBR traffic is measured at different layers of the protocol stack. Bit rate is used to measure packet count at the MAC layer, and the number of packets forwarded at the routing layer is also counted, to have a traffic packet count.
3. **Latency (Average End-to-End Delay):** Latency or end-to-end delay is the average time which is taken by any packet to reach its destination. This delay can be due to the delay in the route discovery process, processing delay, and delay from retransmission from multi-hops. This metric has wide applicability in real-time applications like VoIP and video conferencing, where time delay is not negotiable.

$$\text{Latency}\left(\text{Average end} - \text{to} - \text{end delay}\right) = \frac{\sum\left(\text{CBR sent time} - \text{CBR receive time}\right)}{\sum \text{CBR received packets}} \tag{17}$$

$where CBR = constant\, bit\, rate$

4. **Routing Traffic Count:** Total number of packets, which are related to routing traffic of various protocols like AODV, DSR and OLSR, are counted at different layers of the IP stack.
5. **Data Packet Loss:** Data packet loss can be possible on both the MAC layer as well as on the Network layer. When the packet is received by any node then it can be lost if the buffer of the receiving node is full or the TTL time of packet expires. Packet loss performance can have a significant effect on real-time applications.

$$\textbf{Packet loss} = \left(\textbf{total packet sent by source} - \textbf{packet receive by sink}\right) \tag{18}$$

6. **Throughput of Received Packets:** Throughput gives the fraction of the channel capacity used for useful transmission and the total number of packets received by the destination within a specified time frame.

$$\textbf{Throughput } \% = \frac{\sum_{n}^{1} \textbf{Total received packets (in bits)}}{\sum_{n}^{1} \textbf{Simulation time interval (in seconds)}} \times 100 \quad (19)$$

7. **Routing Overhead:** The total number of routing control packets sent by any node during data transmission is known as routing overhead. This metric is important to check the efficiency of any routing protocol.

$$\text{Normalized routing load\%} = \frac{\sum_{n}^{1} \text{Routing packets sent at MAC layer}}{\sum_{n}^{1} \text{CBR traffic received at Agent layer (AGT)}} \times 100 \quad (20)$$

where CBR = constant bit rate

8. **Normalized Routing Load:** Normalized routing overload is the total number of packets which are transmitted per data packet delivered to the destination. For each hop-to-hop delivery, the transmission of routing is supposed to be one complete transmission. This is the total summation of all control packets sent by all nodes for route discovery. This routing load is also calculated in terms of bytes rather than packets.

$$\text{Normalized routing load (bytes)\%} = \frac{\sum_{n}^{1} \text{Routing bytes sent at routing layer + source header overhead}}{\sum_{n}^{1} \text{CBR traffic received at Agent layer}} \quad (21)$$

9. **Normalized MAC Load:** It is measured as the total number of packets including the routing packet, Address Resolution Protocol (ARP) packet, and MAC control packets such as RTS, CTS, and ACK sent at the MAC layer for each data packet delivered to the destination. Normalized MAC load is also calculated using hop-to-hop transmission of routing and gives the measurement of both routing and MAC overhead associated with each data packet delivered. This is an indicator for media access utilization in a wireless medium. The lower the value of MAC load the better will be the protocol performance.

$$\text{Normalized MAC load} = \frac{\sum_{n}^{1} \text{MAC and routing packets sent at MAC layer}}{\sum_{n}^{1} \text{CBR received packet at destination}} \quad (22)$$

where CBR = constant bit rate

10. **Average Hop Count:** It is an estimation of path length from source to destination for packet delivery. The average hop count can be measured by dividing the total number of packets with constant bit rate transmitted over the MAC layer by the total number of packets with constant bit rate received at the destination node. It is calculated independently from routing protocol simulation.

$$\text{Average hop count} = \frac{\sum_{n}^{1} \text{CBR message sent at MAC layer}}{\sum_{n}^{1} \text{CBR message received at destination} \left(\text{Agent layer}\right)} \quad (23)$$

These are some quantitative performance metrics, used to analyse the performance of any routing protocol within specific environments. In addition, this bandwidth, throughput, and reliability are the other metrics, which are equally important for performance measurement of routing protocols. In the case of real-time applications such as multimedia transmission, it demands high end-to-end delay, and high jitterness values, whereas applications like VoIP and video conferencing demands low jitterness with higher delay values, for providing the desired results. Hence, real-time application requirements differentiate based on the specific scenario of each application.

Performance Metrics for MANET Routing Protocols

Variable numbers of applications have emerged in the past decade to fulfill the various needs of the network and end users. Each application has its own specific characteristics, hence its demands too. With the specific scenario of MANET, various properties such as vulnerable wireless links due to node mobility, and Signals interference produces challenges in providing reliable communication. Hence, each routing protocol must guarantee reliable data transfer, with delay sensitivity while delivering data from source to destination in a real-time environment. The Internet Engineering Task Force (IETF) (Macker & Corson, 1998) suggested some recommendations in RFC 2501 for evaluating the performance of routing protocols. Certain base criteria and performance metrics are provided to evaluate the performance of routing protocols. Performance metrics are divided into two parts as qualitative and quantitative metrics (Patil, 2012).

Qualitative Metrics

RFC 2501 suggests fundamental qualitative metrics for MANET routing protocols (Corson & Macker, 1999):

1. **Loop Freedom:** MANET, having fixed bandwidth, interference with neighbouring nodes, and with high probability of packet collision, hence it is necessary to prevent packet looping before its delivery to a destination node. TTL values are indicating the packet's dropping time if it reaches to its maximum hop count.
2. **Distributed Operation:** It suggests the way of interconnecting nodes under various distributed environments.

3. **Demand-Based Operation:** Bandwidth constraints in MANET or in wireless networks can be facilitated by the use of reactive-based or on-demand routing protocols to minimize the control packets in the network and preserves the power of mobile nodes, hence bandwidth utilization too.
4. **Proactive Operation:** Proactive behaviour of the routing protocols provides availability of routes to any destination at any time. In case of any link failure no extra latency is required for establishing a new route to the destination.
5. **Sleep Mode Operation:** Power management is an important aspect in MANET as each node is operated with a limited battery power source. To make the protocol more efficient it must be capable of working even when some nodes are in sleep mode for short periods without affecting its performance.
6. **Unidirectional Support:** Nodes in MANET must be capable of communicating with unidirectional links too. The routing protocol must ensure that they must support unidirectional and bidirectional links.

Quantitative Metrics

The quantitative metrics are defined in RFC2501 and can be used to compare and evaluate the performance of a routing protocol. The metrics are (Patil, 2012):

1. **Route Acquisition Time:** It indicates the total time taken by a protocol to discover a route. The higher the route discovery time, the higher the latency in the network.
2. **Out of Order Delivery:** The ratio of packets which are delivered out of order, also affects the performance of higher layer protocols, such as when TCP is used.
3. **Efficiency:** It can be measured by the total overhead used by a protocol to route the packet.
4. **End-to-End Delay Data Throughput:** It can be measured as the total time taken in the delivery of a packet from source to destination.

Internal Efficiency Metrics for the Routing Protocol

Various ratios need to be analysed for tracking the internal efficiency of a routing protocol. Some of the ratios are (Corson & Macker, 1999):

1. **Average Number of Data Bits Transmitted for Each Data Bit Delivered:** It is used for measuring the bit efficiency for delivering the data within a network. Hence, it also gives the average hop count to deliver the packet.
2. **Average Number of Control Bits Transmitted for Each Data Bit Delivered:** It is an indication of the bit efficiency of the protocol while expending the control overhead within a network to deliver data. This should not only include bits in the control packet but also the bits in the header of the data packet.
3. **Average Number of Both Control Bits and Data Packets Transmitted for Each Data Packet Delivered:** It is the measure of bandwidth efficiency in contention based link layers.

These quantitative metrics are based on same network parameters. Such metrics can be useful to evaluate the bandwidth utilization, memory and power consumption. Evaluating the routing protocol efficiency offered to higher layer protocols is also discussed in RFC 2501, and are summarized in the next subsection.

Network Performance Parameters

Various network parameters also have a significant impact while studying the behaviour of various routing protocols. Some network parameters such as network size, mobility, and dynamic topology need to be taken into consideration. The main network parameters are (Hsu, et al,, 2004):

1. **Network Size:** MANET is a self-configuring network, so the nodes can join or leave the network. The number of nodes in the network can have a significant effect on routing protocol performance, so network size needs to be taken into consideration while analysing any routing protocol.
2. **Dynamic Topology:** High node mobility in the network leads MANET into a highly unstable state. Therefore, node mobility has a significant impact on routing protocol performance.
3. **Network Connectivity:** Network connectivity is the average of neighbours with whom a node is communicating in one hop communication.
4. **Link Capacity:** This is the data rate between two nodes, remaining after packet loss due to MAC access and data encoding. Data rate has a significant effect on link capacity.
5. **Unidirectional Link:** In a routing protocol for a full duplex communication, if unidirectional links were used, then it may affect routing protocol efficiency.
6. **Mobility:** Use of different mobility models by a routing protocol for node mobility can affect the performance of the routing protocol.
7. **Sleeping Nodes Frequency:** Numbers of sleeping nodes to live nodes also have a significant impact on routing protocol performance.

Definition of Mobility Terms

Functioning of any routing protocol also depends on various factors like node mobility, connectivity of nodes, and node density. Such factors make the network dynamic in nature. Some of the mobility terms are (Macker & Corson, 1998):

1. **Nodes Density:** It is defined as the total number of nodes within a specified region. Node density for a simulation region can be obtained as:

$$Node\,density = \frac{Number\,of\,nodes}{Simulation\,area} \tag{24}$$

2. **Nodes Distance:** Node distance is the measurement of the total distance between two nodes. The distance between two nodes n_1 and n_2 within a network is given by a Euclidean distance.

$$Node\,distance\left(n_1, n_2\right) = \sqrt{\left(x_{n_1} - x_{n_2}\right)^2 + \left(y_{n_1} - y_{n_2}\right)^2} \tag{25}$$

Performance of AODV Routing Parameters

The main advantage of the AODV protocol is its loop-free nature generating less traffic overhead in the network. As for the reactive nature of AODV, the node itself starts a route discovery process to a destination when needed. AODV is designed to minimize the flooding in the network, therefore it uses the shortest possible route to the destination node. Furthermore, AODV reuses an already known updated route to the destination by using the TTL increment features. Each node starts to generate a Route Request (RREQ) packet with an initial value of TTL in the IP header. If the source node does not receive a Route Reply (RREP) after a certain time interval, it resends the RREQ packet and increases the TTL value and waiting time, unless and until the source node gets a RREP or reaches its maximum number of retries. Therefore, this mechanism takes a sufficient amount of time or a considerable delay for the route discovery process to the farthest nodes. Hence, in such a scenario, this protocol can raise issues like time delays for the real-time application. In multimedia applications, the waiting time during the route discovery process runs smoothly at the start. However, in case of link failure, a rerouting process is generated using TTL values. Hence, in the case of the farthest nodes, this process takes a longer time. Therefore, the route discovery process that uses TTL has a significant effect on AODV performance and can degrade the quality of application for end users. The Route Discovery Time (RDT) is the total time taken by the AODV protocol to discover the route, and is very important for protocol performance. During the route discovery process, when any node receives a RREQ, it first checks in its routing table whether a route to the destination exists or not. If it does not exist then it buffers the message and rebroadcasts the RREQ. Only after receiving a RREP message can the original message be sent. Hence as the number of nodes increases, the value of RDT will also be changed. Previous research also suggests that the value of RDT is not changing linearly (Albero, et al., 2012). This is because with each extra intermediate node in the route, the IP address has to be recorded at those intermediate nodes, hence the RREP will be larger in size. Hence, the size of RREP message will be larger for the farthest node in comparison to nearby nodes, and changes the RDT value too.

Further prior studies suggest that the use of HELLO messages by the AODV routing protocol has a significant effect on its performance while using with real-time applications (Javed & Prakash, 2012) (Albero, et al., 2012). HELLO messages are used by the AODV routing protocol for knowing the link status of its neighbours. Thus, each node in AODV has a link table, which maintains nodes' active communication by receiving constant HELLO messages from each neighbouring node. Each node analyses whether the links with its neighbouring nodes are stable or lost. Noise in the medium plays an important role in communicating within any application. If a HELLO message is received by any node after the waiting time due to noise, impairment in the network or due to distance to farthest node, then the node will analyse that the link is lost and it will update the entries in its routing table. In case the deleted node entry belongs to any active route, the protocol needs to restart the route discovery process and produces higher packet loss and latency in the network. This can affect the continuity of any multimedia application streaming as there will be random delay in the AODV protocol, which is not a favourable condition for real-time applications. Prior research also suggests that the value of random delay in the AODV protocol is higher than in other routing protocols (Albero, et al., 2012). With further studies, authors have also analysed that the average routing overhead of the AODV routing protocol during communication has higher values in comparison to protocols like OLSR (Goswami, Joardar, & Das, 2014) (Narayan S., et al., 2010). Another issue is reaction time in AODV. While dealing with the nodes' mobility during implementation of the AODV routing protocol for real-time applications and the nodes' mobility is on the higher side, the reaction time is also on the higher side. This high reaction time in AODV is not acceptable as it leads to communication termination, and become an unavoidable condition for real-time applications. With the increased number of nodes in the AODV routing protocol, its performance is becoming more critical. This is because the increasing number of nodes that will increase the routing overhead in the network. Therefore, the routing performance will be decreased.

Prior studies suggest that the AODV routing protocol has the lowest average latency and lowest jitter while comparing it with other protocols like OLSR, DSR, DSDV, and others (Samara, Karapistoli, & Economides, 2012), whereas AODV shows the lowest packet loss rate. For delay sensitive applications like multimedia applications, a protocol with low value of average end-to-end delay and low packet delay will be best suited, like DSR (Hosek & Molnar, 2011) . Thus, it is concluded that AODV can represent good performance, while having higher routing overheads. The AODV routing protocol has a higher packet delivery ratio due to its reactive nature. In general, it is concluded that the DSR and OLSR protocols outperform AODV within real-time applications (Hosek & Molnar, 2011). Hence depending upon various situations, the AODV protocol varies in its QoS metric performance.

Performance of OLSR Routing Parameters

As OLSR is a proactive routing protocol, it maintains routing tables to provide instant routes and updates it periodically, as explained in previous section. OLSR uses the Multipoint Relay (MPR) to minimize the rebroadcasting of HELLO messages in the network. Since each node in the OLSR routing protocol keeps a routing table, therefore a shortest path to destination is provided by this protocol. Because of the availability of various routes to the destination there will be no route discovery delay in case of any link failure. The OLSR protocol utilizes the topology control messages with information about link status after a fixed time interval and maintains a route to the destination.

OLSR uses various messages like the HELLO message and topology control messages in the network for the efficient working of the protocol. HELLO_INTERVAL and TC_INTERVAL time slots are used by HELLO and Topology Control (TC) messages, while updating the table and deleting the entries if timeout occurs. It means HELLO_MESSAGE and TC_MESSAGE are used for maintaining each nodes' state and are directly proportional to HELLO_INTERVAL and TC_INTERVAL simultaneously. Therefore, any change in HELLO_INTERVAL and TC_INTERVAL can change the performance of the OLSR protocol (Sundaram, S. Palani, & A. Babu, 2013). Within a network topology, there are various nodes, out of which some nodes are selected as MPR. In the case of any node, which is itself a MPR, and it moves away from the range of its neighbourhood nodes, then for a certain time period some of those neighbourhood nodes have stale information, unless these nodes get updated from some other new neighbouring nodes of that MPR node. It will take some multiple times of HELLO_INTERVAL. The performance of real-time applications will be affected more for higher time intervals in comparison to low time interval value.

Previous research also suggests that after any link failure, like an MPR left the network, new routes are available in the routing table but cannot be actually used until the topology control waiting time (Abolhasan, Hagelstein, & Wang, 2009). It can also degrade the performance of the applications. In the case where the value of the HELLO_INTERVAL is increased, it will improve the performance of the OLSR protocol, but it will increase the overhead in the network, which is not suitable for real-time applications. Prior research concluded that the OLSR protocol is better than AODV and DSR for real-time applications like the multimedia video streaming application, in the case of the route discovery process after the link failure (Abolhasan, Hagelstein, & Wang, 2009).

Packet loss and erroneous packets are not acceptable in real-time applications such as VoIP, where multimedia streaming causes the quality degradation. In the OLSR protocol, neighbourhood hold time interval is used for link expiry. In case of any node leaving from the transmission range of the neighbourhood node, the numbers of packet loss will increase hence the performance of the real-time applications will decrease. With further studies, it can be concluded that HELLO_INTERVAL and TC_INTERVAL are used for route stability, but the DEFAULT values of both can be ineffective for route discovery as a two-hop link cannot be established with frequent node mobility and frequent route changes (Ghosh, 2013). It means in this case the packet loss ratio will be higher and the delays for a new route discovery process will be higher, which needs to be addressed for reliability of real-time applications.

With the increase in the size of the network, the OLSR protocol outperforms other routing protocols in real-time scenarios, as OLSR has a lower ratio of packet drop because OLSR is a table-driven proactive multipath routing protocol. The average delay for the network is significantly small and almost constant with the variation of the network size. Since each node in OLSR has a routing table, therefore end-to-end delays in the OLSR protocol are almost constant. The packet delivery variation and jitter for the OLSR routing protocol is observed to be slightly higher. Therefore, high quality video and audio streaming in the OLSR protocol does not provide best results. The OLSR routing protocol has lower performance in terms of packet delivery ratio and jitter (Samara, Karapistoli, & Economides, 2012). Even the proactive nature of the protocol does not guarantee lowest values of jitter. Therefore, it can be argued that OLSR cannot be a better choice for hard real-time applications broadly.

CONCLUSION

The SIP-based VoIP applications are affected by the SIP signaling performance, voice quality, and the routing performance. The SIP end-to-end performance metrics of RFC 6076 are considered as the most efficient approach to evaluate SIP signaling. However, there are no benchmarking values proposed for these performance metrics. In this research study the RRD, SRD, and SDD will be used for the evaluation efforts for SIP signaling for the registration, call setup, and termination processes. In addition, the study considered the throughput, packet loss rate, end-to-end delays, and jitter for voice quality evaluation. Furthermore, the research study considered routing performance metrics of VoIP application processes to evaluate the routing efficiency. The study of these performance parameters will help to improve the overall performance for SIP-based VoIP applications by controlling the values of these parameters to be within the acceptable range (Jayakumar & Ganapathi, 2008). Further considerations regarding the performance metrics of MANET routing protocols in general have been discussed in this chapter. In addition, a review of related performance parameters for both AODV and OLSR have been covered as well.

REFERENCES

Abolhasan, M., Hagelstein, B., & Wang, J. C.-P. (2009). Real-world Performance of Current Proactive Multi-hop Mesh Protocols. *Communications, 2009. APCC 2009, 15th Asia-Pacific Conference on*. doi:10.1109/APCC.2009.5375690

Agbinya, J. I. (2010). *IP Communications and Services for NGN*. CRC Press.

Albero T., et al. (2012). *AODV Performance Evaluation and Proposal of Parameters Modification for Multimedia Traffic on Wireless Ad Hoc Networks*. Academic Press.

Ali, M., Liang, L., Sun, Z., & Cruickshank, H. (2011). Optimisation of SIP Session Setup for VoIP over DVB-RCS Satellite Networks. *International Journal of Satellite Communications Policy and Management, 1*(1), 55–76. doi:10.1504/IJSCPM.2011.039741

Babar, U., & Malliah. (2011). *Call Setup Delay Analysis of H. 323 and SIP*. Academic Press.

Boumezzough, M., Idboufker, N., & Ouahman, A. (2013). Evaluation of SIP Call Setup Delay for VoIP in IMS. In *Advanced Infocomm Technology* (pp. 16–24). Springer Berlin Heidelberg. doi:10.1007/978-3-642-38227-7_6

Brajdic, A., Suznjevic, M., & Matijaševic, M. (2009). Measurement of SIP Signaling Performance for Advanced Multimedia Services. *Proceedings of the 10th International Conference on Telecommunications*, 381-388.

Chang, L., Sung, C., Chiu, S., & Lin, Y. (2010). Design and Realization of Ad Hoc VoIP with Embedded p-SIP Server. *Journal of Systems and Software, 83*(12), 2536–2555. doi:10.1016/j.jss.2010.07.053

CISCO. (2009). *Voice over IP – Per Call Bandwidth Consumption*. Available: http://www.cisco.com/en/US/tech/tk652/tk698/technologies_tech_note09Goswami,a0080094ae2.shtml

Corson, S., & Macker. (1999). *Mobile Ad Hoc Networking (MANET): Routing Protocol Performance Issues and Evaluation Considerations*. RFC 2501.

Cortes, M., Ensor, J. R., & Esteban, J. O. (2004). On SIP performance. *Bell Labs Technical Journal, 9*(3), 155–172. doi:10.1002/bltj.20048

De Rango, F., Tropea, M., Fazio, P., & Marano, S. (2006). Overview on VoIP: Subjective and Objective Measurement Methods. *International Journal of Computer Science and Network Security, 6*(1), 140–153.

Eyers, T., & Schulzrinne, H. (2000). Predicting Internet Telephony Call Setup Delay. *Proc. 1st IP-Telephony Workshop*. 2000.

Fabini, J., Hirschbichler, M., Kuthan, J., & Wiedermann, W. (2013). Mobile SIP: An Empirical Study on SIP Retransmission Timers in HSPA 3G Networks. In Advances in Communication Networking, (pp. 78-89). Springer Berlin Heidelberg. doi:10.1007/978-3-642-40552-5_8

Fathi, Chakraborty, & Prasad. (2009). *Voice over IP in Wireless Heterogeneous Networks: Signaling, Mobility, and Security*. Springer Science Business Media B.V.

Fathi, H., Chakraborty, S., & Prasadm, R. (2006). Optimization of SIP Session Setup Delay for VoIP in 3G Wireless Networks. *IEEE Transactions on Mobile Computing, 5*(9), 1121-1132. doi:10.1109/TMC.2006.135

Ganguly, S., & Bhatnagar, S. (2008). VoIP: Wireless, P2P and New Enterprise Voice Over IP. Chichester, UK: Wiley. doi:10.1002/9780470997925

Gannoune & Robert. (2003). *A Survey of QoS Techniques and Enhancements for IEEE 802.11b Wireless LAN's*. Technical report, EIVD-Swisscom. IEEE.

Ghosh, S. (2013). *Comparative Study of QoS Parameters of SIP Protocol in 802.11a and 802.11b Network*. arXiv preprint arXiv: 1311.3184

Goswami, Joardar, & Das. (2014). Reactive and Proactive Routing Protocols Performance Metric Comparison in Mobile Ad hoc Networks NS 2. *Memory, 3*(1).

Grasic & Kos. (2012). Structuring the RCS Services on the IMS Application Layer Part I: Description and Comparison Parameters. *Elektrotehniski vestnik, 79*(3).

Grgic, T., & Matijasevic, M. (2013). Performance Metrics for Context-based Charging in 3GPP Online Charging System. *Telecommunications (ConTEL),201312th International Conference on*, 171-178.

GSM. (2000). *Full Rate Speech; Transcoding (GSM 06.10)*. ETSI std. EN 300 961 v. 8.1.1.

Gurbani, K., Jagadeesan, L., & Mendiratta, V. (2005). Characterizing Session Initiation Protocol (SIP) Network Performance and Reliability. In Service Availability, (pp. 196-211). Springer Berlin Heidelberg. doi:10.1007/11560333_16

Gutkowski, P., & Kaczmarek, S. (2012). The Model of end-to-end Call Setup Time Calculation for Session Initiation Protocol. *Bulletin of the Polish Academy of Sciences: Technical Sciences*, *60*(1), 95–101.

Happenhofer, M., Egger, C., & Reichl, P. (2010). Quality of Signaling: A New Concept for Evaluating the Performance of non-INVITE SIP Transactions. *Proceedings of the 22nd International Teletraffic Congress (ITC),* 1-8.

Hoeher, T., Petraschek, M., Tomic, S., & Hirschbichler, M. (2007). Evaluating Performance Characteristics of SIP over IPv6. *Journal of Networks*, *2*(4), 40–50. doi:10.4304/jnw.2.4.40-50

Hosek, J., & Molnar, K. (2011). Investigation on OLSR Routing Protocol Efficiency. *Proceedings of the 2011 international conference on Computers, digital communications and computing*.

Hsu, J. (2004). Performance of Mobile Ad Hoc Networking Routing Protocols in Large Scale Scenarios. *Military Communications Conference, 2004. MILCOM 2004. 2004 IEEE*, 1. doi:10.1109/MILCOM.2004.1493241

Husic, Hidic, Hadzialic, & Barakovic. (2014). *Simulation-based Optimization of Signaling Procedures in IP Multimedia Subsystem*. Academic Press.

ITU-T. (2004). *TR Q-series Supplements 51 Signaling Requirements for IP-QoS*. ITU-T.

ITU-T Recommendation P.800. (1996). *Methods for Subjective Determination of Transmission Quality*. ITU-T.

Javed, Q., & Prakash, R. (2012). Improving the Performance of Hybrid Wireless Mesh Protocol. *Mobile Ad hoc and Sensor Networks (MSN),2012Eighth International Conference on*. doi:10.1109/MSN.2012.35

Jayakumar, G., & Ganapathi, G. (2008). Reference Point Group Mobility and Random Waypoint Models in Performance Evaluation of MANET Routing Protocols. *Journal of Computer Systems, Networks, and Communications*, 13.

Karapantazis, S., & Stylianos, F. P. (2009). VoIP: A Comprehensive Survey on A Promising Technology. *Computer Networks*, *53*(12), 2050–2090. doi:10.1016/j.comnet.2009.03.010

Kazemitabar, H., Ahmed Ali, S., Nisar, K., Md Said, A., & Hasbullah, H. (2010). A Survey on Voice Over IP Over Wireless LANs. *World Academy of Science, Engineering and Technology*, *71*, 352–358.

Kondoz, A. M. (2004). *Digital Speech* (2nd ed.). John Wiley & Sons, Ltd. doi:10.1002/0470870109

Kulin, M., Kazaz, T., & Mrdovic, S. (2012). SIP Server Security with TLS: Relative Performance Evaluation. *Telecommunications (BIHTEL),2012IX International Symposium on*, 1-6. doi:10.1109/BIHTEL.2012.6412062

Macker, J., & Corson, M. (1998). Mobile Ad Hoc Networking and The IETF. *Mobile Computing and Communications Review*, *2*(1), 9–14. doi:10.1145/584007.584015

Malas, D., & Morton, A. (2011). *Basic Telephony SIP end-to-end Performance Metrics*. Technical Report RFC 6076. Internet Engineering Task Force (IETF). Retrieved from http://tools.ietf.org/html/rfc6076

Metha, P., & Udani, S. (2001, October-November). Voice over IP. *IEEE Potentials*, *20*(4), 36–40. doi:10.1109/45.969596

Minoli, D. (2011). *Voice over IPv6: Architectures for Next Generation VoIP Networks*. Newnes.

Mourtaji, I., Bouhorma, M., Benahmed, M., & Bouhdir, A. (2014). Proposition of a new approach to adapt SIP protocol to Ad hoc Networks. *International Journal of Software Engineering and Its Applications*, *8*(7), 133–148.

Munir, A. (2008). Analysis of SIP-based IMS Session Establishment Signaling for WiMax-3G Networks. *Networking and Services, 2008. ICNS 2008. Fourth International Conference on*, 282-287. doi:10.1109/ICNS.2008.7

Nahum, Tracey, & Wright. (2007). Evaluating SIP Server Performance. ACM SIGMETRICS Performance Evaluation Review, 35(1), 349-350.

Narayan, S. (2010). VoIP Network Performance Evaluation of Operating Systems with IPv4 and IPv6 Network Implementations. *Computer Science and Information Technology (ICCSIT),20103rd IEEE International Conference on*. doi:10.1109/ICCSIT.2010.5564004

Pack, & Lee. (2008). Call Setup Latency Analysis in SIP-based Voice over WLANs. *IEEE Communications Letters, 12* (2), 103-105. doi:10.1109/LCOMM.2008.071230

Pack, S., Jeong, P., & Kim, Y. (2010). Proactive Route Optimization in SIP Mobility Support Protocol. *Consumer Communications and Networking Conference (CCNC), 20107th IEEE*, 1–2. doi:10.1109/CCNC.2010.5421774

Patil, V. (2012). *Qualitative and Quantitative Performance Evaluation of Ad hoc on Demand Routing Protocol in MANET*. Academic Press.

Pirhadi, M., Hemami, S., & Tabrizipoor, A. (2009). Call Set-up Time Modeling for SIP-based Stateless and Stateful Calls in Next Generation Networks. *Advanced Communication Technology, 2009. ICACT 2009. 11th International Conference on, 2,* 1299-1304.

ITU-T Rec. G.723.1. (2006). *Dual Rate Speech Coder for Multimedia Communications Transmitting at 5.3 and 6.3 kbit/s*. ITU-T.

ITU-T Rec. G.728. (2012). *Coding of Speech at 16 Kbit/s Using Low-delay Code Excited Linear Prediction*. ITU-T.

ITU-T Rec. G.107. (2003). *The E Model, A Computational Model for Use in Transmission Planning*. ITU-T.

ITU-T Rec. G.114. (2003). *One-way Transmission Time*. ITU-T.

Rezac, F., Voznak, M., Partila, P., & Tomala, K. (2013). Interactive Video Audio System and Its Performance Evaluation. *Telecommunications and Signal Processing (TSP),201336th International Conference on*, 43-46. doi:10.1109/TSP.2013.6613888

Rozhon, Voznak, Tomala, & Vychodil. (2012). Updated Approach to SIP Benchmarking. *Telecommunications and Signal Processing (TSP),201235th International Conference on*, 251-254.

Rozhon, J., & Voznak, M. (2011). SIP Registration Burst Load Test. In Digital Information Processing and Communications, (pp. 329-336). Springer Berlin Heidelberg. doi:10.1007/978-3-642-22410-2_29

Samara, C., Karapistoli, E., & Economides, A. (2012). Performance Comparison of MANET Routing Protocols Based on Real-life Scenarios. *Ultra Modern Telecommunications and Control Systems and Workshops (ICUMT),20124th International Congress on*. doi:10.1109/ICUMT.2012.6459784

Schulzrinne, Casner, Frederick, & Jacobson. (2003). *RTP: A Transport Protocol for Realtime Applications*. RFC 3550.

Sinnreich, H., & Johnston, A. B. (2012). *Internet Communications Using SIP: Delivering VoIP and Multimedia Services with Session Initiation Protocol* (Vol. 27).Wiley.

Sun, Z. (2005). *Satellite Networking: Principles and Protocols*. John Wiley & Sons. doi:10.1002/047087029X

Sundaram, Palani, & Babu. (2013). Performance Evaluation of AODV, DSR and OLSR Mobile Ad Hoc Network Routing Protocols using OPNET Simulator. *International Journal of Computer Science & Communication Networks, 3*, 54-63.

Tang, J., Davids, C., & Cheng, Y. (2008). A Study of an Open Source IP Multimedia Subsystem Test Bed. *Proceedings of the 5th International ICST Conference on Heterogeneous Networking for Quality, Reliability, Security and Robustness*. doi:10.4108/ICST.QSHINE2008.3952

Tebbani, B., Haddadou, K., & Pujolle, G. (2009). A Session-based Management Architecture for QoS Assurance to VoIP Applications on Wireless Access Networks. *Consumer Communications and Networking Conference, 2009. CCNC 2009. 6th IEEE*, 1–5. doi:10.1109/CCNC.2009.4784757

Thisen, D. J. M., Espinosa, C., & Herpertz, R. (2009). Evaluating the Performance of an IMS/NGN Deployment. *Proceedings of the 2nd Workshop on Services, Platforms, Innovations and Research for new Infrastructures in Telecommunications*, 2561-2573.

Tuong, C., Kuthethoor, S., & Hadynski, K., & Parker. (2010). Performance Analysis for SIP-based VoIP Services over Airborne Tactical Networks. *Aerospace Conference*, 1–8.

Tzvetkov, V. (2008). SIP Registration Optimization in Mobile Environments Using Extended Kalman Filter. *Communications and Networking in China, 2008. ChinaCom 2008. Third International Conference on*, 106-111. doi:10.1109/CHINACOM.2008.4684980

Voznak, M., Kovac, A., & Halas, M. (2012). Effective Packet Loss Estimation on VoIP Jitter Buffer. In *Networking 2012 Workshops* (pp. 157–162). Springer Berlin Heidelberg. doi:10.1007/978-3-642-30039-4_21

Voznak, M., & Rozhon, J. (2010). SIP Back to Back User Benchmarking. *Wireless and Mobile Communications (ICWMC),20106th International Conference on*, 92-96. doi:10.1109/ICWMC.2010.86

Voznak, M., & Rozhon, J. (2012). SIP End To End Performance Metrics. *International Journal of Mathematics and Computers in Simulation*, *6*(3), 315–323.

Voznak, M., & Rozhon, J. (2013). Approach to Stress Tests in SIP Environment Based on Marginal Analysis. *Telecommunication Systems*, *52*(3), 1583–1593. doi:10.1007/s11235-011-9525-1

Voznak, M., & Rozhon, J. (2012). SIP Registration Stress Test. *6th International Conference on Communications and Information Technology*, 95-100.

Voznak, M., & Rozhon, J. (2010). Performance Testing and Benchmarking of B2BUA and SIP Proxy. *Conference Proceedings TSP*, 497-503.

Yahiaoui, S., Belhoul, Y., Nouali-Taboudjemat, N., & Kheddouci, H. (2012). AdSIP: Decentralized SIP for Mobile Ad hoc Networks. *Advanced Information Networking and Applications Workshops (WAINA),201226th International Conference on*, 490–495. doi:10.1109/WAINA.2012.151

Chapter 5
QoS and Performance Evaluation for SIP-Based VoIP Over DMO

Mazin I. Alshamrani
Ministry of Haj and Umra, Saudi Arabia

Ashraf A. Ali
The Hashemite University, Jordan & University of South Wales, UK

ABSTRACT

In this chapter, analyses for the performance metrics that define the quality of service (QoS) of SIP-based VoIP will be introduced. SIP-based VoIP applications over Direct Mode of Operation (DMO), which behaves in a way similar to Mobile Ad-hoc Network (MANET) systems, have three main performance categories related to the QoS. These categories are the SIP signaling, voice data transmission, and MANET routing. The SIP signaling controls the VoIP calls initiation, termination, and modifications. The major QoS parameters of VoIP that are managed by SIP signaling are the registration intervals, call setup time, and call termination time. These QoS parameters are increased in MANET due to the nodes' mobility that affects the routing calculations and the connectivity status. These necessitate mechanisms to reduce the delays in the MANET environment. The voice packets are transferred over the Real Time Protocol (RTP) which is encapsulated in the unreliable transport protocol using the User Datagram Protocol (UDP).

INTRODUCTION

In this chapter, analyses for the performance metrics that define the quality of service (QoS) of SIP-based VoIP will be introduced. SIP-based VoIP applications over Direct Mode of Operation (DMO), which behaves in a way similar to Mobile Ad-hoc Network (MANET) systems, have three main performance categories related to the QoS. These categories are the SIP signaling, voice data transmission, and MANET routing. The SIP signaling controls the VoIP calls initiation, termination, and modifications. The major QoS parameters of VoIP that are managed by SIP signaling are the registration intervals, call

DOI: 10.4018/978-1-5225-2113-6.ch005

setup time, and call termination time. These QoS parameters are increased in MANET due to the nodes' mobility that affects the routing calculations and the connectivity status. These necessitate mechanisms to reduce the delays in the MANET environment. The voice packets are transferred over the Real Time Protocol (RTP) which is encapsulated in the unreliable transport protocol using the User Datagram Protocol (UDP). In addition, the bandwidth consumption, delays, jitter and packet loss are QoS parameters that quantify VoIP performance. The bandwidth is dependent on the codec system used. The delays experienced by voice packets are one-way delays between the two calling ends and are affected by the routing and connectivity delays in the MANET environment. The jitter is related to the variations in the delay and the RTP tries to recover the loss for it, while the packet loss is related to the network congestion and erroneous links. A number of studies have been undertaken for these performance metrics that support the evaluation studies.

In this chapter, the simulation efforts have been carried out using GSM voice codecs to evaluate SIP call processes and QoS parameters together over MANET. The SIP signaling and QoS parameters for VoIP have been assessed on the OPNET® Modeler simulation scenarios. The simulation efforts have not considered other simultaneous applications that could influence the performance of the SIP applications to provide the effort implementations. However, the assumptions considered a background traffic with 30% to 40% of the overall bandwidth.

In this evaluation study, the results for both Static and Uniform mobility models are representing the best effort of the implemented scenarios. The results for the Static and Uniform scenarios are meeting with the evaluation results for similar scenarios over other network systems which support the reliability level of the findings of this research study. Hence, the investigated QoS parameters are considered as the benchmark values for SIP-based VoIP over AODV and OLSR MANET. These benchmarking efforts also considered the Registration Request Delays (RRD), Session Request Delays (SRD), and the Session Disconnect Delay (SDD) of the RFC 6076. These main SIP end-to-end metrics had been implemented in this research study to provide an evaluation for the SIP signaling over VoIP application between the SIP calls' entities as represented in Figure 2. The results show that these parameters are comparable for both IPv4 and IPv6 in AODV and OLSR MANET environments. Furthermore, the simulation efforts in this research study used to design and implement number of important parameters for SIP signaling performance evaluation. These parameters are the RRD, SRD, and RDD from the RFC 6076 together with the parameters of the call registration delay, and the call setup delay.

RELEVANT RESEARCH EFFORTS

In general, limited numbers of researchers have studied and evaluated the performance of real-time applications over MANET. Most of the evaluation efforts considered Constant Bit Rate (CBR) or File Transfer Protocol (FTP) traffic with a different number of MANET nodes. For IPv4 MANET, a performance evaluation with OPNET® Modeler v14.5 for the reactive routing protocols, AODV and DSR, using GSM voice traffic, concluded that AODV has the lowest end-to-end delay and a lower network load compared with DSR (Pandey & Swaroop, 2011). Furthermore, AODV presents higher average throughput and received traffic while DSR does not scale well with large sized networks. Simulation results also showed that AODV reactive routing protocol is the best suited for MANET, while DSR recorded very poor QoS in MANET with high node capacity for GSM voice applications. However, this research did not consider VoIP applications with different mobility models (Skordoulis, et al., 2008).

An evaluation for AODV as a reactive routing protocol and OLSR as a proactive routing protocol in OPNET® Modeler with a variable number of MANET nodes concluded that the performance of routing protocols vary depending on the network type and the selection of accurate routing protocols that affect the application's efficiency. In (Chiang, Dai, & Luo, 2012), three RWP-based mobility models have been implemented for MANET reactive routing protocols over different performance parameters. The results show that AODV in the RWP Waypoint mobility model performs better than TORA and DSR in the RWP walk and RWP direction mobility models. The study also concluded that AODV can be used with intensive mobility models. In (Shrestha & Tekiner, 2009), a performance comparison of selected MANET routing protocols has been performed in varying network sizes with increasing area and numbers of nodes to investigate RWP mobility and scalability of the routing process with high CBR traffic flow. AODV performed very consistently and established quick connection between nodes without delays while TORA conceded high end-to-end delay due to the formation of temporary loops within the network. AODV offers the best efficiency with high traffic applications compared to OLSR and TORA.

In (Kaur & Singh, 2012), an evaluation study used to compare different performance parameters between the most popular routing protocols in MANET: OLSR (proactive), TORA (reactive), and Geographic Routing Protocol (GRP) (hybrid). The study exploited FTP traffic over simulation models with different node capacity. The results show that OLSR offers the best performance in terms of load and throughput. However, it suffers considerable delays and routing overhead compared with other routing protocols as a result of the MPR nodes' selection. In (Palta & Goyal, 2012), a comparison between OLSR and TORA in terms of delay, retransmissions and data drop has shown that OLSR interacts with different nodes' update to reduce the delays and increase the throughput, knowing that the retransmission attempts are considered a real problem in OLSR. A method of proactive MANET routing protocol evaluation has been applied to the OLSR protocol, as proposed in (Thompson, MacKenzie, & DaSilva, 2011). The method analysed the performance of OLSR in medium sized MANET clusters using data from the MANIAC Challenge project specifically for OLSR evaluation. A performance evaluation study about AODV and OLSR routing protocols under realistic radio channel characteristics has already been implemented in ns-2 with Nakagami's fading model in (Khan & A. Qayyum, 2009). The study used CBR traffic with a uniform mobility model with a speed of 40kph. The results show that under realistic channel conditions in both routing protocols the system failed to deliver a good number of data packets to the destination nodes in highly fading environments.

In (Gandhi, Chaubey, Shah, & Sadhwani, 2012), a study has compared DSR, OLSR and ZRP using different mobile scenarios generated by the RWP Mobility model for MANET using CBR (UDP) traffic. The study used ns-2 and has shown that OLSR offers low average jitter and end-to-end delay with high throughput. The study in (Amnai, Fakhri, & Abouchabaka, 2011) has discussed the impact of mobility models and the density of nodes on the performances of OLSR using real-time VBR (MPEG-4) as well as Constant Bit Rate (CBR) traffic. The paper compared the performance in both cases in ns-2 over three mobility models: RWP Waypoint, RWP Direction, and Mobgen Steady State. The simulation results have shown that OLSR behaviour changes according to the traffic and the mobility model used, where RWP Waypoint has the optimal throughput. A proposed QoS extension model for OLSR MANET has been presented and evaluated in OPNET® Modeler for voice applications in (Sondi & Gantsou, 2009) . The simulation result has shown an improvement of the packet's delivery ratio by using the proposed QoS support model for voice communication over MANET compared with native OLSR. The study has focused on voice signaling using PCM voice codec. In (Ivov & Noel, 2004), a study illustrated the performance of real-time streaming media over a mesh OLSR-based network. The study has examined the effect of mobility and background traffic on carried load and jitter for media applications using IEEE 802.11 MAC/PHY with the EMANE software emulator.

On the other hand, few research efforts studied SIP-based applications over IPv6 MANET where the main concern is mobility issues. In (Ivov & Noel, 2004), a proposed solution for the optimization of SIP-based mobility over IPv6 was considered in an 802.11b network with a theoretical evaluation of the actual performance and the proposed work. The study modified the cross-layer triggers and Duplicate Address Detection (DAD) to reduce the handoff delays which improve SIP mobility. In (Chen, Yang, and Hwang, 2007), a design and implementation for a SIP-based MIP6-MANET system was proposed as an integration of Mobile IPv6 and SIP-based MANET. The system focused on providing an efficient handoff mechanism to reduce the handoff time and routing delay for Mobile Nodes (MN) roaming between different MANETs through IP clouds. In (Xiaonan & Shan, 2012), a mobility handover scheme was designed and implemented for IPv6 based AODV MANET by reducing the mobility handover cost and shortening the mobility handover delay.

VOIP CALLS EVALUATION

The simulation works focused on evaluating the SIP signaling performance. Therefore, the study assumptions used to generate many VoIP calls with short period to provide a comprehensive investigation for all sessions of the SIP signaling during the simulations over different MANET models. The results shown in this research study used the average representation method as it provides simple and comparable data representations during the evaluation assessments.

The VoIP calls performance in general can be evaluated from the statistical results provided by OPNET® Modeler from the VoIP calls successful ratio, number of connected calls, and the call durations (Hoeher, Petraschek, Tomic, & Hirschbichler, 2007) . These parameters are related to the performance of the three SIP signaling sessions: registration, initiation, and termination. The total estimated number of the initiated calls between the caller and the callee is 175 calls over all the scenarios where the duration of each call is 10 seconds. However, the simulation results shown that the percentage of the initiated VoIP calls from the total estimated number of the initiated VoIP calls in the best effort scenarios for AODV in the Static mobility model is 95.43% with IPv4 and 93.14% with IPv6, as shown in Table 1, while in the Uniform mobility model the percentage is 92.6% with IPv4 and 90.3% with IPv6. In the RWP mobility model, AODV has a percentage of 76% initiated calls with IPv4 and 69.21% with IPv6. In the RWP-All mobility model, AODV has 56% with IPv4 and only 22.3% with IPv6. On the other hand, OLSR has a percentage of 99.43% with IPv4 and 96% with IPv6 in the Static mobility model, while in the Uniform mobility model the percentage is 94.3% with IPv4 and 89.1% with IPv6, as shown in Table 1. In the RWP mobility model, OLSR has only 28% calls with IPv4 and 22.9% with IPv6. In the RWP-All mobility model, OLSR has 19.4% with IPv4 and only 1.73% with IPv6. Both AODV and OLSR had shown good performance in the Static and Uniform models as it represents the best effort scenarios. OLSR had shown a better percentage of the initiated calls compared to AODV. AODV showed an acceptable number of calls with the RWP mobility model while the RWP-All mobility model showed a low number of calls. OLSR has low numbers with the RWP mobility model and only three successful calls with the RWP-All mobility model. IPv4 in general has shown better performance than IPv6 over both AODV and OLSR.

Table 1. Number of initiated VoIP calls

	AODV IPv4	OLSR IPv4	AODV IPv6	OLSR IPv6
Static	167	174	163	168
Uniform	162	164	158	155
RWP	133	49	121	40
RWP-All	98	34	39	3

Table 2. Number of rejected VoIP calls

	AODV IPv4	OLSR IPv4	AODV IPv6	OLSR IPv6
Static	8	4	12	7
Uniform	12	11	14	16
RWP	42	133	67	144
RWP-All	79	140	125	173

The reduction in the VoIP calls' initiation process can be remarked from the increased number of rejected SIP calls in Table 2 with the RWP mobility models. The calls' rejections are caused by the SIP initiation packet drops between the two ends because of MANET connectivity conditions. In addition, the call rejections happen as a result of the callee being unreachable. The rejection percentage is very low with the Static and Uniform mobility models. The percentage is between 4.6% and 8% for AODV and between 2.3% and 9.14% for OLSR. The rejection percentage is increased with the RWP mobility model and is very high with the RWP-All mobility model where it is between 24% and 71.4% for AODV and between 76% and 98.8% for OLSR.

The average VoIP calls' duration has been investigated for AODV and OLSR over different traffic models, as shown in Figure 1. The call duration for a single VoIP call is determined from the SIP call registration, initiation, and termination processes. The ideal call duration for successful VoIP calls in this simulation work is 10 seconds as implemented and configured in OPNET® Modeler scenarios. The shorter the possible call duration shows the better SIP signaling performance it has. OLSR shows shorter calls' durations compared with AODV for both IPv4 and IPv6 over the Static and Uniform mobility models. The proactive behaviour of OLSR supported the registration and initiation processes for SIP signaling. In general, the optimised VoIP calls' duration for AODV and OLSR are existing over the Static and RWP mobility models. With the RWP mobility model, AODV IPv4 had acceptable call durations, whereas with AODV IPv6 the call durations are between 10.64 and 21.80 seconds for the successful VoIP calls. In the RWP-All mobility model, AODV had call durations of 13.44 to 33.85 seconds for the successful VoIP calls. OLSR had call durations between 15 and 21 seconds with IPv4 and 14 to 36 seconds with IPv6 for its successful VoIP calls over the RWP mobility model. For the RWP-All mobility model, the call durations for the successful VoIP calls were between 28 and 37 seconds with IPv4 and 35 to 55.36 seconds with IPv6. The long call durations for OLSR traffic is related to the long timers for SIP signaling for the initiation or the termination processes. In the simulation configuration steps for OLSR, the SIP termination timers for SIP calls were configured to be unlimited to allow VoIP calls to connect, otherwise no calls were connected over the OLSR RWP mobility models because of the default timers for SIP signaling.

SIP SIGNALING EVALUATION

The evaluation of SIP signaling performance for the simulated scenarios depends on the RFC 6076 SIP end-to-end performance metrics and the call setup time. In addition, the evaluation of SIP signaling considered the B2BUA-based SIP server performance regarding the SIP messages over the SIP registra-

Figure 1. Average call durations in seconds

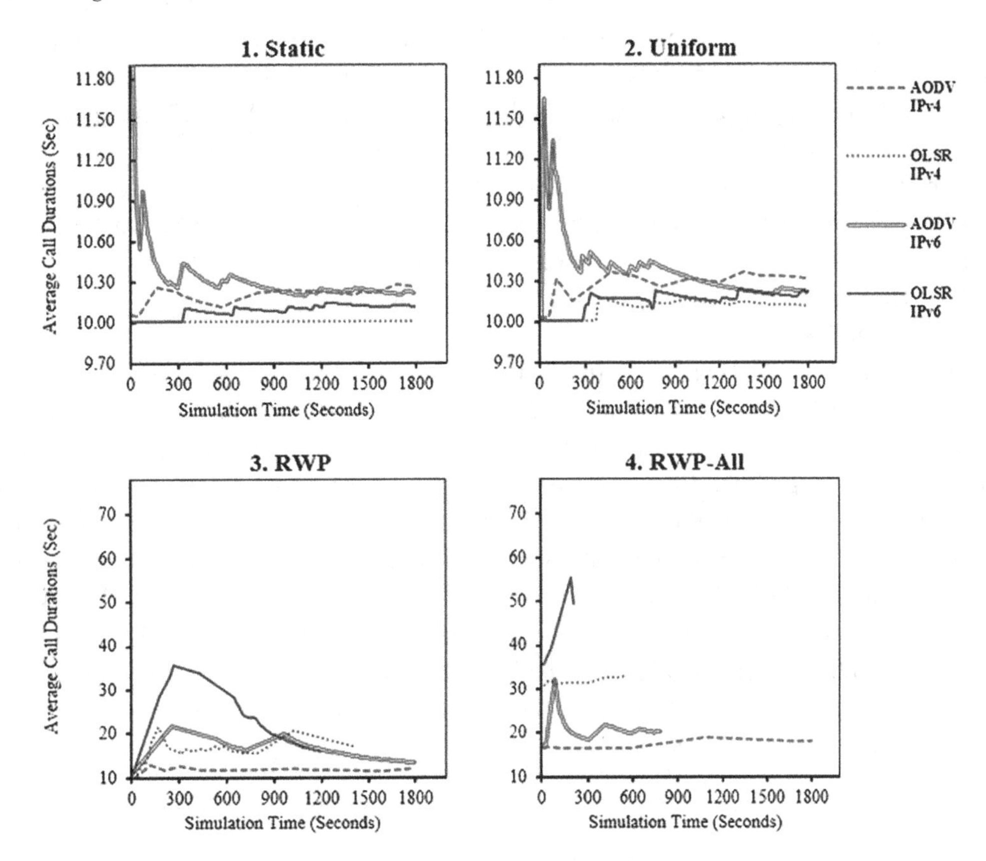

tion, initiation, and termination processes. As the main issue regarding the performance metrics is the benchmarking values, the simulation work shows results using the best effort scenarios over the Static and Uniform mobility models. Then, the collected results were compared with the RWP scenarios to find the related performance differences for SIP signaling. In this research study, the best effort values for SIP signaling over the Static and Uniform mobility scenarios were considered as the reference values for SIP performance metrics to compare them with other scenarios using different mobility models. In addition, the best effort values are used as suggested benchmarking values for SIP end-to-end performance that could be used to compare the SIP signaling performance over MANET for different mobility models.

SIP End-to-End Performance Metrics

The SIP signaling performance had examined three RFC 6076 performance metrics: the RRD, SRD, and SDD.

RRD Values

The RRD values for SIP over MANET for both AODV and OLSR scenarios are shown in Table 3. For AODV, RRD values are in its best effort with the Static mobility model over both IPv4 and IPv6, where the average RRD values are between 2 ms and 2.73 ms and the maximum RRD value was 10.45 ms.

In the Uniform model, the average values increased to be in the range of 14.33 ms to 32.41 ms, where the maximum value was 205.89 ms with IPv6. The RRD values for IPv6 are 25.3% more than IPv4 in the Static AODV scenario, and 55.8% more than IPv4 in the Uniform AODV scenario. For the RWP mobility model, the performance was affected by the random mobility of MANET nodes as the average RRD values are in the range of 578.84 ms to 874.24 ms where the maximum value is 2.84 seconds. The mobility of the SIP server in the RWP-All scenario affected the RRD values to be in the range of 1.41 seconds to 2.11 seconds while the maximum value was 5.69 seconds. The RRD values for IPv6 are 33.8% more than IPv4 in the RWP AODV scenario, and 34.2% more than IPv4 in the RWP-All AODV scenario. The RRD values over AODV IPv4 have lower values compared with AODV IPv6 with different mobility scenarios.

For OLSR, the average RRD values in the Static model are in the range of 1.59 ms to 2.26 ms, and the maximum RRD value was 8.85 ms. In the Uniform mobility models, the average RRD values are in the range of 13.61ms and 27.24 ms where the maximum RRD value was 191.51 ms. The RRD values for IPv6 are 30% longer than IPv4 in the Static OLSR scenario, and 50% longer than IPv4 in the Uniform OLSR scenario. For the RWP mobility model, the performance was influenced by the nodes' random mobility where the average RRD values are in the range of 1.27 seconds to 3.18 seconds where the maximum recorded value is 13.18 seconds. Furthermore, the RRD values for the RWP-All scenario are in the range of 3.49 seconds to 8.36 seconds while the maximum value was 17.9 seconds. The RRD values for IPv6 are 60.2% more than IPv4 in the RWP OLSR scenario, and 58.2% more than IPv4 in the RWP-All OLSR scenario. The RRD values with OLSR IPv4 showed smaller values compared with OLSR IPv6 over different mobility scenarios.

Table 3. RRD for SIP signaling over AODV and OLSR MANET in milliseconds (ms)

	AODV					
	IPv4			**IPv6**		
	Minimum	**Maximum**	**Average**	**Minimum**	**Maximum**	**Average**
Static	1.12	6.56	2.04	2.18	10.45	2.73
Uniform	4.47	182.34	14.33	4.82	205.89	32.41
RWP	108.43	2574.67	578.84	231.11	2835.47	874.24
RWP-All	307.69	4758.98	1405.72	717.82	5688.86	2105.41
	OLSR					
	IPv4			**IPv6**		
	Minimum	**Maximum**	**Average**	**Minimum**	**Maximum**	**Average**
Static	1.03	6.04	1.59	1.74	8.85	2.26
Uniform	3.85	137.43	13.61	3.24	191.51	27.24
RWP	698.93	6126.49	1268.37	1218.47	13176.13	3184.35
RWP-All	2419.59	8418.43	3495.62	4848.51	17870.02	8357.12

SRD Values

The SRD values for SIP over MANET for both AODV and OLSR scenarios are shown in Table 4. For AODV, the best values for SRD are with the Static mobility model where the values are between 16.32 ms and 24.76 ms and the maximum SRD value was 41.03 ms. In the Uniform model, the average value are in the range of 189.48 ms to 265.23 ms where the maximum value was 469.27 ms within IPv6 implementation. The SRD delays for IPv6 are 34.1% more than IPv4 in the Static AODV scenario, and 28.6% more than IPv4 in the Uniform AODV scenario. For the RWP mobility model, the random mobility of MANET nodes increased the SRD average value to be in the range of 1.03 seconds to 1.94 seconds where the maximum value is 3.32 seconds. On the other hand, the mobility of the SIP server in the RWP-All scenario increased the SRD values to be in the range from 2.24 seconds to 3.26 seconds while the maximum value was 12.28 seconds. The SRD values for IPv6 are 53.1% longer than IPv4 in the RWP AODV scenario, and 31.1% longer than IPv4 in the RWP-All AODV scenario. The AODV performance with IPv4 for SRD values is slightly better than IPv6 over different mobility scenarios.

For OLSR, the average SRD value in the Static model are in the range of 15.69 ms to 19.39 ms, and the maximum SRD value was 27.98 ms. In the Uniform mobility models, the average SRD values are in the range of 121.55 ms and 212.73 ms where the maximum SRD value was 581.27 ms. The average SRD values for IPv6 are 19% longer than IPv4 in the Static OLSR scenario, and 42.9% longer than IPv4 in the Uniform OLSR scenario. For the RWP mobility model, the performance of random mobility for MANET nodes is lower than Static and Uniform mobility models as the average SRD values are in the range of 3.59 seconds to 6.78 seconds where the maximum recorded value is 18.88 seconds. Furthermore, the SRD values for the RWP-All scenario are in the range of 6.38 seconds to 11.16 seconds while the maximum value was 34.36 seconds. The SRD values for IPv6 are 47.3% more than IPv4 in the RWP OLSR scenario, and 42.83% more than IPv4 in the RWP-All OLSR scenario. The SRD values over OLSR IPv4 have smaller values compared with OLSR IPv6 over different scenarios.

Table 4. SRD for SIP signaling over AODV and OLSR MANET in milliseconds (ms)

	AODV					
	IPv4			IPv6		
	Minimum	Maximum	Average	Minimum	Maximum	Average
Static	3.08	28.56	16.32	5.11	41.03	24.76
Uniform	5.42	345.67	189.48	7.42	469.27	265.23
RWP	485.04	2141.25	1028.29	684.36	3322.79	1943.57
RWP-All	1179.46	6147.81	2242.74	1408.15	12276.16	3255.35
	OLSR					
	IPv4			IPv6		
	Minimum	Maximum	Average	Minimum	Maximum	Average
Static	2.73	21.43	15.69	4.76	27.98	19.37
Uniform	6.65	417.92	121.55	6.81	581.27	212.73
RWP	1723.17	12804.41	3578.45	2329.71	18875.63	6783.82
RWP-All	4657.41	17876.68	6386.32	9587.54	34358.27	11164.74

SDD Values

The SDD values for SIP over MANET for both AODV and OLSR scenarios are shown in Table 5. For AODV, the SDD values are in its best case in the Static mobility model, where the average SDD values are between 1.48 ms and 2.02 ms and the maximum SDD value was 8.04 ms. For the Uniform model, the average values range increased to be in the range of 11.37 ms to 21.45 ms where the maximum value was 137.45 ms for IPv6. The SDD values for IPv6 are 26.7% more than IPv4 values in the Static AODV scenario, and 47% more than IPv4 in the Uniform AODV scenario. For the RWP mobility model, the average SDD values were affected by the random mobility of MANET nodes where the SDD values are in the range of 453.78 ms to 737.19 ms where the maximum value is 1.19 seconds. Furthermore, the mobility of the SIP server in the RWP-All scenario affected the SDD values to be in the range from 1.19 seconds to 1.79 seconds while the maximum value was 4.05 seconds. The SDD values for IPv6 is 38.4% more than IPv4 in the RWP AODV scenario, and 33.7% more than IPv4 in the RWP-All AODV scenario. The AODV performance for the SDD IPv4 has better performance compared with IPv6 over different mobility scenarios.

For OLSR, the average SDD values in the Static model are in the range of 1.03 ms to 1.95 ms, and the maximum SDD value was 6.79 ms. In the Uniform mobility models, the average SDD values are in the range of 11.08 ms and 19.39 ms where the maximum SDD value was 226.65 ms. The SDD values for IPv6 are mostly 47.2% more than IPv4 in the Static OLSR scenario, and 42.9% more than IPv4 in the Uniform OLSR scenario. For the RWP mobility model, the performance was affected by the nodes' random mobility where the average SDD values are in the range of 1.09 seconds to 1.9 seconds where the maximum recorded value is 6.71 seconds. Furthermore, the SDD values for the RWP-All scenario are in the range of 1.82 seconds to 5.41 seconds while the maximum value was 14.6 seconds. The SDD values for IPv6 are mostly 42.33% longer than IPv4 in the RWP OLSR scenario, and 66.41% longer than IPv4 in the RWP-All OLSR scenario. In general, the OLSR performance for SDD with IPv4 shows lower values and better performance compared with IPv6 over different mobility scenarios.

Table 5. SDD for SIP signaling over AODV and OLSR MANET in milliseconds (ms)

	AODV					
	IPv4			IPv6		
	Minimum	**Maximum**	**Average**	**Minimum**	**Maximum**	**Average**
Static	1.16	4.13	1.48	1.86	8.04	2.02
Uniform	3.86	107.22	11.37	4.68	137.45	21.45
RWP	83.68	1558.64	453.78	178.64	1193.08	737.19
RWP-All	252.36	3871.73	1187.81	613.38	4045.04	1790.23
	OLSR					
	IPv4			**IPv6**		
	Minimum	**Maximum**	**Average**	**Minimum**	**Maximum**	**Average**
Static	0.82	5.18	1.03	1.45	6.79	1.95
Uniform	3.71	146.14	11.08	4.12	226.65	19.39
RWP	489.13	3712.68	1088.14	814.91	6698.92	1886.68
RWP-All	1256.14	6452.16	1818.38	1912.51	14582.47	5412.85

The comparison of the results for RRD, SRD, and SDD have shown that AODV and OLSR are slightly similar with delays over the Static and Uniform scenarios for both IPv4 and IPv6. However, AODV has shorter delays for the RWP and RWP-All scenarios compared with OLSR that showed longer delays. In general, the SDD average values are lower than the RRD values as the registration process has a simple direct signaling system. Furthermore, the nature of session request processes affects the SRD values. The SRD values are representing the most considerable delays for SIP signaling over MANET for both AODV and OLSR.

SIP Registration Interval

The registration process for SIP clients is the first stage of the SIP call initiation. The registration happens once in the beginning of the network setup and is repeated each time the calling nodes use it to change its status by re-joining the network or reconnecting with the SIP server. Table 6 shows the average registration time for both the caller and the callee to register with the SIP server during the provided registration period. For AODV, the average registration time values for SIP sessions over the Static mobility model for both IPv4 and IPv6 are between 3.13 ms and 4.21 ms and the maximum registration time was 17.12 ms. In the Uniform model, the average registration time values increased to be in the range of 37.72 ms to 52.83 ms where the maximum value was 395.15 ms with IPv6. The registration time values for IPv6 are 25.7% more than IPv4 in the Static AODV scenario, and 28.6% more than IPv4 in the Uniform AODV scenario. For the RWP mobility model, the random mobility of MANET nodes increased the average registration time values to be in the range of 812.91 ms to 1.42 seconds where the maximum value is 3.83 seconds. Furthermore, the mobility of the SIP server in the RWP-All scenario added further delays to the registration time values to be in the range from 1.98 seconds to 3.79 seconds while the maximum value was 13.37 seconds. The registration time values for IPv6 are 42.56% more than IPv4 in the RWP AODV scenario, and 47.84% more than IPv4 in the RWP-All AODV scenario. The registration time values for AODV IPv4 have shorter registration times compared with AODV IPv6 over different mobility scenarios.

For OLSR, the average registration time values in the Static models are in the range of 3.32 ms to 5.13 ms, and the maximum value was 15.62 ms. In the Uniform mobility models, the average registration time values are in the range of 29.54 ms and 44.83 ms where the maximum registration time value was 434.45 ms. The average registration time values for IPv6 is 35.28% more than IPv4 in the Static OLSR scenario, and 34.12% more than IPv4 in the Uniform OLSR scenario. For the RWP mobility model, the nodes' random mobility increased the average registration time values to be in the range of 2.49 seconds to 5.33 seconds where the maximum recorded value is 23.17 seconds. Furthermore, the average registration time values for the RWP-All scenario are in the range of 6.02 seconds to 13.82 seconds while the maximum value was 25.77 seconds. The average value of the registration time for IPv6 is 53.29% more than IPv4 in the RWP OLSR scenario, and 56.46% more than IPv4 in the RWP-All OLSR scenario. In general, the average registration time values with OLSR IPv4 have shorter registration times compared with OLSR IPv6 over different mobility scenarios.

SIP Call Setup Interval

Table 7 shows the number of SIP/TCP retransmission attempts during the call setup process. The number of retransmission attempts increased when one or more of the call setup messages had been lost or

Table 6. SIP registration time for SIP clients over AODV and OLSR MANET in milliseconds (ms)

	AODV					
	IPv4			IPv6		
	Minimum	Maximum	Average	Minimum	Maximum	Average
Static	2.05	9.72	3.13	3.07	17.12	4.21
Uniform	9.42	248.18	37.72	7.71	395.15	52.83
RWP	138.31	3112.1	812.91	651.14	3813.21	1415.16
RWP-All	698.26	9177.17	1976.12	1576.22	13369.54	3789.19
	OLSR					
	IPv4			IPv6		
	Minimum	Maximum	Average	Minimum	Maximum	Average
Static	2.65	8.59	3.32	2.27	15.62	5.13
Uniform	8.31	207.17	29.54	6.81	434.45	44.83
RWP	981.36	9235.17	2489.19	1981.48	23174.78	5326.59
RWP-All	2853.24	17276.43	6017.38	5341.66	25769.87	13819.43

delayed. With the Static model, AODV had 4 to 5 SIP/TCP retransmissions, while with the Uniform mobility model, it had 6 to 10 retransmissions. The number of retransmission attempts for IPv6 is 20% more than IPv4 in the Static AODV scenario, and 40% more than IPv4 in the Uniform AODV scenario. Because of the random mobility of MANET nodes, the number of SIP/TCP retransmission attempts increased to be 23 to 42 for the RWP mobility model, and 48 to 87 for the RWP-All mobility model. This also shows that the number of retransmission attempts for IPv6 has 45.24% more than IPv4 in the RWP AODV scenario, and 44.83% more than IPv4 in the RWP-All AODV scenario. On the other hand, OLSR has 14 to 19 SIP/TCP retransmissions, while with the Uniform mobility model, it has 20 to 36 retransmissions. Thus, the number of retransmission attempts for IPv6 is 26.32% more than IPv4 in the Static OLSR scenario, and 44.44% more than IPv4 in the Uniform OLSR scenario. Furthermore, the random mobility of MANET nodes increased the SIP/TCP retransmission attempts to be 49 to 83 for the RWP mobility model, and 80 to 141 for the RWP-All mobility model. Thus, the number of the retransmission attempts for IPv6 are 41% more than IPv4 in the RWP OLSR scenario, and 43.3% more than IPv4 in the RWP-All AODV scenario.

In Figure 2, the average call setup time showed variable delays over the RWP mobility models compared with the static level of the call setup over the Static and Uniform scenarios. The average call setup time is always more than the values of the SRD performance metrics over all SIP calls. The average call setup

Table 7. Number of SIP/TCP retransmission attempts

	AODV IPv4	OLSR IPv4	AODV IPv6	OLSR IPv6
Static	4	14	5	19
Uniform	6	20	10	36
RWP	23	49	42	83
RWP-All	48	80	87	141

Figure 2. Average SIP call setup time in seconds

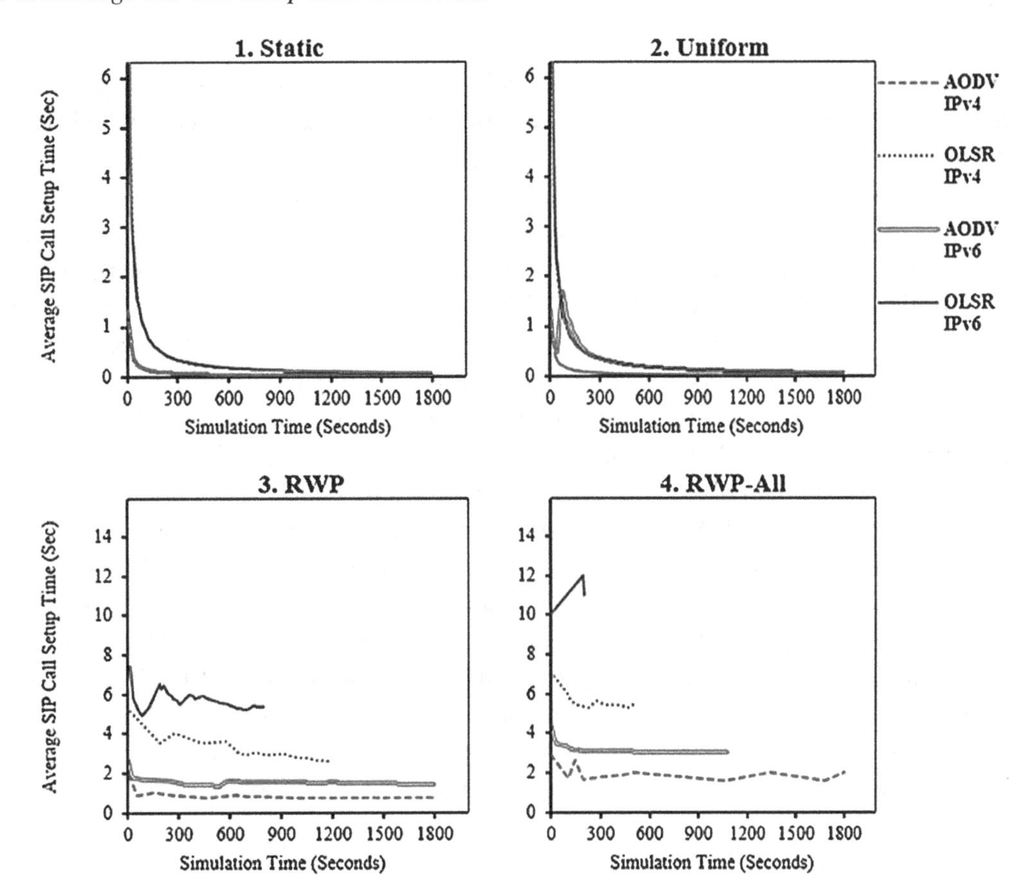

time is similar on AODV and OLSR, as it is in the range of 17.15 ms to 278 ms. Furthermore, OLSR has shown slightly better performance compared with AODV for the Static and Uniform scenarios. In the RWP scenarios, the average call setup time for AODV is in the range of 0.93 seconds to 2.24 seconds, while in the RWP-All model the average call setup time is in the range of 1.56 seconds to 4.43 seconds. The average call setup time for IPv6 is 46.1% more than IPv4 for the RWP AODV scenario, and 39.3% more than IPv4 for the RWP-All AODV scenario. For OLSR, the average call setup time in the RWP scenarios is in the range of 2.93 seconds to 7.18 seconds, while in the RWP-All model, the average call setup time is in the range of 5.14 seconds to 12.19 seconds. The average call setup time for IPv6 is 47.8% more than IPv4 for the RWP OLSR scenario, and 44.1% more than IPv4 for the RWP-All OLSR scenario. In general, IPv4 showed better performance over all scenarios compared with IPv6 for both AODV and OLSR. These results support the research findings in (Boumezzough, Idboufker, and Ouahman, 2013 and (Tebbani, Haddadou, and Pujolle, 2009) which shows that IPv6-based SIP has longer delays, and size compared with IPv4-based SIP. In addition, the call setup process has the longest delays compared with other SIP call processes over both AODV and OLSR.

SIP Call Termination Interval

The call termination process between two callers in a MANET environment has a lower affect over the SIP signaling performance compared with the registration and the call setup processes. This is because of the call termination process usage for a simple direct form of SIP messages as represented in Figure 2. In addition, the SIP call termination time has the same values of the SDD performance metric, which makes it simpler to detect the delays and refer it to the termination performance for the benchmarking values. According to (Tebbani, Haddadou, & Pujolle, 2009), the maximum recommended delays for SIP call termination should not exceed 1 second. Thus, based on the SDD values for the investigated scenarios, the termination delays in both the Static and Uniform scenarios are at the accepted level of delays for both AODV and OLSR. However, for the RWP and RWP-All scenarios, the termination time had exceeded 1 Second which reflects on the termination performance over the overall QoS. Therefore, the average termination time for AODV over the RWP mobility model is in the range of 453.78 ms to 737.19 ms where the maximum time is 1.19 seconds as shown in Table 5. For the RWP-All, the average termination time for AODV RWP-All scenarios is in the range from 1.19 seconds to 1.79 seconds while the maximum time is 4.05 seconds. Thus, the average termination time for IPv6 is 38.4% more than IPv4 in the RWP AODV scenario, and 33.7% more than IPv4 in the RWP-All AODV scenario. For the OLSR RWP mobility model, the average termination time is in the range of 1.09 seconds to 1.9 seconds where the maximum time is 6.71 seconds. Furthermore, for the OLSR RWP-All scenario the average termination time is in the range of 1.82 seconds to 5.41 seconds while the maximum time was 14.6 seconds. Thus, the average termination time for IPv6 is mostly 42.33% more than IPv4 in the RWP OLSR scenario, and 66.41% more than IPv4 in the RWP-All OLSR scenario.

SIP Server Efficiency

The SIP server efficiency during the registration, call setup, and termination processes has an important role over the overall performance of the SIP call. Figure 3 shows the average number of active SIP/TCP connections each 5 seconds over the B2BUA-based SIP server during the simulation time. In the Static and Uniform scenarios, the average number of active SIP/TCP connections each 5 seconds on the SIP server is 6 to 8 per 5 seconds for both AODV and OLSR, whereas the SIP server performance for IPv4 scenarios is slightly better compared with IPv6. In the RWP scenarios, AODV scenarios have 5 to 7 average number of active SIP/TCP connections per 5 seconds in the SIP server, while the OLSR scenarios have an active number of SIP/TCP connections between 4 and 5 active SIP/TCP connections per 5 seconds in the SIP server. In the RWP-All scenarios, the IPv4 AODV has a performance of 4 to 5 average number of active SIP/TCP connections per 5 seconds in the SIP server, while in the IPv6 AODV and IPv4 OLSR scenarios, the number of active connections at the SIP server had dropped from the range of 4 to 5 per 5 seconds, to the range of 1 to 2 active connections per 5 seconds as most of the connections between MANET nodes had been lost after a while because of the nodes' movement start. For IPv6 OLSR over the RWP-All, the average of active SIP/TCP connections start with 2 to 3, and then it dropped down to 1 active connection per 5 seconds. In general, the number of active SIP/TCP connections reflects the status of the SIP signaling performance over different SIP processes. As the number of active SIP/TCP connections increased, the SIP signaling performs better.

Figure 3. Average number of active SIP/TCP connection per 5 seconds in the SIP server

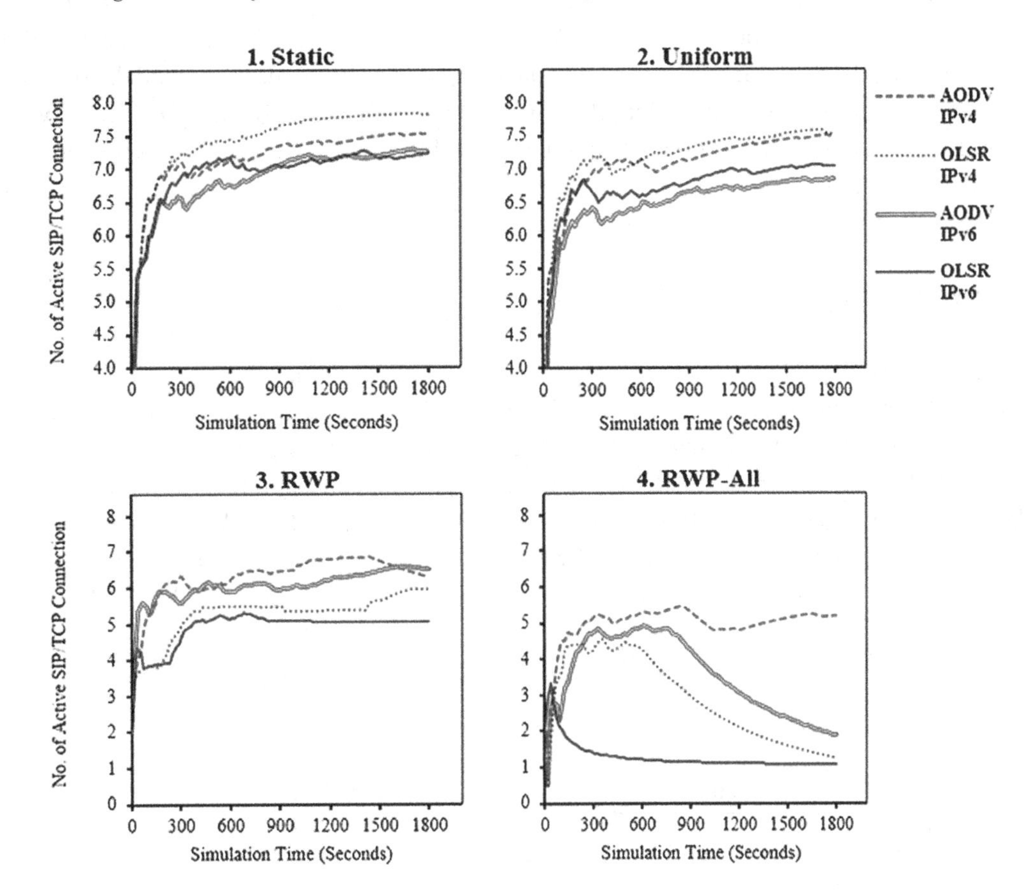

VOICE QOS

When the caller receives the call initiation acknowledgment from the callee, the voice RTP/UDP traffic starts flooding between each end through different MANET nodes. The voice performance metrics had investigated the voice traffic from the caller (node 1) to the callee (node 24) over four mobility models. The voice performance metrics had investigated five measures, as detailed below.

Throughput

In OPNET® Modeler, voice streams can be analysed using RTP statistics to determine the total volume of sent and received voice traffic per second to find out the actual voice throughput transferred from the caller to the callee. The average number of sent voice traffic from the caller (node 1) for the investigated MANET protocols is shown in Figure 4. On the other hand, the received voice traffic by the callee (node 24) is shown in Figure 5. The average received voice traffic by the receiver needs to be as close as possible to the average sent by the sender with a low percentage of packet drops. The best effort for the total sent and received voice traffic for both AODV and OLSR is with Static and Uniform mobility models.

For the RWP mobility models, in general, AODV has better throughput compared with OLSR. The average of voice traffic throughput for IPv4 AODV is 27.23% more than the IPv6 AODV average throughput. On the other hand, OLSR showed a very low throughput over RWP mobility models. Furthermore, the average of voice traffic throughput for IPv4 AODV is 48.52% more than the IPv6 AODV average throughput. In general, the results showed that the bandwidth consumption of IPv6 is higher than IPv4 because of its larger header size. However, for the RWP mobility models, IPv4 has higher consumptions than IPv6 because it has more initiated VoIP calls.

RTP One-way Delay

The one-way RTP/UDP packet delays for voice streams had been examined between the caller and the callee for VoIP applications. It was used to find the time difference between the received RTP/UDP packets in the receiver side depending on the RTP packet parameters, which are Timestamps, Unique Synchronization Source identifiers (SSRC), port numbers, and sequence numbers. The average RTP one-way delays had been measured for all the generated VoIP calls over both IPv4 and IPv6 as represented in Figure 6. The delay range for AODV and OLSR with Static and Uniform mobility are acceptable as it is from 143 ms to 170 ms. OLSR showed lower one-way delays compared to AODV for Static and

Figure 4. Average voice traffic sent from the caller (Node 1) to the callee (Node 24)

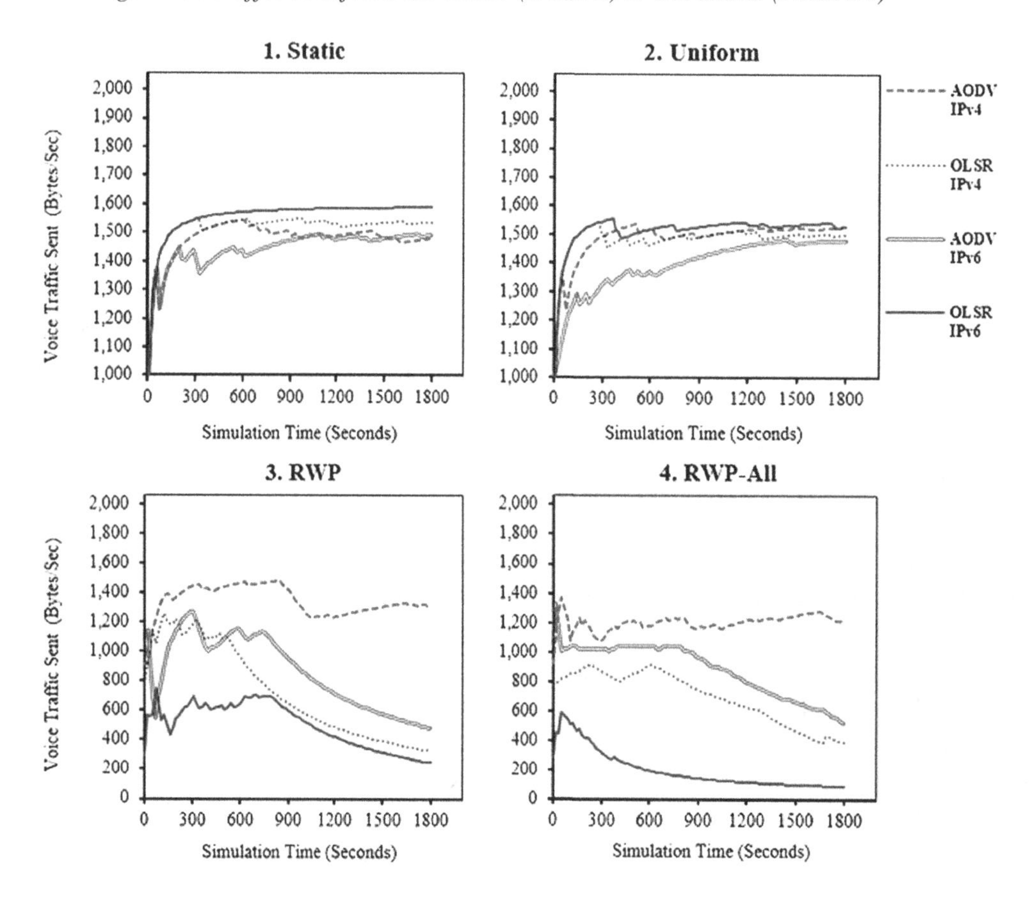

Figure 5. Average voice traffic received by the callee (Node 24) from the caller (Node 1)

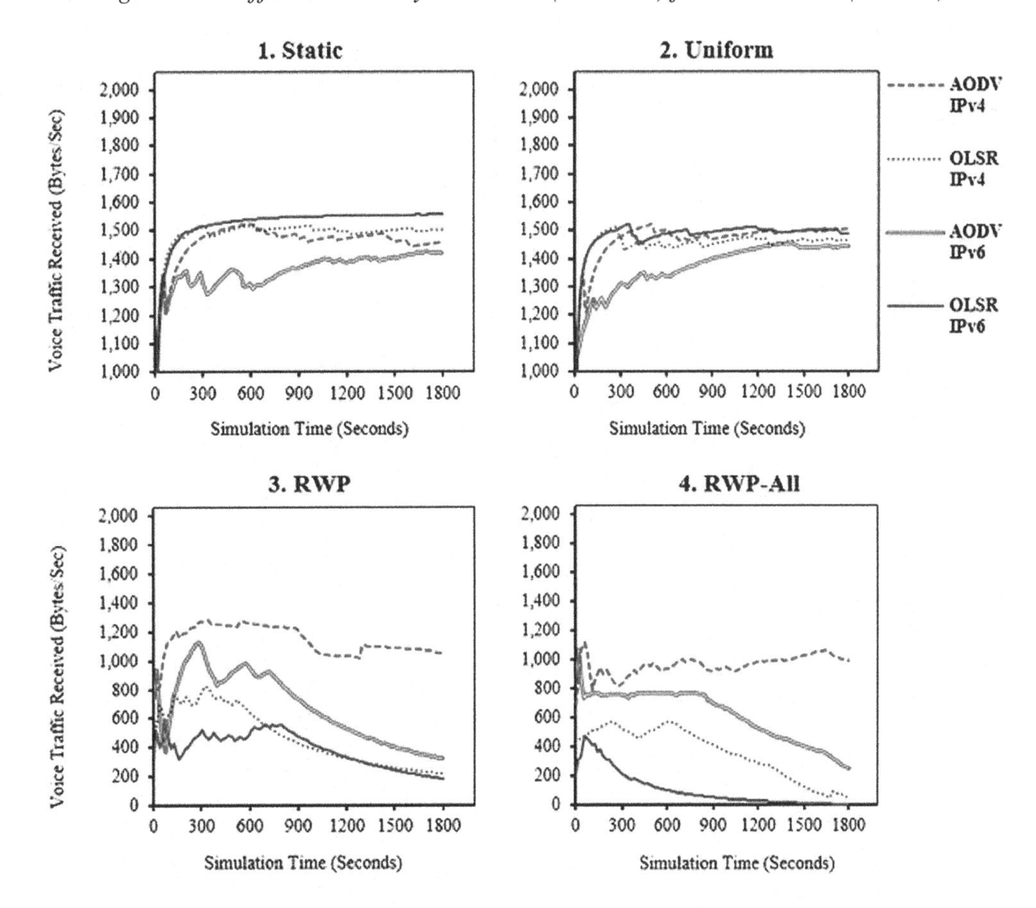

Uniform mobility models because of the proactive routing nature for OLSR that provides more reachability choices for RTP/UDP packets with lower delays (Khan & A. Qayyum, 2009) .

On other hand, RWP mobility models showed longer delays for those successfully received RTP packets. For AODV, the delays are in the range of 156 ms to 166 ms for RWP mobility while it is in the range of 175 ms to 208 ms for the RWP-All mobility model. OLSR has lower packets delivered with end-to-end delays between 216 ms and 223 for RWP mobility, and between 196 ms and 235 ms for the RWP-All mobility model. The header processing for IPv4 and IPv6 RTP packets affects the one-way delays because of the header sizes. However, the overall average of end-to-end delays for RTP/UDP packets that were successfully received are within the acceptable delay range for VoIP applications. Therefore, there is no significant deference between IPv4 and IPv6 in the RTP one-way delays.

Jitter

OPNET® Modeler implements the jitter determination equation to calculate the maximum jitter for voice packets of the investigated VoIP applications over MANET as shown in Table 8. In the Static and Uniform mobility models, the maximum jitter for both AODV and OLSR varies between 2.86 ms and 5.23 ms which is an acceptable variation range for VoIP applications (Xiaonan and Shan, 2012) . AODV in both RWP mobility models has an acceptable jitter range for both IPv4 and IPv6, except with

Figure 6. Average voice data end-to-end delay for traffic from (Node 1) to (Node 2)

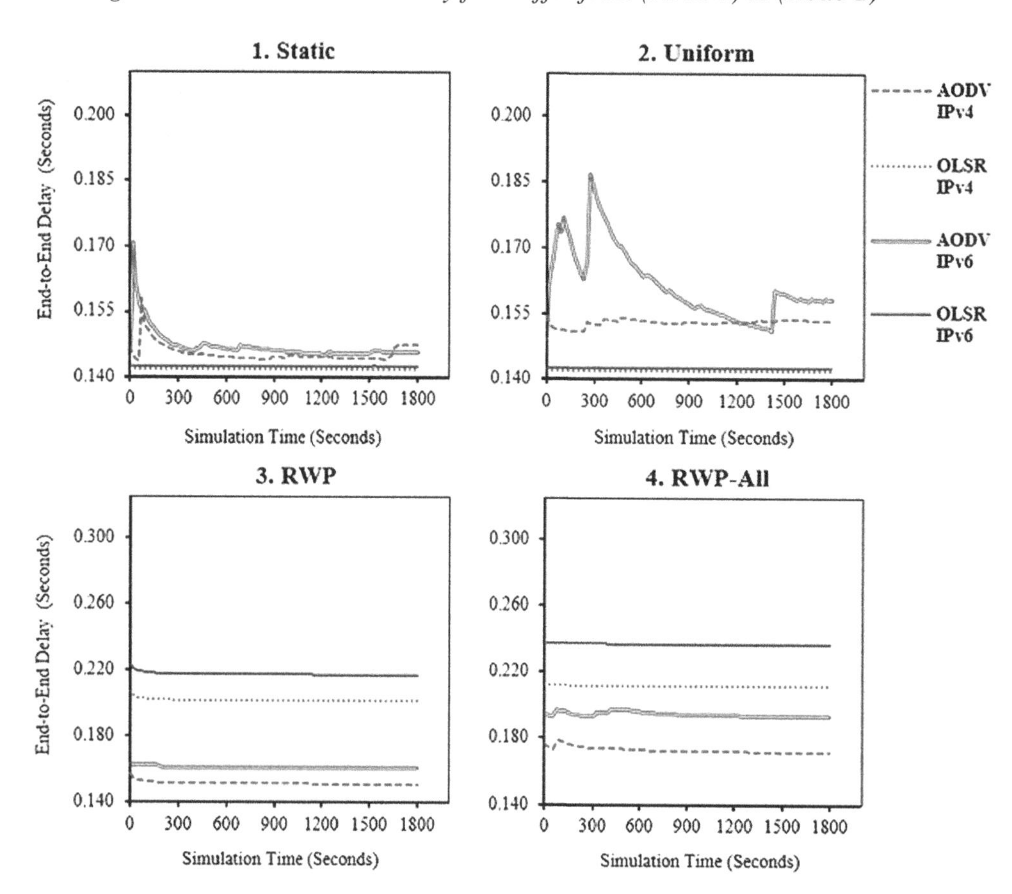

RWP-All where AODV IPv6 exceeds the acceptable range of the maximum jitter. On the other hand, OLSR IPv4 showed an acceptable maximum jitter for both mobility models while OLSR IPv6 had an acceptable maximum jitter variation. The difference between IPv4 and IPv6 in the jitter variation is not related to the header size. It is related to the delay variation of the received consecutive voice packets.

Packet Loss

The calculation of packet loss percentage for RTP traffic depends on the difference between the average number of voice traffic sent from the caller (node 1) in Figure 4, and the average number of received

Table 8. Maximum jitter variation for voice packets flow in milliseconds

	AODV IPv4	OLSR IPv4	AODV IPv6	OLSR IPv6
Static	2.86	2.23	3.81	3.12
Uniform	4.37	4.78	5.58	5.23
RWP	23.52	37.33	45.89	53.41
RWP-All	38.64	42.75	54.18	62.42

voice traffic by the callee (node 24) in Figure 5. Static and Uniform mobility models had the lowest packet loss percentage of the total sent and received voice traffic for both AODV and OLSR. For AODV, the average sent traffic is between 1305 and 1610 Bytes per second, and the average received traffic is between 1280 and 1520 Bytes per second, with a traffic loss percentage between 0.62% and 1.11% for the overall initiated voice traffic for IPv4 and IPv6 VoIP calls. The organized movement and the fixed number of hops between the source and destination nodes reduced the number of lost packets between the source and the destination nodes. However, OLSR showed a slightly better traffic compared with AODV where packet loss is lower, as it is between 0.56% and 1.02% of the overall VoIP calls.

For RWP mobility models, AODV has higher percentage of packet loss for the successful calls compared with OLSR. For AODV, the average sent traffic is between 1090 and 1490 Bytes per second for the first 900 seconds, and the average received traffic is between 770 and 1290 Bytes per second, with a packet loss percentage of 3.15% to 7.79%, where the percentage increased with the RWP-All mobility model that reached to 14.75% of the total initiated VoIP calls. On the other hand, the average sent traffic is between 600 and 1150 Bytes per second for the first 735 seconds, and the average received traffic is between 100 and 540 Bytes per second, with a packet loss percentage of 5.18% to 14.75%. The percentage of the packet loss increased with the RWP-All mobility model. It reached to 17.13% of the total generated voice traffic for the very limited number of successful VoIP calls.

These increases in the packet loss percentage happen because of the variable increments in hop numbers between source and destination. In addition, the MANET nodes' reachability during voice traffic transmission was reduced in the RWP mobility models which has another influence on the packet delivery ratio. Table 9 summarizes the RTP packet loss percentage for the evaluated voice traffic. The packet loss percentage of the voice calls in the Static mobility mode is less than 1%, which is an acceptable packet loss percentage in VoIP applications. With the Uniform mobility model, the packet loss is still acceptable as it is around 1%, which is the highest recommended percentage of packet loss. However, with the RWP mobility models the packet loss percentage increased more than the acceptable packet loss ratio, which reflects on the voice QoS for both AODV and OLSR. The mobility nature and limited destination reachability are responsible for this increased percentage of packet loss (Ivov and Noel, 2004) .

Mean Opinion Score

The subjective MOS performance metric for VoIP calls was evaluated using the E-model equation over the simulation scenarios in OPNET®. The ITU-T determined that the average MOS values for GSM codec system is 3.5 out of 5. In Figure 7, the best MOS values are with the Static model over all scenarios as the results swing between 3 and 3.5. For Uniform mobility, OLSR scenarios are slightly better when compared with AODV, and IPv4 performs better with both routing protocols. However, the MOS voice

Table 9. RTP/UDP packet loss percentage for connected VoIP calls

	AODV IPv4	OLSR IPv4	AODV IPv6	OLSR IPv6
Static	0.62%	0.56%	0.94%	0.73%
Uniform	1.03%	0.94%	1.11%	1.02%
RWP	3.15%	5.18%	7.79%	14.75%
RWP-All	7.81%	9.07%	14.64%	17.13%

quality values for OLSR scenarios dropped to Fair with the RWP model and to Poor with the RWP-All mobility model while OLSR IPv6 had Poor results compared with OLSR IPv4. For AODV, the MOS values for IPv4 performs Fair with RWP and the RWP-All Mobility model while AODV IPv6 had a little poor MOS value over the RWP-All mobility model. This is because of the delays and the increased jitter over both RWP mobility models that affect the R factor of MOS parameters. In general, the MOS values for OLSR are better with the Static and Uniform mobility models while AODV mostly have acceptable MOS values with both RWP mobility models.

GENERAL ROUTING PERFORMANCE

A relation between the applications' performance and MANET routing parameters had been shown. In this section, the considered performance metrics for the general routing performance are the consumed bandwidth, sent/received routing traffic, and hops number between the correspondent nodes. Figure 8 shows the average consumed bandwidth for the routing processes in the MANET during the simulation time of the investigated scenarios. In the Static and Uniform scenarios, the average consumed bandwidth for routing data in the MANET is in the range of 25 to 34.6 Kbits/s for AODV, and in the range of 31 to 40.5 Kbits/s for OLSR. The average of consumed bandwidth for IPv4 scenarios is slightly lower com-

Figure 7. Average voice MOS value for VoIP applications

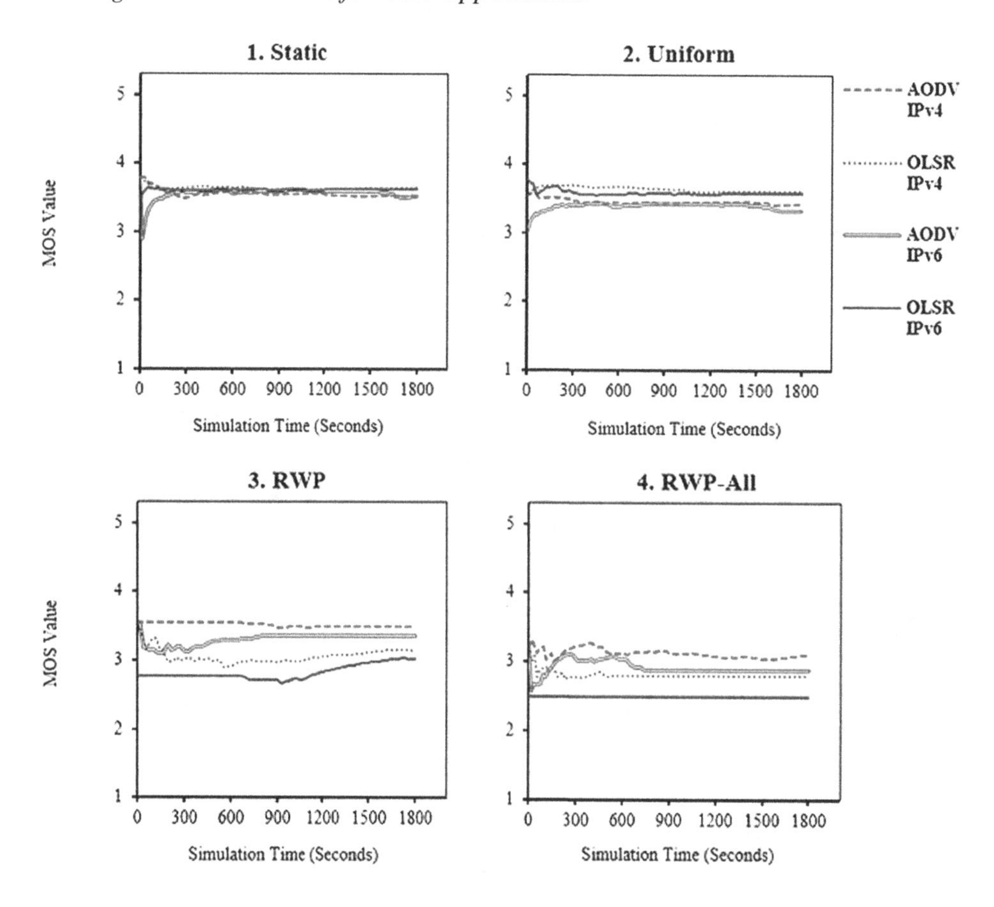

pared with the consumed bandwidth in IPv6. This is because of the packet overhead of IPv6 traffic that increased the amount of consumed traffic. OLSR has higher bandwidth consumptions compared with AODV because of the proactive nature of OLSR. On the other hand, the RWP scenarios showed that the average of total consumed bandwidth for routing messages IPv4 AODV is around 18 to 24 Kbits/s while IPv6 AODV has lower consumption with 13 to 16 Kbits/s. For IPv4 OLSR scenarios, the total consumed bandwidth for routing messages were in the range of 21 to 43 Kbits/s and for IPv6 OLSR in the range of 5 to 10 Kbits/s. For the RWP-All scenarios, the total consumed bandwidth for routing messages in IPv4 AODV is in the range between 20 and 24 Kbits/s while in IPv6 AODV it is in the range between 10 and 21 Kbits/s. While with IPv4 OLSR scenarios, the total consumed bandwidth for routing messages were in the range of 10.5 to 29.7 Kbits/s and for IPv6 OLSR in the range of 1.8 to 14.7 Kbits/s. This variance in the routing traffic bandwidth for RWP and RWP-All scenarios are related to the successful reachability and active calls over MANET nodes during VoIP applications. The routing performance in OLSR with RWP and RWP-All scenarios has considerable delays that affect the connectivity for VoIP applications and reduced the number of the established VoIP calls, thus the voice traffic had reduced as well.

Figure 8. Average consumed bandwidth for routing data in (Bits/Sec) for AODV-based and OLSR-based MANET

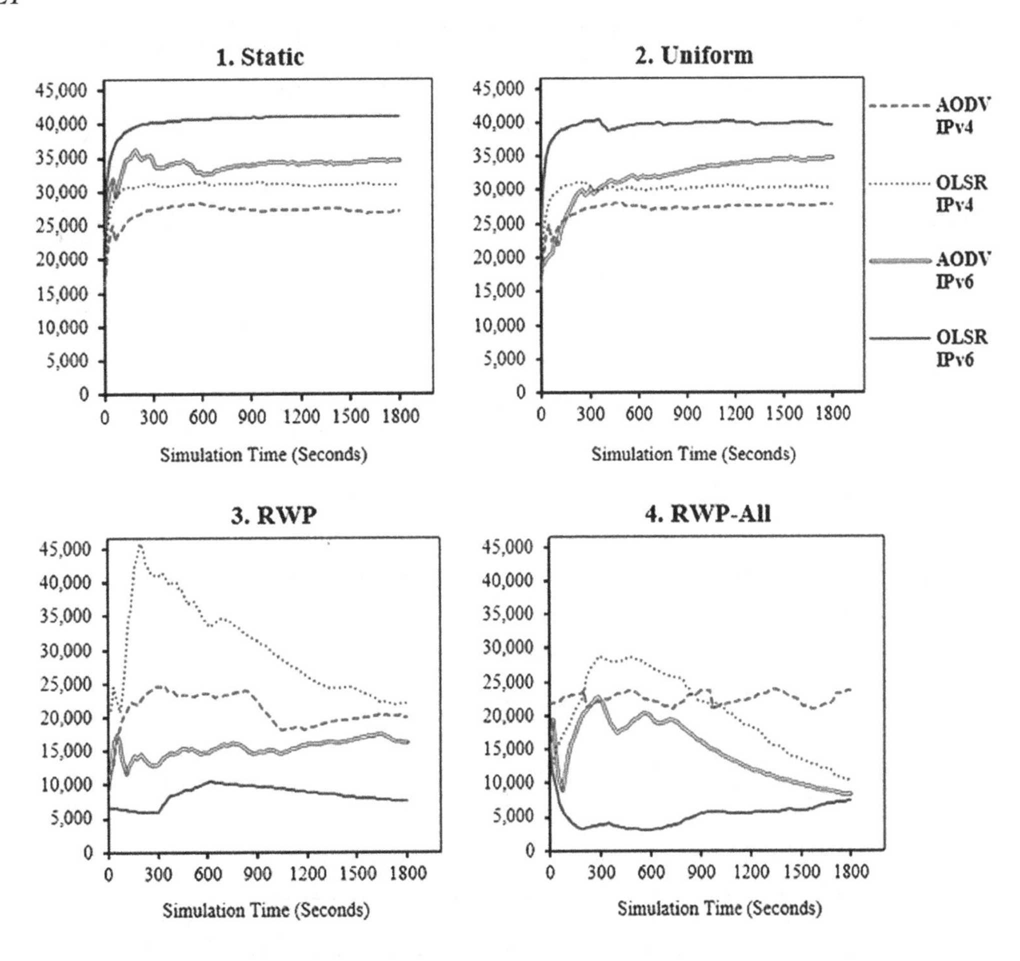

In Figure 9, the average routing traffic sent from the caller node regarding the established VoIP applications is represented. For the Static and Uniform scenarios, the average of sent routing traffic from the caller node is in the range of 450 to 800 bits/s for AODV, and in the range of 100 to 430 bits/s for OLSR. While in the RWP scenarios, the average of sent routing traffic is in the range of 480 to 600 bits/s for AODV, and in the range of 500 to 2000 bits/s for OLSR. For the RWP-All scenarios, the average of sent routing traffic is in the range of 220 to 680 bits/s for AODV, and in the range of 100 to 2700 bits/s for OLSR. The low sent routing traffic in both RWP and RWP-All scenarios is reflecting the fact that the voice traffic has not sent and that most of the sent routing requests are for the messages of the SIP call setup attempts.

On the other hand, Figure 10 shows the average routing traffic received by the caller node during the simulation time. For the Static and Uniform scenarios, the average received routing traffic by the caller node is between 1 and 2 Kbits/s for AODV, and in the range of 2 to 4 Kbits/s for OLSR. In the RWP scenarios, the average of received routing traffic is between 1 and 3 Kbits/s for AODV, and in the range of 2 to 13 Kbits/s for OLSR. For the RWP-All scenarios, the average is between 1 and 2.5 Kbits/s for AODV, and in the range of 3 to 16 Kbits/s for OLSR. The results showed that the received routing traffic for both scenarios are mostly double or triple the sent routing traffic. The RWP and RWP-All scenarios

Figure 9. Average routing traffic sent by the caller node (Bits/Sec)

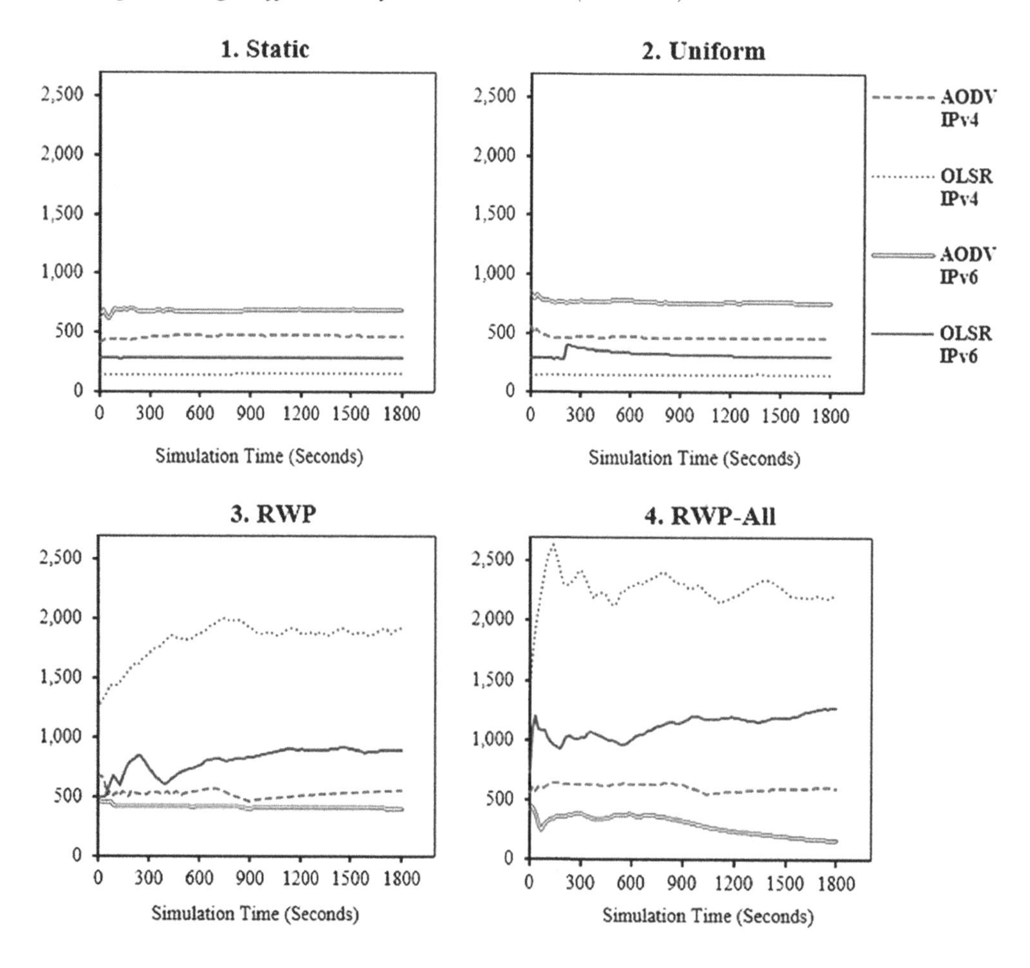

showed a high number of received routing traffic compared with the sent routing traffic. This happens as a result of the high number of generated routing messages that try to find routes for the requested connections during the variable movements of MANET nodes.

In Figure 11, the average number of hops between the caller node and the callee node during the generated VoIP calls is represented. For the Static and Uniform scenarios, the average number of hops is between 7 and 8 hops for both AODV and OLSR, while for the RWP scenarios, the average number of hops is between 3 and 7 for AODV, and 2 to 4 for OLSR. In the RWP-All scenarios, the average number of hops is between 3 and 5 for AODV, and 4 to 5 for IPv4 OLSR, while IPv6 OLSR has registered only two successful call setups at the beginning of the simulation. For the RWP and RWP-All scenarios, successful connections mostly happen when the hops number between the source and destination is lower than 4. Thus, both communicated nodes need to have a low number of hops between themselves during the nodes' mobility to be able to initiate the SIP-based VoIP calls over the RWP mobility models.

Figure 10. Average routing traffic received by the caller node (Bits/Sec)

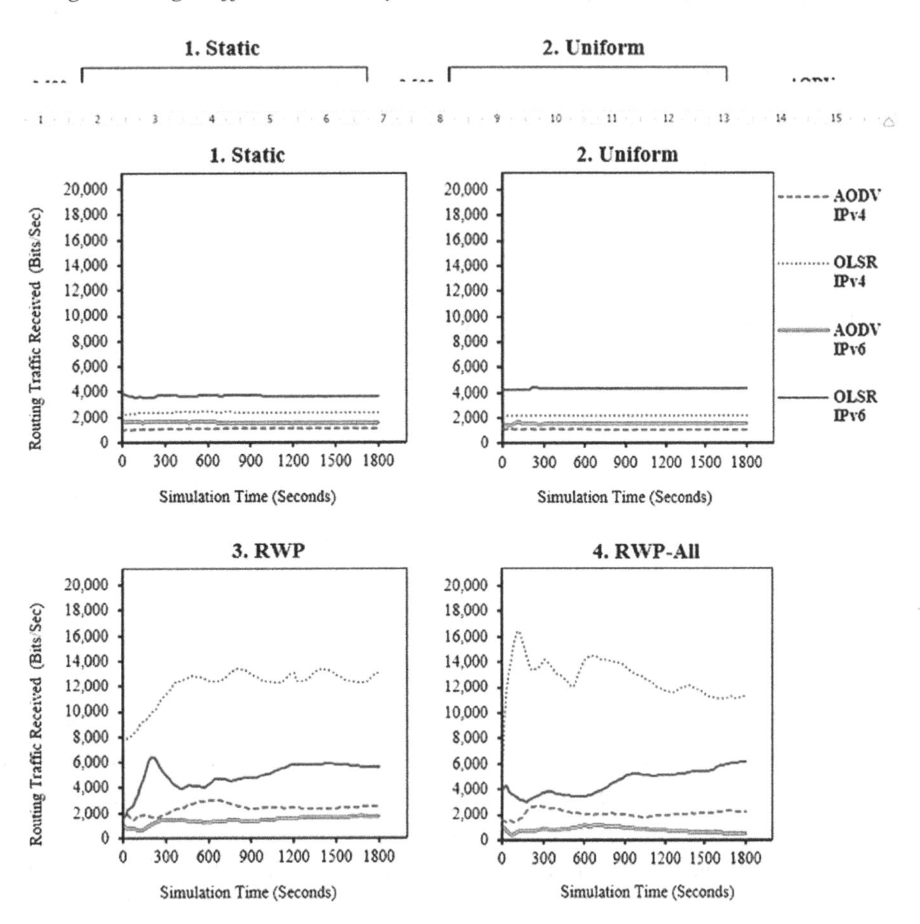

Figure 11. Average number of hops between caller (Node 1) and callee (Node 24)

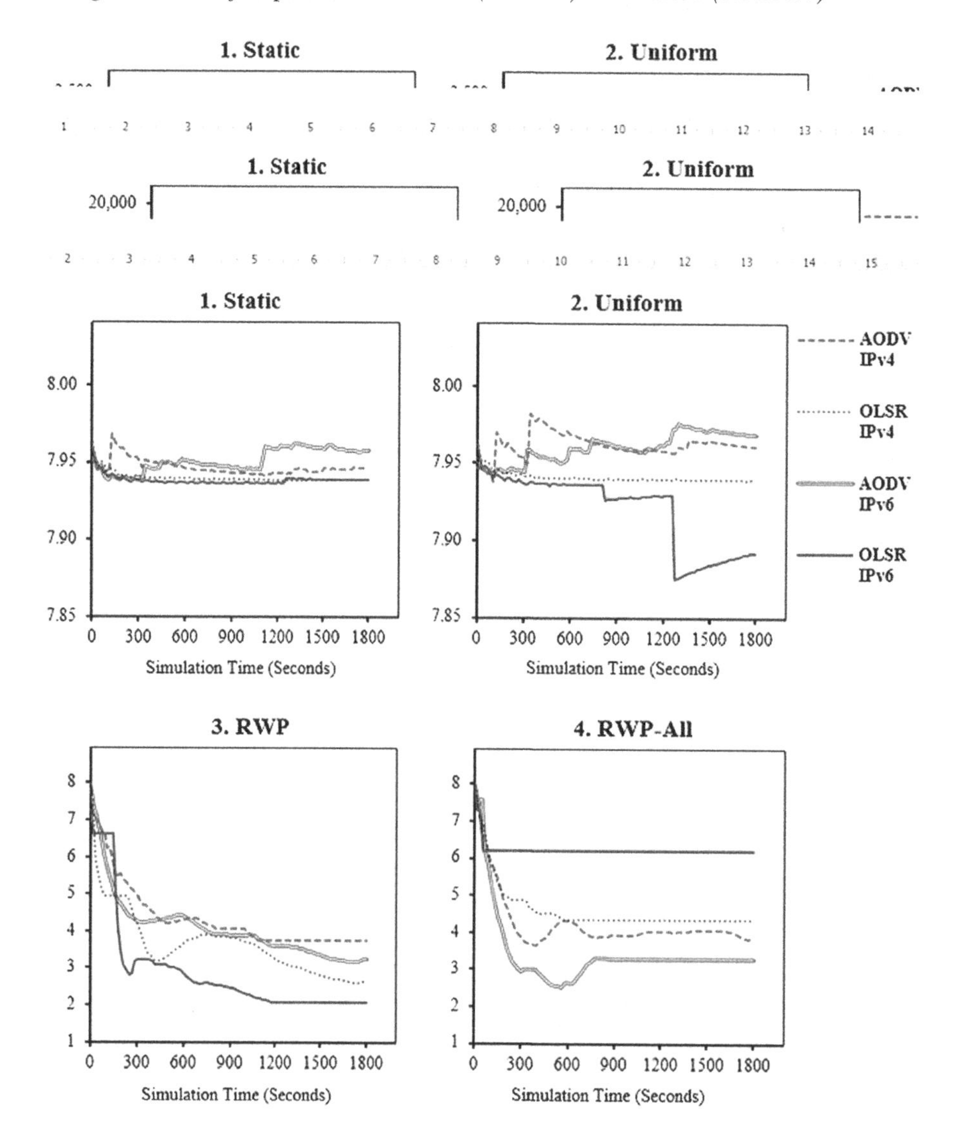

AODV ROUTING PERFORMANCE

In this section, a simple investigation is reported for related AODV routing parameters that affect the SIP-based VoIP performance over AODV MANET. The routing performance for related parameters will be evaluated throughout the examinations for the route discovery time, number of route requests sent by the caller, and the packet queue size on the caller side.

Figure 12 shows the average Route Discovery Time (RDT) for the caller node over AODV routing processes during the simulation time for the implemented scenarios, while Table 10 shows the maximum registered route discovery time for AODV throughout the simulations for both IPv4 and IPv6 traffic. For the Static and Uniform scenarios, the average route discovery time is up to 210 ms. The maximum registered value was 980 ms for IPv6 AODV. Hence, these values are meeting with the best effort conditions for AODV route discovery time that provide the best performance for real-time applications. In the RWP scenarios, the average route discovery time increased to be in the range of 0.93 to 2.4 seconds where the maximum registered value was 24.25 seconds for IPv6 AODV, while in RWP-All, the average was between 0.96 and 6.92 seconds and the maximum value was 47.24 seconds for IPv6 AODV.

Figure 13 shows the average number of Route Requests (RREQ) per 5 seconds that is sent by the caller node over AODV for the implemented scenarios. For the Static and Uniform scenarios, the average number of route requests sent by the caller is between 4 and 9 requests per 5 seconds, while for the RWP scenarios, the average number has reduced to 3 to 7 requests per 5 seconds. In the RWP-All scenarios, the average number of route requests sent by the caller is between 2 and 6 per 5 seconds. The IPv6 AODV showed a slightly higher number of sent route requests compared with IPv4.

Figure 12. Average route discovery time for the caller (Node 1)

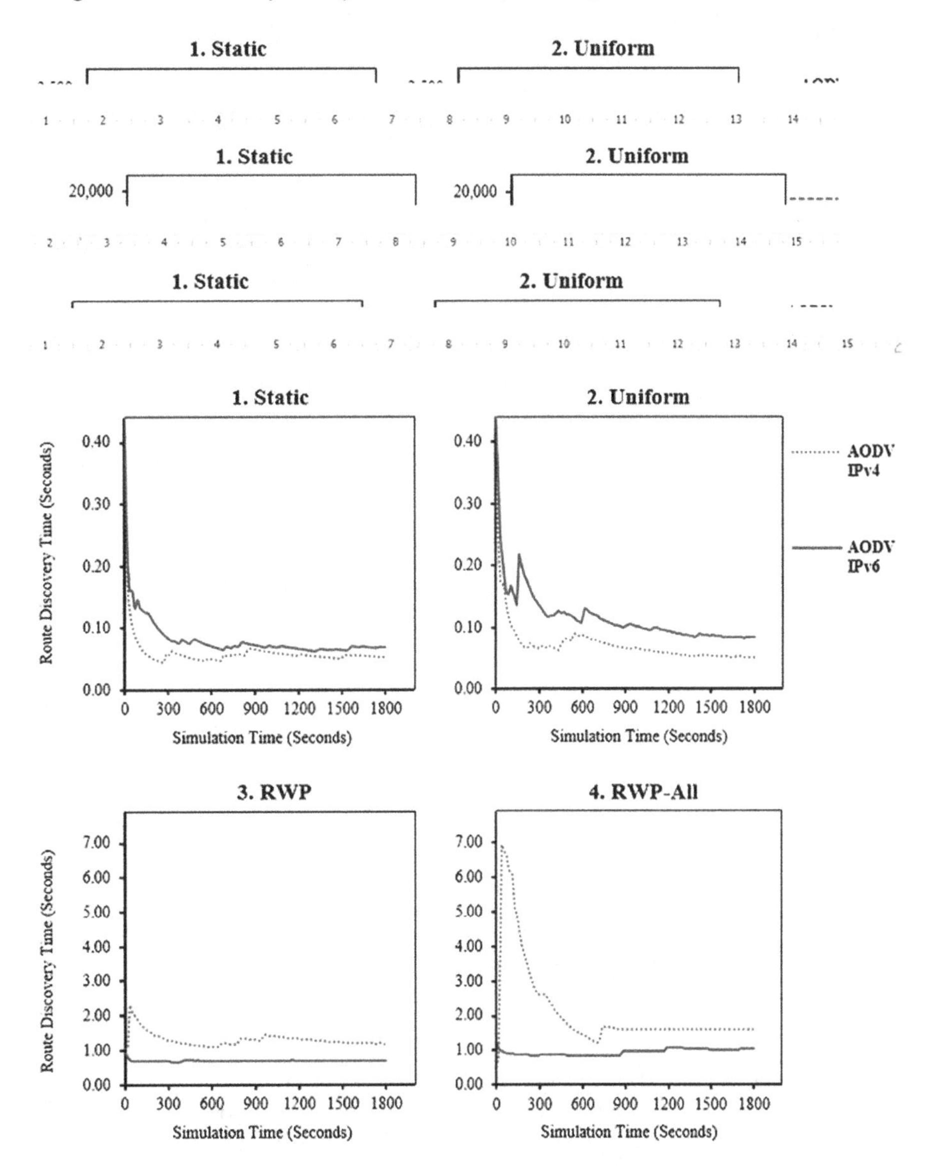

Table 10. Maximum route discovery time for AODV in seconds

	AODV IPv4	AODV IPv6
Static	0.48	0.66
Uniform	0.59	0.98
RWP	17.35	24.25
RWP-All	19.45	47.24

Figure 13. Average number of route requests sent by the caller (Node 1)

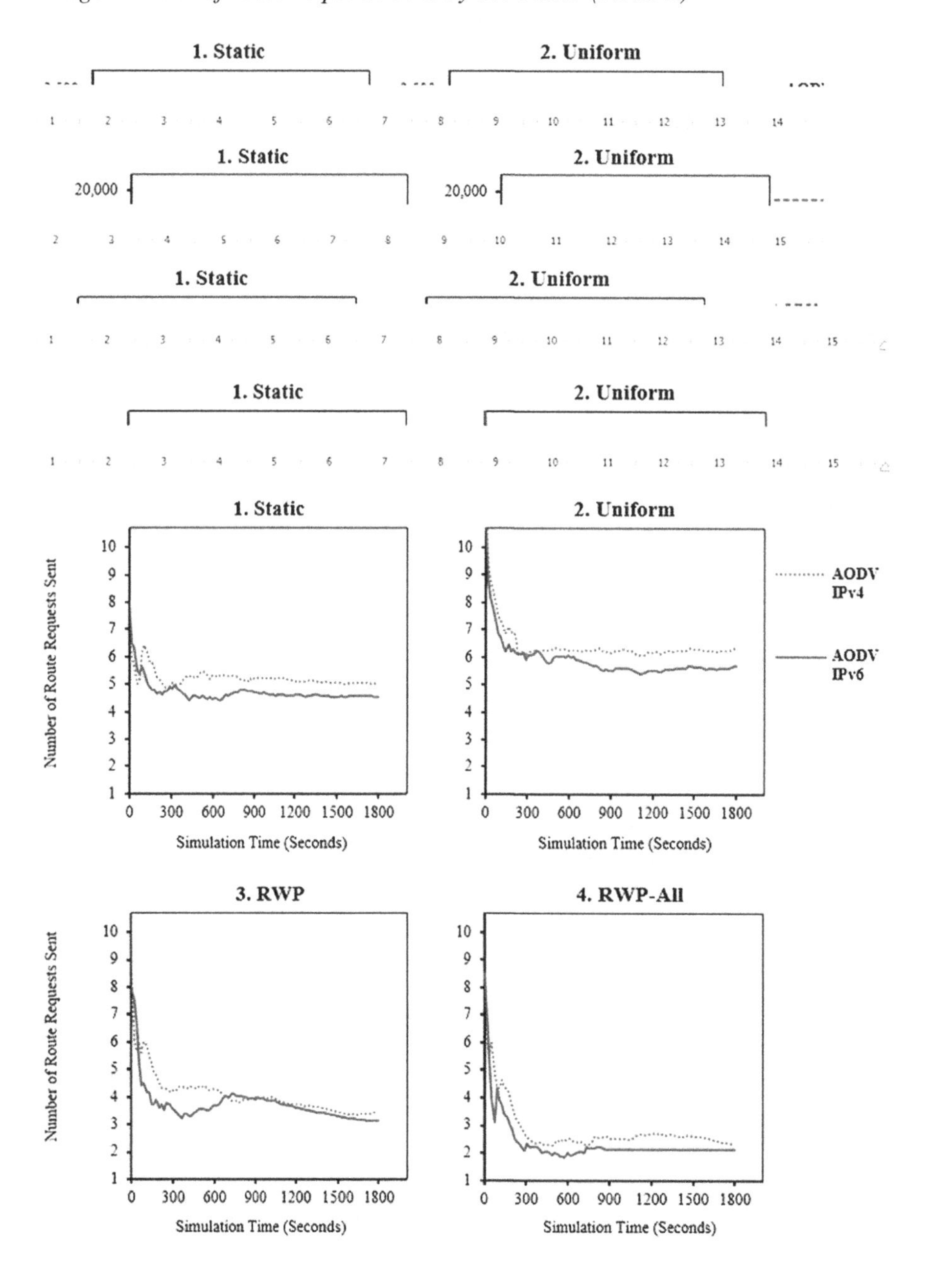

Figure 14 shows the average number of queued packets by the middle nodes between the caller and callee for data sent from the caller side during SIP-based VoIP implementations over AODV MANET. The average number of queued packets by middle nodes shows its related affects over the RDT and the RREQ performance for AODV routing that affect the SIP-based VoIP applications over MANET. As much as the queue size increases, the RDT time increases and the number of RREQ messages increase which reduce the performance of the VoIP calls. For the Static scenarios, the average number of queued packets is between 9 and 21 packets per second, while in the Uniform scenarios, the average number is between 7 and 42 packets per second. In the RWP scenarios, the average number is between 12 and 83 packets per second. With the RWP-All scenarios, the average number of queued packets is between 14 and 273 packets per second. In general, IPv6 AODV representations have larger numbers of queued packets compared with IPv4.

In general, the performance of AODV routing parameters showed related effects over SIP-based VoIP performance. The increased number of RREQ messages caused an increase in the number of RREP messages. The TTL could enhance the routing performance if its values were controlled to enhance the routing performance depending on the current status of the routing table, number of lost RREQ messages, and number of RREP messages. Furthermore, the RDT values increased with the RWP and RWP-All scenarios as a result of the increased impacts of the nodes' mobility and nodes' lost connectivity over routing parameters.

Figure 14. Average packet queue size on the caller Side (Node 1)

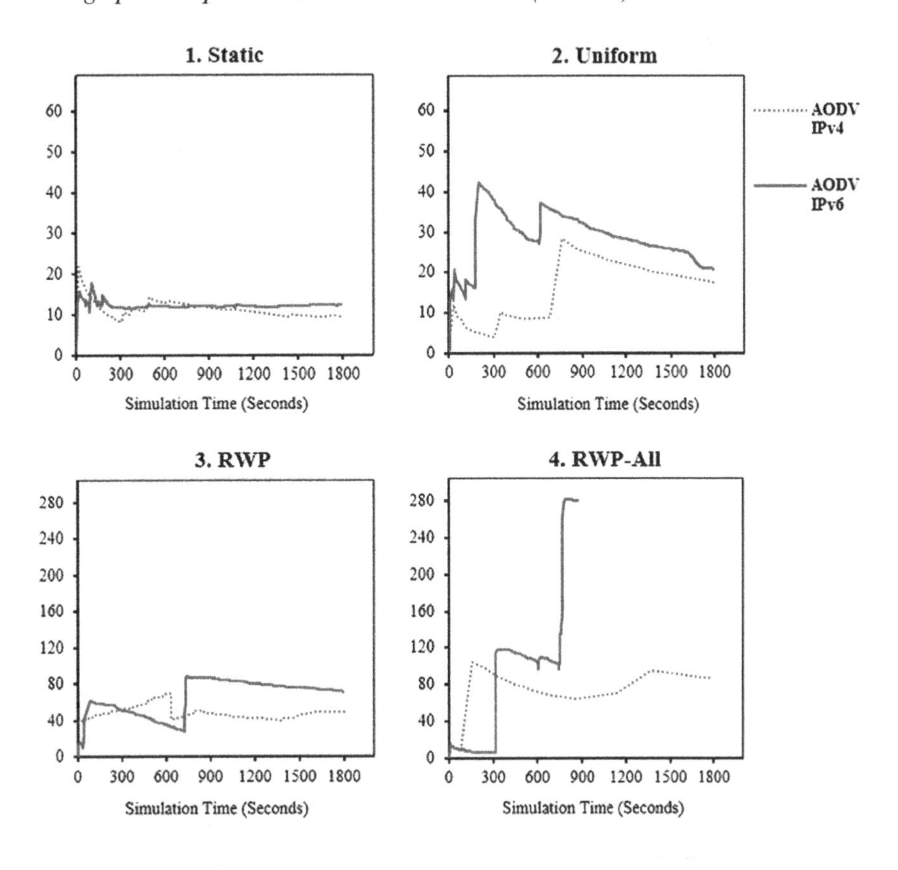

OLSR ROUTING PERFORMANCE

In this section, a simple investigation is reported for related OLSR routing parameters that affect the SIP-based VoIP performance over OLSR MANET. The related OLSR routing parameters had been evaluated throughout the simulation results by examining the traffic of HELLO messages sent, the number of Multipoint Relay (MPR) messages sent, and the number of OLSR Topology Control (TC) messages that were forwarded in MANET. Figure 15 shows the average number of HELLO traffic sent over OLSR routing processes for MANET during the simulation time for the implemented scenarios. The traffic amount of HELLO messages represents the actual status of route discovery for the required connectivity between different nodes. As the amount of HELLO traffic increases, so the routing overhead increases in the MANET. For the Static and Uniform scenarios, the average HELLO traffic sent in MANET is between 4.3 Kbits/s and 5.1 Kbits/s where these values are meeting with the best effort conditions for OLSR HELLO traffic that provide the best performance for real-time applications. In the RWP scenarios, the average HELLO traffic sent increased to be in the range of 4.4 Kbits/s to 5.9 Kbits/s for IPv4 OLSR, and in the range of 9.2 Kbits/s to 13.6 Kbits/s for IPv6 OLSR, while in the RWP-All, the average HELLO traffic sent is between 6.5 Kbits/s and 7.8 Kbits/s for IPv4 OLSR, and in the range of 9.6 Kbits/s to 14.4 Kbits/s for IPv6 OLSR.

Figure 15. Average HELLO traffic sent in MANET

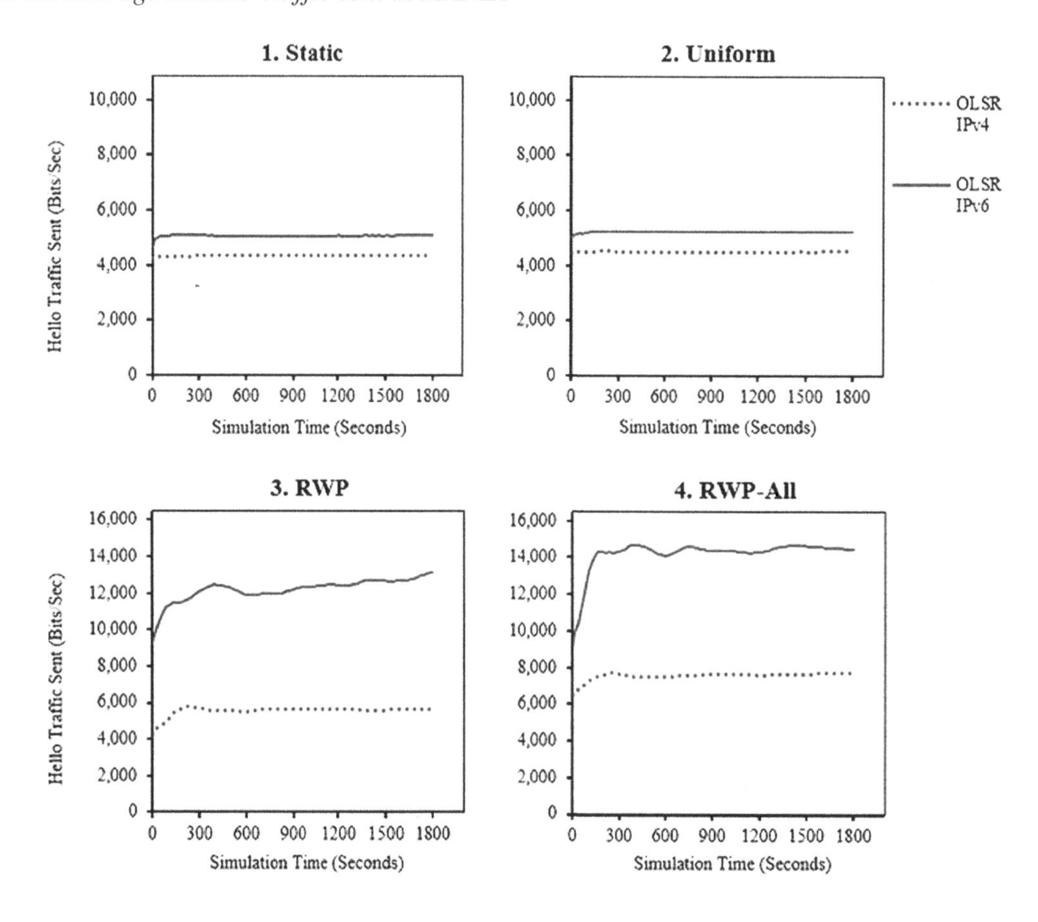

Figure 16 shows the average number of MPR messages sent over the OLSR routing process during the simulation time for the implemented scenarios. The functionality of MPR is to minimise the rebroadcasting of HELLO messages in the network. Hence, the decreased number of MPR values indicate that high numbers of HELLO messages are generated from MANET nodes and the MPR mechanism is not able to control the high number of generated HELLO messages. For the Static and Uniform scenarios, the average number of MPR messages sent is between 12 and 17 message per second, while for the RWP scenarios, the average number has reduced to be between 10 and 12 messages per second. In the RWP-All scenarios, the average number of MPR messages sent in MANET is between 6 and 10 messages per second.

Figure 17 shows the average number of OLSR Topology Control (TC) messages that were forwarded in OLSR MANET during the SIP-based VoIP implementations. The number of TC messages sent indicates the route status of MANET. The increased number of TC messages reflects the status of the correct routing data that provide the best route between two nodes in the MANET. The TC values decrease with the increase of the nodes' mobility factor. For the Static and Uniform scenarios, the average number of TC messages is between 1240 and 1330 messages every 10 seconds, while in the RWP scenarios, the average number of TC messages is between 750 and 1350 messages every 10 seconds. With the RWP-All scenarios, the average number of TC messages is between 520 and 1230 messages every 10 seconds. Furthermore, IPv6 OLSR representations showed larger numbers of TC messages compared with IPv4.

Figure 16. Average number of MPR in MANET

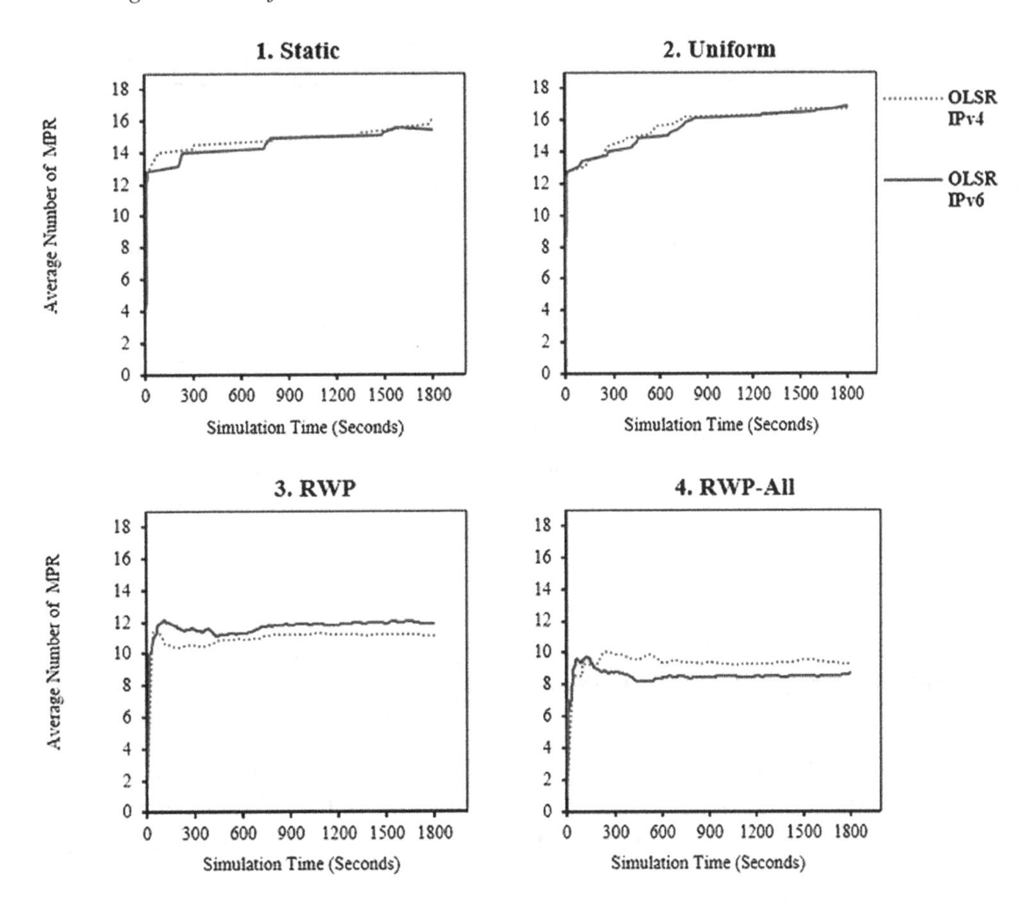

Figure 17. Average number of OLSR Topology Control (TC) messages forwarded in MANET

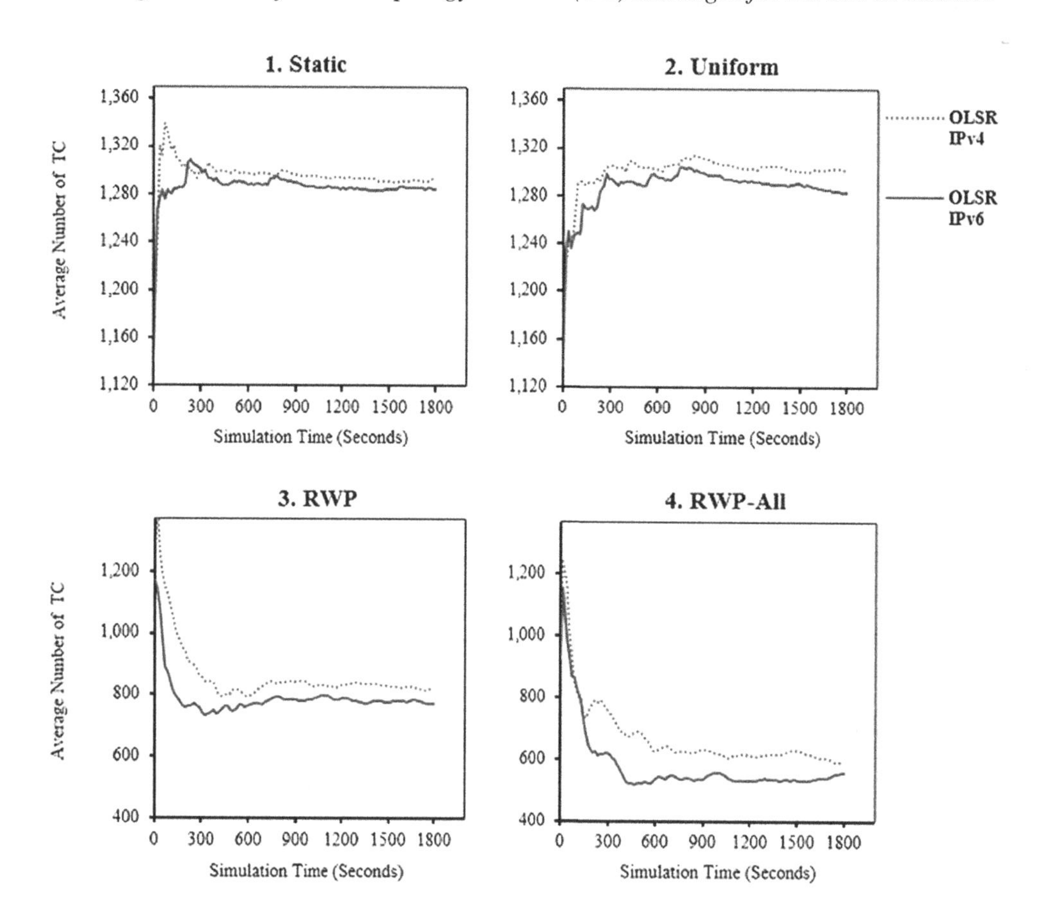

The performance of OLSR routing parameters showed a considerable level of effect over the performance of SIP-based VoIP implementations. The increased number of HELLO messages increased the routing overhead for route discovery in the network. In general, the number of HELLO messages increased with the RWP and RWP-All scenarios because of the increased impact of the nodes' mobility and route updates of OLSR routing tables. Furthermore, the RWP scenarios had shown lower numbers of TC messages and weak performance for the MPR mechanism compared with the Static and Random scenarios. Controlling the HELLO_INTERVAL and TC_INTERVAL time slots that are used by HELLO and TC messages can improve the MPR mechanism performance with frequent updates for the routing table and reducing the routing overhead.

RESULTS DISCUSSION FOR THE BENCHMARKING AND EVALUATION EFFORTS OF THE SIP PROCESSES

The delays in the SIP signaling processes and voice end-to-end delays for collected statistics in this evaluation chapter for both Static and Uniform scenarios for MANET VoIP implementations are in the same range as the results of the implementations of SIP-based applications over other network systems, such as in (Fabini, Hirschbichler, Kuthan, & Wiedermann, 2013) . The similarity between the findings of SIP-based VoIP performance in this chapter with other research efforts supports the accuracy level of the simulation efforts and the reliability of the research implementations. The evaluation results showed that IPv6 SIP has longer delays, call setup time and throughput compared with IPv4 SIP for both RWP and RWP-All scenarios, which support the results' findings in (Yasinovskyy, Wijesinha, Karne, & Khaksari, 2009) .

For the SIP signaling performance, the time intervals of the registrations, calls setup, and calls termination processes are quite long within the RWP and RWP-All MANET. These long-time intervals are related to the nodes' mobility which affects the general routing performance. For example, the call setup time can be reduced with the dynamic modifications for SIP timers using the optimisation algorithms for SIP messages' retransmission processes depending on the routing status. Further improvements in terms of call setup delays can be achieved based on these modifications. In general, a set of QoS parameters from the upper three layers of the Open Systems Interconnection (OSI) could be considered to enhance the SIP signaling performance over MANET. This is because the routing processes during the nodes' mobility had been shown to have a direct effect on the SIP signaling performance over MANET. Therefore, the SIP performance is only as strong as the weakest link at the MANET and performs well where there are no hardware problems. The performance evaluation of the examined systems revealed the following results regarding the SIP processes:

- As part of the benchmarking efforts, the values of the investigated performance metrics of the RFC 6076 have been determined from the evaluation efforts of the Static and Uniform mobility models for the best effort scenarios. The RRD values are up to 35 ms, the SRD values are up to 270 ms, and the SDD values are up to 25 ms for both IPv4 and IPv6 representations.
- The average registration time values for AODV IPv4 have shorter registration, initiation and termination times when compared with AODV IPv6 over different mobility scenarios as shown before. The registration time value for IPv6 is about 43% longer than IPv4 in the RWP AODV-based MANET. In addition, the registration time for IPv6 is about 48% more than for IPv4 over the RWP-All AODV-based. In the OLSR-based MANET, the average value for the registration intervals for IPv6 is 54% more than IPv4 in the RWP OLSR scenario and 57% more than IPv4 in the RWP-All scenario. In general, the average values of the registration time for OLSR-based IPv4 have shorter registration times compared with OLSR-based IPv6 over different mobility scenarios.
- The call setup process has the longest delays compared with other SIP call processes over both AODV and OLSR. The average call initiation time for an IPv6-based SIP-based call is about 46.1% longer than IPv4 for the RWP AODV scenario, and 39.3% longer than IPv4 for the RWP-All AODV scenario. For OLSR, the average call setup time for IPv6 is 48% more than IPv4 for the RWP OLSR scenario and 44% more than IPv4 for the RWP-All OLSR-based scenario.
- The average termination time for IPv6 is about 38% more than IPv4 in the RWP scenario and about 34% more than IPv4 in the RWP-All scenario for AODV-based MANET. For OLSR, the average termination time for IPv6 is about 42% more than IPv4 in the RWP scenario and 66% more than IPv4 in the RWP-All for OLSR-based scenario.
- In general, IPv4 implementations have better performance over all scenarios when compared with IPv6 implementations for both AODV and OLSR. The findings of the evaluation efforts support the research findings in (Boumezzough, Idboufker, and Ouahman, 2013 and (Tuong, et al., 2010) that showed that IPv6-based SIP has longer delays and traffic size when compared with IPv4-

based SIP. Both IPv4 and IPv6 implementations had been considered in the investigated scenarios without the mobility support, which clearly influenced the performance of the IPv6 traffic when compared with IPv4 traffic.

The differences for the VoIP QoS for SIP-based VoIP applications over IPv4 and IPv6 are in different ranges depending on the nature of the network system, mobility, bandwidth, and the connectivity statues. The VoIP metrics such as end-to-end delays, jitter, throughput, and packet loss are quite comparable for both IPv4 and IPv6 over RWP and RWP-All MANET scenarios. The IP networks still cannot meet the required level of QoS for VoIP applications; however, the VoIP QoS can be improved by controlling the values of the VoIP performance parameters to be within the acceptable range as declared in (Pandey & Swaroop, 2011; Shrestha & Tekiner, 2009). As this research considered the GSM as the voice codec for the implemented VoIP applications, the voice end-to-end performance metrics could differ with the usage of other voice codecs. From the evaluation efforts, the majority of successful VoIP calls within the random mobility scenarios happened in the first half of the simulations because of the initial positions of the nodes that provided the best possible connectivity and reachability level until the nodes' distribution increased the percentage of the changed routing values. In general, MANET mobility characteristics are considered unfriendly with the VoIP nature, as shown in the RWP and RWP-All mobility models. The comparison results of different QoS parameters for SIP-based VoIP in the MANET for IPv4 and IPv6 are as follows:

- The average bandwidth consumption of IPv6 is higher than IPv4 because of its larger header size in IPv6. In addition, IPv6 in the implemented systems of this research effort does not support the mobility features. However, for the RWP mobility models, IPv4 has higher consumption than IPv6 because it has more initiated VoIP calls. In addition, the VoIP calls are in their best effort conditions over the Static and Uniform scenarios, where the RWP mobility models consumed a convergent amount of bandwidth related to the identified and calculated bandwidth in Table 1. In general, OLSR implementations have higher bandwidth consumption compared with AODV implementations because of the proactive nature of OLSR.
- The average one-way RTP/UDP packet delays for voice streams between the caller and the callee for VoIP applications have longer delays for those successfully received voice data. The header processing for IPv4 and IPv6 RTP packets affects the one-way delays; however, the overall average of the end-to-end delays for RTP/UDP packets that were successfully received are within the acceptable delay range for VoIP applications. Therefore, there is no significant difference between IPv4 and IPv6 for the RTP one-way delays.
- The maximum jitter variations for both AODV-based and OLSR-based MANETs for the Static and Uniform scenarios are in the acceptable variation range for VoIP applications (Xiaonan and Shan, 2012) . For the RWP mobility models, the jitter range for both IPv4 and IPv6 are mostly in the acceptable range. The difference between IPv4 and IPv6 in the jitter variations is related to the delay variation of the received consecutive voice packets not to the header size.
- The average packet loss ratio for Static and Uniform scenarios is between 0.5% and 1% of the overall voice data for VoIP calls over both AODV and OLSR implementations. For RWP mobility models, the implementations of AODV-based MANET has a higher percentage of packet loss for the successful calls compared with OLSR-based MANET. For AODV, the average packet loss ratio is between 3% to 15% of the overall voice data for the VoIP calls, and between 5% to 17% for

OLSR implementations. The mobility nature and limited destination reachability are responsible for the increased percentage of the packet loss (Ivov and Noel, 2004) .

- The average Mean Opinion Score (MOS) value is Good (3.5 out of 5) for VoIP calls over both Static and Uniform mobility models. The MOS value dropped to Fair with the RWP model and to Poor with the RWP-All mobility model. IPv6 implementations have Poor MOS results compared with IPv4. In general, the MOS values for OLSR implementations are better with the Static and Uniform mobility models while AODV implementations mostly have acceptable MOS values with both RWP mobility models.

This evaluation in this chapter is focused on the SIP process delays and VoIP QoS. The related RFC 6076 performance metrics in the evaluation of the SIP processes are employed. The performance metrics values precisely indicate the performance level of the SIP signaling of the SIP processes. In (Voznak, & Rozhon, 2010), the RRD values are in the range of (10ms to 30ms) for the registration processes with the B2BUA SIP server. Furthermore, the SRD values are in the range of (25ms to 100ms) for the call setup processes with the same resources. The results of the RRD and SRD representations concur with the evaluation results in this chapter. The evaluation methods for the SIP signaling in (Voznak, & Rozhon, 2010), (Rozhon, Jan, Voznak, Tomala, & Vychodil, 2012), (Voznak, & Rozhon, 2012), and (Voznak, & Rozhon, 2010) do not consider a large number of hops between the nodes for the generated traffic as the main concern in this research was the performance of the SIP server. In addition, the research efforts do not cover all of the SIP processes in the simulation work as the termination process was not considered. However, the evaluation and benchmarking efforts in this chapter do consider the performance of all of the SIP processes for the callers and also the SIP server. In addition, the benchmarked values are clearly identified and used in this research study. The SDD values in this chapter are in the range of (2ms to 21ms), which is in the acceptable range based on the related delay values for both the RRD and SRD performance metrics in this study and in (Voznak & Rozhon, 2010) .

Compared with the other performance evaluation methods for the SIP signaling processes in (Gupta, Sadawarti, & Verma, 2011), the evaluation and benchmarking method considered in this thesis provide more reliable results for the registration, initiation and termination processes of the SIP-based VoIP. In addition, these benchmarked values can be used over different network platforms based on the identified equations for the RRD, SRD, and SDD. Furthermore, the implementation of these performance metrics is considered a standardised method using a reliable simulation tool (OPNET). On the other hand, the values of the investigated SIP processes within the RWP and RWP-All mobility models cannot be used for the benchmarking results as the node connectivity is affected by the routing nature in MANET. This issue is considered the main performance problem for the SIP-based VoIP implementations over MANET (Bah, Glitho, & Dssouli, 2012)

SUMMARY

This chapter represented an evaluation and comparison study for SIP signaling and VoIP performance metrics over AODV and OLSR for both IPv4 and IPv6 MANET. As IP networks still cannot meet the required QoS of VoIP, VoIP QoS is improved by controlling the values of these parameters to be within the acceptable range as declared in (Pandey & Swaroop, 2011). The differences for the call setup delays for SIP-based VoIP applications over IPv4 and IPv6 are in different ranges depend on network system, bandwidth, and the connectivity statue. The VoIP metrics such as end-to-end delays, jitter, throughput, and packet loss are quite comparable for both IPv4 and IPv6 over RWP and RWP-All MANET scenarios. Most of the successful VoIP calls during MANET mobility scenarios occur in the first half of the simulation as the nodes' initial positions provide better connectivity and reachability before they began moving. Both Static and Uniform mobility models had shown the best effort for SIP end-to-end performance metrics and VoIP QoS over AODV and OLSR MANET. In general, MANET mobility characteristics are considered unfriendly for VoIP nature as shown in RWP and RWP-All mobility models. The effects of such characteristics on different voice metrics have been studied in this chapter. In terms of SIP signaling performance, the call setup time is quite large in RWP and RWP-All MANET which relates to the nodes' mobility. The call setup time can be reduced with the dynamic modifications for SIP timers using optimisation algorithms for SIP messages' retransmission processes depending on the routing status. Further improvements in terms of call setup delays can be achieved based on these modifications.

The results for both Static and Uniform mobility models are representing the best effort of the implemented scenarios that meets with the evaluation results for similar scenarios over other network systems. The evaluation results showed that these performance parameters are comparable for both IPv4 and IPv6 over MANET environments. With the comparisons of the evaluation results for RRD, SRD, and SDD, the implementations of the SIP-based VoIP over AODV-based and OLSR-based MANET are slightly similar in terms of the delays over the Static and Uniform scenarios for both IPv4 and IPv6. However, the implementations of the AODV-based MANET have shorter delays over the RWP and RWP-All scenarios compared with the OLSR-based that showed longer delays. In general, the average values of the SDD are lower than the average values of RRD because the SDD processes have a simple and direct signaling representation system. On the other hand, the nature of the session initiation requests affects the SRD values. Therefore, the session initiation processes represent the most considerable delays for the SIP signaling over MANET for both AODV-based and OLSR-based MANET representations. Furthermore, the values of the evaluated RFC 6076 metrics in this chapter could be used as reference values to evaluate the SIP signaling performance over the registration, call initiation, and termination processes for SIP-based VoIP applications over MANET.

REFERENCES

Amnai, M., Fakhri, Y., & Abouchabaka, J. (2011). *QoS Routing and Performance Evaluation for Mobile Ad Hoc Networks Using OLSR Protocol.* arXiv preprint arXiv: 1107.3656

Bah, S., Glitho, R., & Dssouli, R. (2012). A SIP Servlets-based Framework for Service Provisioning in Stand-Alone MANETs. *Journal of Network and Computer Applications.*

Boumezzough, M., Idboufker, N., & Ouahman, A. (2013). Evaluation of SIP Call Setup Delay for VoIP in IMS. In *Advanced Infocomm Technology* (pp. 16–24). Springer Berlin Heidelberg. doi:10.1007/978-3-642-38227-7_6

Chen, Y., Yang, Y., & Hwang, R. (2007). SIP-based MIP6-MANET: Design and Implementation of Mobile IPv6 and SIP-based Mobile Ad Hoc Networks. *Computer Communications*, *29*(8), 1226–1240. doi:10.1016/j.comcom.2005.08.015

Chiang, W., Dai, H., & Luo, C. (2012). Cross-layer Handover for SIP Applications Based on Media-Independent Pre-Authentication with Redirect Tunneling. *Digital Information and Communication Technology and it's Applications (DICTAP), 2012Second International Conference on*, 348-353. doi:10.1109/DICTAP.2012.6215373

Gandhi, S., Chaubey, N., Shah, P., & Sadhwani, M. (2012). Performance Evaluation of DSR, OLSR and ZRP Protocols in MANETs. *Computer Communication and Informatics (ICCCI),2012International Conference on*, 1–5. doi:10.1109/ICCCI.2012.6158841

Gupta, Sadawarti, & Verma. (2011). Review of Various Routing Protocols for MANETs. *International Journal of Information and Electronics Engineering*, *1*(3), 99.

Fabini, H. Kuthan, & Wiedermann. (2013). Mobile SIP: An Empirical Study on SIP Retransmission Timers in HSPA 3G Networks. In Advances in Communication Networking (pp. 78-89). Springer Berlin Heidelberg.

Hoeher, T., Petraschek, M., Tomic, S., & Hirschbichler, M. (2007, August). Evaluating Performance Characteristics of SIP over IPv6. *Journal of Networks*, *2*(4), 40–49. doi:10.4304/jnw.2.4.40-50

Jain, Somasundaram, Wang, Baras, & Roy-Chowdhury. (2010). *Study of OLSR for Real-time Media Streaming over 802.11 Wireless Network in Software Emulation Environment*. Academic Press.

Kaur & Singh. (2012). Performance Comparison of OLSR, GRP and TORA Using OPNET. *International Journal of Advanced Research in Computer Science and Software Engineering*, *2*(10), 1-7.

Khan, I., & Qayyum, A. (2009). Performance Evaluation of AODV and OLSR in Highly Fading Vehicular Ad Hoc Network Environments. *Multitopic Conference, 2009. INMIC 2009. IEEE 13th International*, 1–5. doi:10.1109/INMIC.2009.5383121

Pandey & Swaroop. (2011). A Comprehensive Performance Analysis of Proactive, Reactive and Hybrid MANETs Routing Protocols. *International Journal of Computer Science Issues*, *8*(6).

Palta & Goyal. (2012). Comparison of OLSR and TORA Routing Protocols Using OPNET Modeler. *International Journal of Engineering Research and Technology*, *1*(5).

Rozhon, Voznak, Tomala, & Vychodil. (2012). Updated Approach to SIP Benchmarking. *Telecommunications and Signal Processing (TSP),201235th International Conference on*, 251-254.

Skordoulis, D., Ni, Q., Chen, H.-H., Stephens, A. P., Liu, C., & Jamalipour, A. (2008). IEEE 802.11 n MAC Frame Aggregation Mechanisms for Next-Generation High-Throughput WLANs. *Wireless Communications, IEEE*, *15*(1), 40–47. doi:10.1109/MWC.2008.4454703

Shrestha, A., & Tekiner, F. (2009). On MANET Routing Protocols for Mobility and Scalability. *International Conference on Parallel and Distributed Computing, Applications and Technologies*, 451-456. doi:10.1109/PDCAT.2009.88

Sondi, P., & Gantsou, D. (2009). Voice communication over Mobile Ad Hoc Networks: Evaluation of A QoS Extension of OLSR Using OPNET. *Asian Internet Engineering Conference*, 61–68. doi:10.1145/1711113.1711125

Shrestha, A., & Tekiner, F. (2009). On MANET Routing Protocols for Mobility and Scalability. *International Conference on Parallel and Distributed Computing, Applications and Technologies*, 451-456. doi:10.1109/PDCAT.2009.88

Tebbani, B., Haddadou, K., & Pujolle, G. (2009). A Session-based Management Architecture for QoS Assurance to VoIP Applications on Wireless Access Networks. *Consumer Communications and Networking Conference, 2009. CCNC 2009. 6th IEEE*, 1–5. doi:10.1109/CCNC.2009.4784757

Thompson, M. S., MacKenzie, A. B., & DaSilva, L. A. (2011). A Method of Proactive MANET Routing Protocol Evaluation Applied to The OLSR Protocol. *Proceedings of the 6th ACM international workshop on Wireless network testbeds, experimental evaluation and characterization*, 27–34. doi:10.1145/2030718.2030726

Tuong, L., Cook, S., Kuthethoor, G., Sesha, P., Hadynski, G., Kiwior, D., & Parker, D. (2010). Performance Analysis for SIP-based VoIP Services over Airborne Tactical Networks. *Aerospace Conference*, 1-8.

Voznak, M., & Rozhon, J. (2010). SIP Back to Back User Benchmarking. *Wireless and Mobile Communications (ICWMC),20106th International Conference on*, 92-96. doi:10.1109/ICWMC.2010.86

Voznak, M., & Rozhon, J. (2012). SIP End To End Performance Metrics. *International Journal of Mathematics and Computers in Simulation*, *6*(3), 315–323.

Voznak, M., & Rozhon, J. (2010). Performance Testing and Benchmarking of B2BUA and SIP Proxy. *Conference Proceedings TSP*, 497-503.

Xiaonan, W., & Shan, Z. (2012). Research on Mobility Handover for IPv6 Based MANET. *Transactions on Emerging Telecommunications Technologies*.

Yasinovskyy, R., Wijesinha, A., Karne, R., & Khaksari, G. (2009). A Comparison of VoIP Performance on IPv6 and IPv4 Networks. *Computer Systems and Applications, 2009. AICCSA 2009. IEEE/ACS International Conference on*, 603-609. doi:10.1109/AICCSA.2009.5069388

Chapter 6
Ultra-High-Definition Video Transmission for Mission-Critical Communication Systems Applications:
Challenges and Solutions

Anthony Olufemi Tesimi Adeyemi-Ejeye
Kingston University, UK

Mohammed Abdulrahman Alreshoodi
Qassim University, Saudi Arabia

Geza Koczian
University of Essex, UK

Michael C. Parker
University of Essex, UK

Stuart D. Walker
University of Essex, UK

ABSTRACT

With the standardization of ultra-high-definition formats and their increasing adoption within the multimedia industry, it has become vital to investigate how such a resolution could impact the quality of experience with respect to mission-critical communication systems. While this standardization enables improved perceptual quality of video content, how it can be used in mission-critical communications remains a challenge, with the main challenge being processing. This chapter discusses the challenges and potential solutions for the deployment of ultra-high-definition video transmission for mission-critical applications. In addition, it examines the state-of-the-art solutions for video processing and explores potential solutions. Finally, the authors predict future research directions in this area.

INTRODUCTION

Video content consumption over Internet Protocol (IP) has become ubiquitous, especially since it can be accessed anywhere, anytime, and on an expanding variety of user devices, such as laptops, tablets, smartphones, and PCs. With current advancements in both processing and networking capabilities, a

DOI: 10.4018/978-1-5225-2113-6.ch006

demand in higher quality video has also emerged. Consequently, recent years have seen the rise of video applications, such as streaming and video conferencing. Another area that would benefit immensely from the advancement of video is Public Safety Communication (PSC).

Traditional public safety communication currently uses Plesiochronous Digital Hierarchy (PDH) and/or Synchronous Digital Hierarchy (SDH)-/Synchronous Optical Network (SONET)-based Time-Division Multiplexing (TDM) technologies. IP-based video systems have evolved and now provide superior performance over traditional approaches to mission-critical systems. Indeed, many PSC networks are evolving to broadband solutions that use IP Wide Area Networks (WAN) for video surveillance and better integration with growing Information Technology (IT)-based applications (Alcatel-Lucent, 2013).

Video surveillance systems utilize video content to enable personnel to monitor an area, or even a collection of areas, from a central location, 24 hours, 7 days a week, and therefore reduces the need to delegate a large number of people to patrol the area. Surveillance systems are not new; however, there have been changes in the deployment of such systems in recent years. For Instance, a basic system might simply save the content on a storage device for further or future analysis, while more advanced systems can enable real-time analysis and can perform tasks such as recognizing changes in critical areas or the identification of certain objects. The video surveillance market is expected to grow by 14.8% by 2018 (Hsu & Chen, 2015).

Despite remarkable advances in the technical specifications of video surveillance, especially in the commercial domain, there has been an increasing need for higher quality content. In addition, most operators of video surveillance systems struggle both with the needed storage and computational complexity imposed by video applications of increasing quality and complexity. This is because most devices used for video surveillance have intrinsic constraints in storage and processing. The issue of processing can be partially solved with parallel computing (Wen-Mei, 2011), thereby providing new opportunities for future resolutions, while that of storage can be solved with the use of cloud computing resources (Zhu, Luo, Wang, & Li, 2011). Currently, resolutions up to full high definition (1080p) are available and can be deployed sparsely for Mission-Critical Communications (MCC). While 1080p resolution provides improved video quality, the use of this resolution is limited, especially with capture distance (approximately 120m), thus limiting the number of objects.

In recent years in the video domain, there have been significant improvements in video resolutions, with a major shift towards Ultra-High-Definition (UHD) TV (Nakasu, 2012), aimed at increasing the overall viewing experience. Currently, UHDTV standards allows two resolutions, namely 3840 x 2160 (4kUHD) and 7680 x 4320 (8kUHD), with 4kUHD now gaining ground in sectors such as broadcasting, cinemas, and video gaming. UHDTV is thus gaining increasing momentum. According to Cisco (Webster, 2015), IP video traffic is expected to account for 80% of all traffic by the year 2019, especially with the increasing consumption of IP-based UHD content. Mission-critical communication systems can therefore benefit enormously from the adoption of resolutions above 1080p, which will give better video quality, image analysis, and a wider area of coverage. UHD technology increases situational awareness and significantly enhances the ability of the operator, since finely detailed imagery is received. For example, covering a large environment will be much easier with the increased resolution. In addition, it will also allow the operator to zoom into the picture without sacrificing image quality, since more pixels are available, thereby making it optimal for mission-critical communication.

In this book chapter, the authors focus on UHD video transmission for mission-critical communications. The building blocks of UHD video transmission systems are illustrated in Figure 1. The first process involves the capture of UHD content using either a camera or a file-based source. In the second layer,

Figure 1. Building blocks of UHD transmission

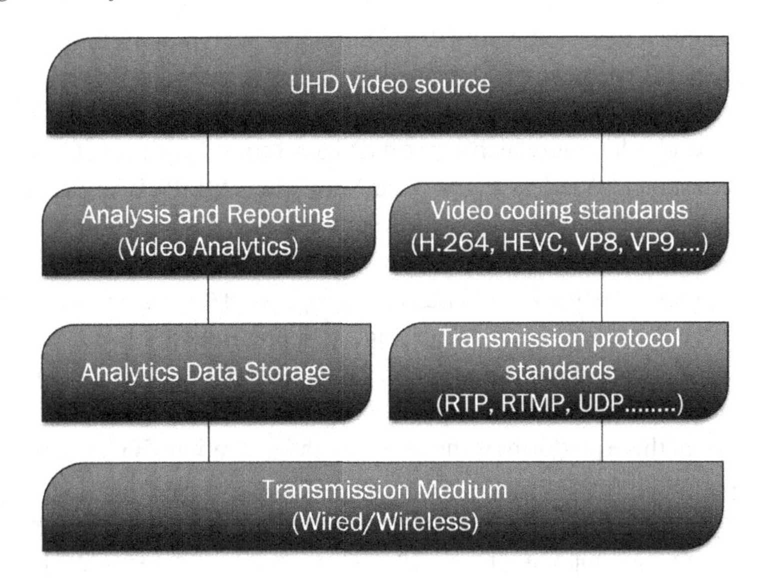

two processes can be implemented to support UHD transmission in mission-critical communications. For bandwidth limited resources, a video coding standard can be used to compress the input video to meet desired requirements, while analysis can also be performed on the input source for certain events (event detection) or objects (object detection). In the third layer, the transport protocols are used to encapsulate the received video and transmit the content, while the fourth layer is dependent on the physical medium available for transmission (wired/wireless).

The remaining sections of this chapter are organized as follows. The background provides a summary of the evolution of ultra-high definition and video transmission. The subsequent four sections provide more insights with regards to the building blocks of UHD transmission for mission-critical communication and current solutions, while the future trends section presents the future directions in the area in terms of future research topics. Finally, concluding remarks are provided.

Background

Video applications over mission-critical communication systems is a cross-layer, multidisciplinary topic that covers video (resolution and processing) standards and techniques, networking standards, transport and resource management, and the quality of both service and experience. There has been rapid development of both video and networking standards in order to meet with the demands of higher quality video applications (Grecos & Wang, 2011).

Video resolution has played a big role in improving the quality received at the end user, and has evolved from Standard-Definition Television (SDTV) to a more recent standard called Ultra-High-Definition Television (UHDTV) (ITU, 2012). UHDTV has been adopted in areas such as broadcast, scientific research, and gaming. At the same time, video coding standards have evolved. Built on the past successes of previous standards, the latest generation is High-Efficiency Video Coding (HEVC) (Benjamin Bross, Han, Ohm, Sullivan, & Wiegand, 2012), which is used side-by-side with its immediate predecessor H.264/MPEG-4 Advanced Video Coding (ITU-T&ISO/IEC, 2003). Further functionality extensions such as Scalable Video Coding to HEVC have also been proposed.

From a network standards perspective, a range of standards have evolved over the years. Traditional mission-critical networks use PDH/SDH/SONET-based TDM technologies. Many of these networks are evolving towards broadband solutions, which support IP-based communication. Many IP-based solutions have also evolved. For instance, Wireless Local Area Networks (WiFi/WLAN) have been standardized in the IEEE 802.11 series and are able to provide data rates above 300Mbps (Matsuura, 2014). More recent 802.11 standards, such as 802.11ac and 802.11ad (X. Zhu & Kocak, 2011), provide gigabit data rates. At the same time, mobile network standards have progressed towards higher data rates. Currently, Fourth Generation (4G) can achieve up to 1Gbps, and is represented by Long-Term Evolution (LTE) advanced and wide-area cellular systems and metropolitan area (WiMAX) 2 standards. It is believed that these existing standards in coverage and capacity will enable a convergence into a common IP-based integrated platform. One of these platforms is the future Fifth Generation (5G) systems, which will enable the collaborative support of legacy and new applications at improved data rates (Thompson et al., 2014).

Through the above technologies, video can be transmitted uncompressed or compressed and stored for future use. One of the major applications of video in mission-critical systems is video surveillance. Figure 2 illustrates the typical overview of video transmission (for surveillance purposes).

Figure 2 illustrates an overview of a typical video surveillance system, which can also be used in a mission-critical communication system scenario. From the user's perspective, video must be on par with typical video applications. Therefore, the overall Quality of Experience (QoE) of the user has to have acceptable quality of service parameters, such as bandwidth, delay, and jitter. In addition, the perceived

Figure 2. Typical video surveillance system

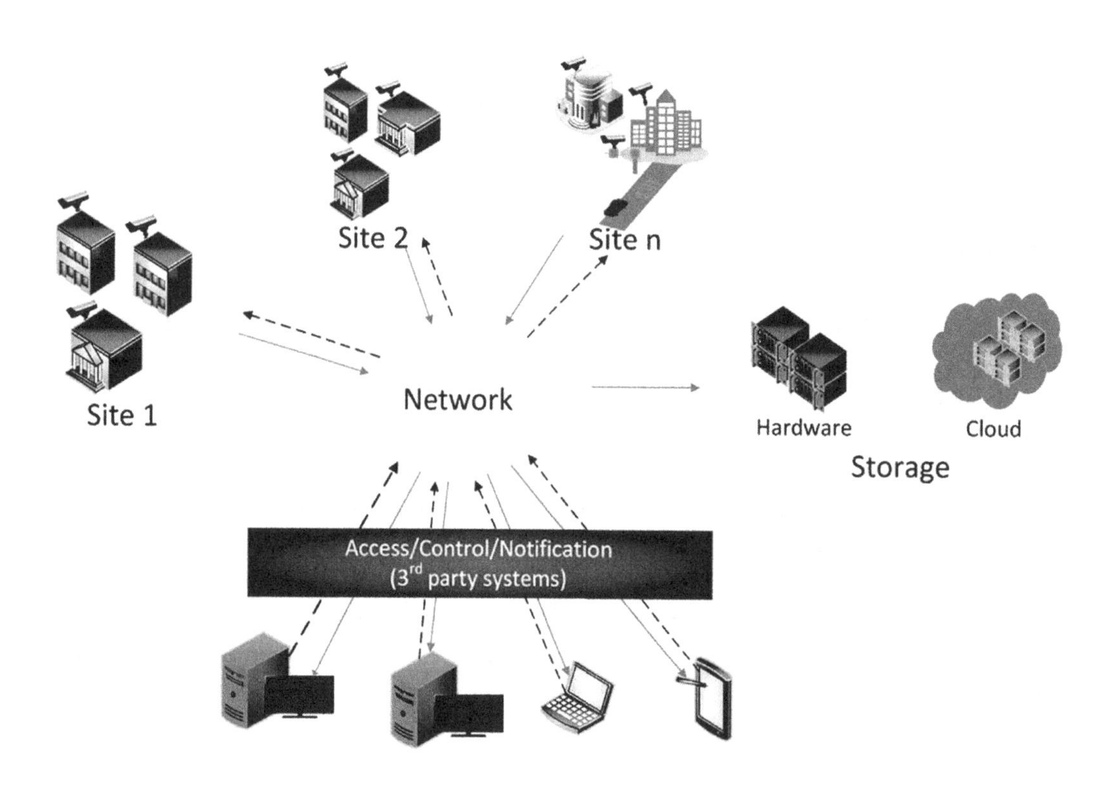

visual quality will also have an impact on the QoE. Video surveillance applications for mission-critical systems can currently transmit up to full-high definition (Oh et al., 2011). Obviously, high-detail scenes increase both perceptual quality and accuracy and can be used for other purposes, such as video analytics. However, full-high-definition cameras have a maximum distance of approximately 120m for effective video capture (SYNESIS, 2010). Therefore, more hardware would need to be provided, especially in larger areas.

ULTRA-HIGH DEFINITION

Spatial Resolution

UHD standardization enables two categories, namely 4K UHD (3840 x 2160) and 8K UHD (7680 x 4320), which are 4 times (4x) and 16 times (16x), respectively, more definitive than the popular full-high-definition standard (1080p). With the proliferation of these standards into the multimedia industry, user-based video applications will find success in surveillance, as previous video standards have.

The transportation and delivery of ultra-high-definition video poses a problem for live transmission since current compression schemes do not provide a robust platform for this application. This is because encoding and channel errors can damage a compressed video frame and also propagate across dependent video frames. Furthermore, the computational load required for encoding 4kUHD is quite high due to the increase in spatial resolution, thus making it more tasking to encode in real-time for a live transmission.

Uncompressed UHD Transmission

The transmission of uncompressed video is mostly inhibited by network bandwidth. Where network bandwidth is sufficient, it can be inhibited by interface connectivity and bandwidth. The bandwidth requirement for transmission of UHD uncompressed video is normally from 4.78 Gb/s for 4kUHD and 19.11 Gb/s for 8K UHD when using the full-RGB color depth, based on the equation provided by (Poynton, 2002), as specified in both Equation 1 and Table 1.

$$B = R * F_{HZ} * C \tag{1}$$

Where B is the resultant bit-rate, R is video image resolution per frame, F_{HZ} is the refresh/frame rate, and C is the color depth. Based on the above Equation 1, the resultant bit-rates of UHD 24Hz RGB video using different bit-depths are shown in Table 1.

Due to the enormous bandwidth required for uncompressed video, it is widely accepted that the chrominance can be reduced since the human visual system has lower acuity for color differences as

Table 1. Uncompressed UHD bit rate at physical layer for video frames moving at 24 frames/s

	8-bit (Gb/s)	10-bit (Gb/s)	12-bit (Gb/s)	16-bit (Gb/s)
4K UHD	4.78	5.67	7.17	9.56
8K UHD	19.11	23.89	28.67	38.22

Table 2. 4kUHD Chroma subsampling bit-rates at 24Hz

	8-bit 4:1:1 (Gb/s)	8-bit 4:2:0 (Gb/s)	8-bit 4:2:2 (Gb/s)	8-bit 4:4:4 (Gb/s)
4K UHD	2.39	2.39	3.19	4.78
4K UHD	9.56	9.59	12.74	19.11

compared to luminance (Pearson, 1975; Poynton, 2002; Winkler, van den Branden Lambrecht, & Kunt, 2001) to optimize bandwidth allocation. This form of compression is only based on color information and not the redundancies between or within the frames. Therefore, the minimum bandwidth requirements for a 4kUHD 24Hz 8-bit video can be recalculated as:

$$B = R * F_{HZ} * S \tag{2}$$

Where S is the subsampling notation used. Table 2 shows the different bit rates of UHD video using different chroma subsampling (YCbCr) notations using 8 bits per color space per pixel.

Wired Transmission

Data rates for Ultra-High Definition (UHD) with 8-bit color chroma-subsampling of 4:2:0 begins at 2.39 Gb/s for 4kUHD and 9.59 Gb/s for 8kUHD. The transmission of beyond FHD uncompressed video format has already being demonstrated in (Halák, Krsek, Ubik, Žejdl, & Nevřela, 2011; Kataoka, 2009; Shirai, Kawano, & Fujii, 2007; Shirai et al., 2009; Shirai, Shimizu, Sameshima, & Takahashi, 2007), which were all conducted over wired networks and over long distances. They all transmitted 4K (4096 x 2160) DCI (Digital Cinema Initiatives) resolution with 10-bit color chroma-subsampling of 4:2:2, thereby making the transmitted bandwidth (from 5.31 Gb/s) even higher, however this amount of bandwidth is only applicable to cinema-based video applications.

In all cases, the video content is streamed over the network using four different FHD sources. At the receiver, all four streams are converted from IP into Four FHD signals. All FHD signals are then synchronized to produce a master 4k image. With the evolution of technology, 4kUHD video capture cameras have become available in the commercial domain.

Wireless Transmission

In terms of transmission over wireless networks, currently the 60 GHz spectrum has been favoured in providing data rates with standardization of IEEE 802.11 and WLAN (X. Zhu & Kocak, 2011), which enables multi-gigabit data rates of up to 7 Gb/s. Currently, the possibility of transmitting uncompressed 4kUHD over 60 GHz networks is being explored, with research presented in (Ejeye & Walker, 2012) demonstrating its possibility, where 4 FHD video streams were transmitted over different 60 GHz channels and then stitched together to give one 4kUHD video with a frame rate of 24Hz. While the 60GHz spectrum works, there are concerns with wireless transmission using this spectrum due to its characteristic properties, such as transmission range and attenuation. For instance, for the same range in comparison to radio signals operating in the 5 GHz spectrum, the Free Space Loss (FSL) is 21dB more (e.g., for 1m range free space loss is 68dB at 60 GHz, and 47dB at 5.5 GHz) when considering the general rule of thumb, which is that for every 6dB increase in propagation loss it halves the coverage distance. To compensate for oxygen attenuation, most regulators now allow fairly high-power transmissions and antenna gains. Even though high power enables more range, some of that power will still be absorbed.

UHD Video Compression

Compression codecs are generally used for the reduction in file size of video sequences. For 4kUHD video sequences three codecs standout based on their specifications, namely Motion JPEG 2000 (Rec, 2007), H.264/MPEG-AVC (Advanced Video Coding), and HEVC. It should also be noted that open standards such as VP9 and its predecessor VP8 are also competitive alternatives (Bankoski et al., 2013).

Motion JPEG 2000

Motion JPEG 2000 establishes the use of the JPEG 2000 (Marcellin, Gormish, Bilgin, & Boliek, 2000) format for motion sequences. Motion JPEG2000 seems to be the preferred codec amongst all three in terms of encoding latency. The major reason is that interframe compression techniques increase the encoding latency since they compress a group of frames, as discussed in (Fujii, Kitamura, Murooka, Shirai, & Takahara, 2009; Shimizu et al., 2006) where the authors discuss the possibilities of 4K DCI video transmission at frame rates from 30Hz. The study provides a solution for higher refresh rates using only JPEG 2000 with a bit-rate of 500 Mb/s. In addition, (Fujii et al., 2009) also takes into consideration the scenario where there is a multicast stream with a lower resolution (2560 x 2048) video.

H.264/AVC

H.264/AVC provides both intra- and inter-frame compression. This video codec is able to offer increased compression efficiency compared to JPEG2000, and is widely used across the spectrum of consumer products and services. H.264/AVC codecs have been implemented both as software and hardware solutions.

The encoder is split into layers, namely video coding and network abstraction layers. The compression algorithms, which include motion compensation, entropy encoding, and discrete cosine transformation, are provided by the Video Coding Layer (VCL). At the VCL, the intra-frame compression is applied to the initial frame (I-frame) in a GOP while subsequent frames (either P or B frames) of that group are compressed based on temporal redundancies with respect to the I-frame.

For transmission or storage, the VCL representation of the video content is partitioned to the Network Abstraction Layer (NAL) units, which in turn can be packetized into packets (with transport protocol headers). H.264 video coding is used in a wide range of applications, but its specifications can only allow up to 4kUHD video input at a frame rate of 60Hz.

H.264/AVC codec features has been well investigated over the last few years with immense improvements and is still widely used alongside newer standards.

High-Efficiency Video Coding

High-Efficiency Video Coding (HEVC), also known as H.265, is a successor to H.264/MPEG-4 AVC. Like H.264, it is a joint effort of both ISO MPEG and ITUT VCEG. HEVC proposes to improve video quality at increased data compression when compared to its predecessor. It also supports up to 8k resolution as input and improved parallel processing methods (Gary J Sullivan & Ohm, 2010; G. J. Sullivan, Ohm, Han, & Wiegand, 2012).

The replacement for its predecessor codec (H264/AVC) began in January 2010 with the partnership of the ITU-T, ISO, and JCT-VC with the aim of designing a standard that can provide at least 50% compression without a reduction in visual quality. The HEVC codec standard had its breakthrough with the draft specification in February 2012 (Benjamin Bross et al., 2012) with both VCL and NAL similar to those of its predecessor coding standard. At the inception of the standardization of HEVC, it was projected to improve compression over its predecessor codec by at least 50%. However, results provided from experiments performed by the authors in (Ohm, Sullivan, Schwarz, Tan, & Wiegand, 2014) show that there was a variation in HEVC Main Profile (MP) bit-rate savings after considering two scenarios (interaction and entertainment).

The major difference between the VCL in HEVC and that of H.264/AVC is the coding structure being used in each picture. In H.264/AVC, each frame is divided into macroblocks with each block containing 16 x 16 luma samples. These macroblocks in turn can be further divided into smaller blocks with 4 x 4 being the smallest unit; other sizes are 16 x 8, 8 x 16, 8 x 8, 8 x 4, and 4 x 8. On the other hand, pictures in HEVC are divided into Coding Tree Units (CTUs), which can be 16 x 16, 32 x 32, or 64 x 64 pixels, with a larger size usually improving coding efficiency (McCann, Bross, Kim, Sekiguchi, & Han, 2011). The largest tree unit specified at input is called the Largest Coding Unit (LCU) and is then recursively sub-divided using quad-tree segmentation into smaller Coding Units (CUs); therefore, HEVC allows CUs of 8 x 8, 16 x 16, 32 x 32, and 64 x 64. The use of larger coding units can lead to an increase in compression gain especially on homogeneous regions within a frame (intra-prediction) or with minimal motion between adjacent frames (inter-prediction). The smallest CU is determined by the depth and number of levels within the quad-tree structure. At any level of the tree subdivision, the four CUs present can either be encoded as a single block or sub-divided even further to form the next level.

Interactive applications show an average of 40% in bit-rate savings, while entertainment applications show an average bit-rate savings of 35.4% for objective measurement (PSNR) and 49.3% average bit-rate savings for perceived video quality, all in comparison to the H264/AVC High Profile (HP) using the HEVC reference codec. This was achieved based on the maximum coding unit size for luma permitted in HEVC MP of 64 x 64 in the VCL and is also beneficial for higher resolution videos as well as video sequences with sparse contents. A comparison between HEVC and H.264 can be seen in Figure 3, which suggests that at each bit-rate HEVC provides improved quality compared to H.264/AVC using different bit-rates.

Transmission

The transmission of compressed UHD video poses less challenges in comparison to its uncompressed formats, especially since a reduction in bandwidth is obtainable. Current studies have shown that the 4kUHD can be transmitted at 20 Mb/s over packet-switched networks using best-effort protocols (Adeyemi-Ejeye & Walker, 2014; Adeyemi-Ejeye & Walker, 2013), such as Real-Time Transport Protocol (RTP) (Schulzrinne, Casner, Frederick, & Jacobson, 2003) and User Datagram Protocol (Protocol, 1980). When using reliable protocols, 4kUHD can be transmitted at an even lower data rate (Clift, Adeyemi-Ejeye, Koczian, Walker, & Clarke, 2014) of 8 Mb/s. These transmissions have been done using both IEEE 802.11 n and ac compliant equipment, which operate using 5 GHz and enable increased bandwidth through Multiple in Multiple out (MIMO) and data link layer frame aggregation. 5 GHz spectrum WLAN are currently available to purchase in the market.

Figure 3. Comparison between HEVC and H.264/AVC using different bit-rates for a 4kUHD high temporal index sequence (Sintel) (Adeyemi-Ejeye, 2015)

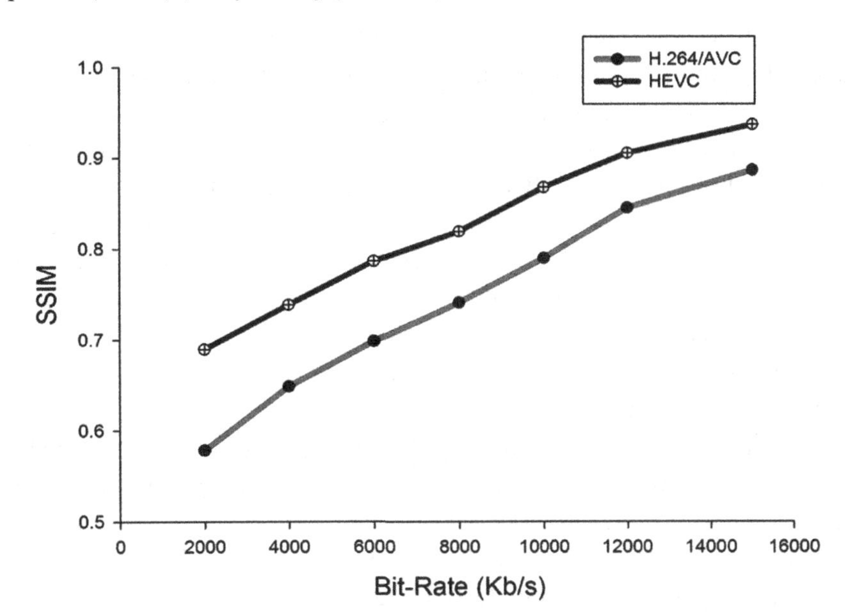

Exploring Video Coding Standards for Mission-Critical Communication Systems

In choosing video coding standards for transmitting UHD over MCC systems, target bandwidth, encoding latency, protocol stack, and QoS parameters must all be taken into consideration to ensure acceptable QoE.

For instance, although JPEG 2000 is an intra-frame codec and therefore encoding latency is minimal, the resulting bit-rate might be considered unacceptable for MCC, especially since bit-rates between 200-400 Mb/s would be obtainable per camera (Shirai, Yamaguchi, Shimizu, Murooka, & Fujii, 2006). This would also be the case for uncompressed UHD video transmission. However, with the evolution of networks enabling increased bandwidth, video transmission at higher data rates can be exploited.

Hybrid video codecs such as H.264/AVC and HEVC enable significant data rate reduction in comparison to JPEG 2000 and uncompressed UHD video. However, there is always trade-off in encoding with regards to compression efficiency, as the semantics vary with intended application. For example, compression ratio should be increased if its intended application is only for storage or VOD. In the latter, encoding latency and transmission medium are not a constraint even for video resolutions above FHD.

Several other features can be exploited for other purposes such as error resilience, random access, etc. Since prerecorded video is preprocessed by appropriate spatial and temporal preprocessing, video bit-rates can be well matched with the network environment. However, this changes when certain parameters are varied, such as resolution and refresh rate. Live encoding and streaming already presents a significant problem as encoding latency for H.264/AVC and HEVC encoders are high due to computational complexity (Saponara, Denolf, Lafruit, Blanch, & Bormans, 2004; Zrida, Abid, Ammri, & Jemai, 2008). The computational complexity arises from data dependencies that exist in the prediction process. For H.264/AVC the data dependencies exist between the Macroblocks (MB), while in HEVC they exist between the Coding Units (CU). These dependencies also vary with increases in both the spatial and temporal indices of any video content.

In terms of target bit-rate and transmission, HEVC video coding would be an obvious choice, since there are indications that up to 50% bit-rate reduction is obtainable when compared with H.264/AVC. Although it has been demonstrated that 4kUHD can be transmitted at 8 Mb/s using both H.264/AVC and the Real-Time Messaging Protocol (RTMP) (Clift et al., 2014) with an end-to-end latency of four seconds, this might not be suitable for MCC applications that need low-latency video transmission. Therefore, the obvious choice of transmission protocols would be the one with the best effort delivery: UDP. Based on the choice of delivery, the data rates have to be increased since the parameters, such as the size of the Group of Pictures (GOP), would be reduced. Shortening the GOP is also considered a way of mitigating the effects of packet loss during transmission.

The computational intensity of encoding UHD both with HEVC and H.264/AVC lends itself the processing advantage available with parallel processing. In recent years, video codecs have been mapped onto different parallel processing architectures, with H.264/AVC being more popular in implementation. In Ge, Tian, & Chen, 2003, H.264/AVC is mapped on an Intel hyper-threading architecture, while a more efficient multi-level implementation was proposed in (Chen, Tian, Ge, & Girkar, 2004). In both studies, a frame is divided into several slices and these slices bring huge overheads due to the slice headers and their impairments to data dependencies among MBs. Furthermore, their implementation trades off coding efficiency for parallel threading. On the other hand, a more efficient algorithm by (Zhuo & Ping, 2006) exploits the spatial and temporal data dependencies for optimal coding efficiency. In all three cases, the reduction in encoding latency is done without any consideration for streaming protocols. This in itself poses a problem of latency bottlenecks during the packetization process for video streaming, as CPU thread resources are limited during the encoding/transmission process. To solve this problem, the use of a GPU or an external system that deals with encoding can be used for the implementation parallel encoding. The use of an external hardware system for encoding, as done with JPEG 2000 for 4K videos, provides reduced latency since the other processes, such as transmission, can be shared effectively across different devices. HEVC is still new in the video compression industry, and due to its specifications (such as the inter-prediction), it is computationally complex and intensive in terms of process performance when compared to H.264/AVC. For example, Table 3 shows the average frame rates achieved while encoding on a 16-core processing server.

The results in Table 3 suggests that the processing power needed for HEVC encoding would be much higher. Furthermore, it can be estimated that for a 64-core encoding machine, choosing a maximum CTU of 64 x 64 will enable a frame rate of up to 16 Hz, while 32 x 32 and 16 x 16 will enable up to 32Hz and 48Hz, respectively, when using such a system as a dedicated encoder. Therefore, parallel solutions for video encoding using HEVC have to be investigated.

Table 3. Coding Tree Unit (CTU) vs achievable average frame rate (Adeyemi-Ejeye, 2015)

CTU	Average (Hz)
64 x 64	4
32 x 32	8
16 x 16	12

SOLUTIONS AND RECOMMENDATIONS

Processing Performance

Processing performance is a key metric in determining the overall QoE since it could lead to an increase in encoding latency. A popular platform used for parallel encoding solutions is the Graphics Processing Unit (GPU). The very popular Compute Unified Device Architecture (CUDA) (Nvidia, 2011) supports a rigorous implementation of this, though it is only available to nVidia GPUs. For H.264/AVC, there has been enormous progress in porting all encoding processes for GPU processing both in the research and commercial domains. For instance, in (Wu, Wen, Su, Ren, & Zhang, 2012), four major processes (interprediction, intraprediction, entropy encoding, and deblocking) of H.264/AVC encoding were mapped onto CUDA. They use data localization to organize data and threads to work efficiently on the GPU. However, this approach is constrained to available GPU resources, as their approach eliminates the use of higher than HD resolutions, especially on GPUs with constrained local memory. Furthermore, their implementation allows a latency bottleneck for critical video applications, such as live streaming, as the data in their proposed architecture has to be transferred back to the system memory before it can be accessed by other processes. This research was extended to accommodate 4kUHD live encoding by enabling zero-copy between the GPU and CPU memory, dynamic parallelism for inter-prediction, and the horizontal mode for intra-prediction (Adeyemi-Ejeye & Walker, 2014). The entire implementation reducing the transfer latency incurred by the GPU-CPU and could also encode 4kUHD at an average frame rate of 15 Hz, but there are issues such as an increase in target bit-rate since the intra-prediction algorithm reduced compression ratio and therefore increased the I-frame by approximately 4%.

In terms of HEVC CUDA parallelization, the process currently being optimized is the inter-prediction process. In Radicke, Hahn, Grecos, & Wang, 2013, a motion estimation algorithm, which is up to 42% faster when compared to the HM9 reference encoder (Kim, 2012) with negligible impact on the target bit-rate and objective video quality, is tested. An average of 70% in time savings was achieved for FHD sequences (Juncheng, Falei, Shanshe, & Siwei, 2014) when compared to encoding them with HM 10.0 (B Bross, Han, Sullivan, Ohm, & Wiegand, 2013), while a multiple layer parallel inter-prediction algorithm for both Prediction Unit (PU) and CTU saves over 60% of encoding time in comparison with HM10.

Table 4 shows other implementations of H.264/AVC encoding using CUDA and their performance comparisons. It should also be noted that the GPU architectures used in the research stated above are mostly first generation GPUs and have some limitations due to their compute capabilities. Furthermore, they used an open source commercial implementation of the H.264/MPEG-AVC encoder called x264 (Aimar et al., 2005) that has proven to be approximately 50 times faster than the widely used JM (Sühring, Heising, & Marpe, 2009) reference software and also provides better video quality for the same target bit-rate (Merritt & Vanam, 2007) due to its improved process algorithms (rate control, motion

Table 4. CUDA parallel H.264/AVC encoder implementations

	Platform	Reference code	Target Resolution	Average Performance (fps)
Wu et al., 2012	GTX 260	x264	720p	25.2
Su, Wen, Wu, Ren, & Zhang, 2014	Tesla C2050	x264	720p	32.3
Adeyemi-Ejeye & Walker, 2014	NVS510	x264	2160p	16

estimation, etc.). Based on the previous progress with commercial implementations, it is expected that the HEVC version x265 (Angelini, 2013) would be faster than the reference software (HM) and will also have more flexible features, such as allowing live encoding for transportation.

Transmission

At the IP packet level, video and data use the same technology (TCP/IP) to ensure large-scale and fast transmission. As the number of video capture devices increases, an increase in the need to transmit more and more video data over the same medium also arises. UHD video transmission could also benefit from current practices to reduce bandwidth being used by multiple cameras in terms of video streams. One such solution is multicast, which means that each camera only needs to send one stream at a time while the switches reproduce and forward the same video stream to multiple receivers. The multicast approach in comparison to the unicast helps to reduce bandwidth enormously. For example, if there are 400 HD cameras and 20 clients, a unicast transmission would incur bandwidth costs of up to 46 Gb/s, while a multicast configuration reduced the bandwidth consumption to about 2 Gb/s (Hsu & Chen, 2015).

FUTURE RESEARCH DIRECTIONS

HEVC and Beyond

While HEVC reached the Final Draft International Standard status in January 2013 and is now being deployed in most video application sectors, the major issue is still its computational complexity. With the open source version (x265) now available, it would be of great interest to investigate the parallelization of this tool on GPUs, especially since UHD MCC cameras would have to have a discrete design and a number of solutions for embedded GPUs such as nVidia jetson TK1 (Nvidia, 2015) are now available.

5G Networks

With the advent of 5G networks, more bandwidth will become available. One of the proposed underlining technologies for transmission is the 802.11ad WLAN, which will provide multi-gigabit data rates. Due to this development, it would be worth investigating the impact of very low data rates for transmission over wireless networks for MCC systems. For instance, the 802.11ad specifies a new mac layer that enables packet aggregation. A problem might arise when viewing important data, with a factor such as packet loss affecting the QoE, during the transmission of live events. In addition, the transmission of video within the 5G ecosystem to see how processes such as network virtualization, device-2-device networking, or cloud-based storage. Each process would have its limitations and therefore this must be taken into consideration when designing a 5G MCC system.

Furthermore, it would also be worth investigating solutions for uncompressed UHD video transmission, since data rates above 4.6 Gb/s will become available.

Video Analytics

Video analytics (SYNESIS, 2010) increases the effectiveness of surveillance systems through the implementation of image-processing algorithms. These enhancements enable real-time event/object detection, post-event analysis, and the extraction of statistical data. By defining the set of events the surveillance system should alerted to, video analytics continues to analyze the video and provides immediate alerts upon detection of a relevant even, while at the same time saving costs on manpower. UHD MMC systems can benefit from the use of video analytics since the resolution is increased and therefore it will make the detection of smaller objects and events easier, especially at a range beyond 120 meters.

CONCLUSION

With the evolution and in particular the latest advancements in spatial resolution of video alongside processing and network transmission, ultra-high-definition video transmission is gaining increasing popularity in the multimedia industry. Mission communication systems can benefit enormously from the amount of video data available in terms of range and wider angle of view. Key technology enablers include video coding parallelization, video analytics, and improved transmission mediums. With further development in these areas, we should expect the deployment of UHD MCC systems in the future.

ACKNOWLEDGMENT

This work has been partially funded by the European Union's Horizon 2020 iCIRRUS project (grant no. 644526)

REFERENCES

Adeyemi-Ejeye, A. O. (2015). *Ultra-High Defnition Wireless Video Transmission (PhD Thesis)*. University of Essex.

Adeyemi-Ejeye, A. O., & Walker, S. (2014). 4kUHD H264 wireless live video streaming using CUDA. *Journal of Electrical and Computer Engineering*.

Adeyemi-Ejeye, A. O., & Walker, S. D. (2013). *Ultra-high definition Wireless Video transmission using H. 264 over 802.11 n WLAN: Challenges and performance evaluation*. Paper presented at the Telecommunications (ConTEL), 2013 12th International Conference on.

Aimar, L., Merritt, L., Petit, E., Chen, M., Clay, J., Rullgrd, M., . . . Izvorski, A. (2005). *x264-a free h264/AVC encoder*. Retrieved from http://www. videolan. org/developers/x264. html

Alcatel-Lucent. (2013). *Mission-critical Communications Networks for Public Safety*. Retrieved from http://www.tmcnet.com/tmc/whitepapers/documents/whitepapers/2013/9270-mission-critical-communications-networks-public-safet.pdf

Angelini, C. (2013). Next-Gen Video Encoding: x265 Tackles HEVC/H. 265. *Tom's Hardware Online Magazine*.

Bankoski, J., Bultje, R. S., Grange, A., Gu, Q., Han, J., Koleszar, J., . . . Xu, Y. (2013). *Towards a next generation open-source video codec*. Paper presented at the IS&T/SPIE Electronic Imaging. doi:10.1117/12.2009777

Bross, B., Han, W.-J., Ohm, J.-R., Sullivan, G. J., & Wiegand, T. (2012). *High efficiency video coding (HEVC) text specification draft 8*. JCTVC-J1003.

Bross, B., Han, W., Sullivan, G., Ohm, J., & Wiegand, T. (2013). *High efficiency video coding (HEVC) text specification draft 10 (JCTVCL1003)*. Paper presented at the JCT-VC Meeting (Joint Collaborative Team of ISO/IEC MPEG & ITU-T VCEG).

Chen, Y.-K., Tian, X., Ge, S., & Girkar, M. (2004). *Towards efficient multi-level threading of H. 264 encoder on Intel hyper-threading architectures*. Paper presented at the Parallel and Distributed Processing Symposium.

Clift, L., Adeyemi-Ejeye, A. O., Koczian, G., Walker, S. D., & Clarke, A. (2014). Delivering Live 4K Broadcasting Using Today's Technology.*International Broadcast Convention (IBC 2014)*.

Ejeye, A. O., & Walker, S. D. (2012). *Uncompressed Quad-1080p Wireless Video Streaming*. Paper presented at the 4th Computer science and Electronic Engineering Conference.

Fujii, T., Kitamura, M., Murooka, T., Shirai, D., & Takahara, A. (2009, 2009). *4K& 2K multi-resolution video communication with 60 fps over IP networks using JPEG2000*. Paper presented at the Intelligent Signal Processing and Communication Systems, 2009. ISPACS 2009. International Symposium on.

Ge, S., Tian, X., & Chen, Y.-K. (2003). *Efficient multithreading implementation of H. 264 encoder on Intel hyper-threading architectures*. Paper presented at the Information, Communications and Signal Processing, 2003 and Fourth Pacific Rim Conference on Multimedia.

Grecos, C., & Wang, Q. (2011). Advances in video networking: Standards and applications. *International Journal of Pervasive Computing and Communications*, *7*(1), 22–43. doi:10.1108/17427371111123676

Halák, J., Krsek, M., Ubik, S., Žejdl, P., & Nevřela, F. (2011). Real-time long-distance transfer of uncompressed 4K video for remote collaboration. *Future Generation Computer Systems*, *27*(7), 886–892. doi:10.1016/j.future.2010.11.014

Hsu, R., & Chen, A. (2015). *Ensure Nonstop IP Surveillance with an Optimized Industrial Ethernet Network*. Retrieved from http://www.remotemagazine.com/main/articles/ensure-nonstop-ip-surveillance-with-an-optimized-industrial-ethernet-network/

ITU-T & ISO/IEC. (2003). *Recommendation and Final Draft International Standard of Joint Specification. ITU-T Rec. H. 264/IEC 14496-10 AVC*. Author.

ITU. (2012). *Parameter values for ultra-high definition television systems for production and international programme exchange*. Geneva: International Telecommunication Union.

Juncheng, M., Falei, L., Shanshe, W., & Siwei, M. (2014). *Flexible CTU-level parallel motion estimation by CPU and GPU pipeline for HEVC.* Paper presented at the Visual Communications and Image Processing Conference.

Kataoka, N. (2009). *4K Uncompressed Streaming over Colored Optical Packet Switching Network.* Paper presented at the OptoElectronics and Communications Conference.

Matsuura, N. (2014). *Wireless LAN device and controlling method thereof.* Google Patents.

McCann, K., Bross, B., Kim, I.-K., Sekiguchi, S.-I., & Han, W.-J. (2011). *HM5: High Efficiency Video Coding (HEVC) Test Model 5 Encoder Description.* Paper presented at the JCTVC-G1102, JCT-VC Meeting, Geneva, Switzerland.

Merritt, L., & Vanam, R. (2007). *Improved rate control and motion estimation for H. 264 encoder.* Paper presented at the Image Processing, 2007. ICIP 2007. IEEE International Conference on. doi:10.1109/ICIP.2007.4379827

Nakasu, E. (2012). Super Hi-Vision on the Horizon: A Future TV System That Conveys an Enhanced Sense of Reality and Presence. *Consumer Electronics Magazine, IEEE, 1*(2), 36–42. doi:10.1109/MCE.2011.2179821

Nvidia. (2015). *JETSON TK1*. Author.

Nvidia, C. (2011). *Compute unified device architecture programming guide*. Academic Press.

Oh, S., Hoogs, A., Perera, A., Cuntoor, N., Chen, C.-C., Lee, J. T., . . . Davis, L. (2011). *A large-scale benchmark dataset for event recognition in surveillance video.* Paper presented at the Computer Vision and Pattern Recognition (CVPR), 2011 IEEE Conference on. doi:10.1109/CVPR.2011.5995586

Ohm, J., Sullivan, G. J., Schwarz, H., Tan, T. K., & Wiegand, T. (2014). Comparison of the Coding Efficiency of Video Coding Standards 2014; Including High Efficiency Video Coding (HEVC). *Circuits and Systems for Video Technology. IEEE Transactions on, 12*(12), 1669–1684.

Pearson, D. E. (1975). *Transmission and display of pictorial information*. Pentech Press Limited.

Poynton, C. (2002). Chroma subsampling notation. In Digital Video and HDTV: Algorithms and Interfaces. Morgan Kaufmann.

Protocol, U. D. (1980). RFC 768 J. Postel ISI 28 August 1980. *Isi.*

Radicke, S., Hahn, J., Grecos, C., & Wang, Q. (2013). *Highly-parallel HVEC motion estimation with CUDA.* Paper presented at the Visual Information Processing (EUVIP), 2013 4th European Workshop on.

Rec, I. (2007). T. 800| ISO/IEC 15444-3: 2007. *Information technology -- JPEG 2000 image coding system -- Part 3: Motion JPEG 2000.*

Saponara, S., Denolf, K., Lafruit, G., Blanch, C., & Bormans, J. (2004). Performance and complexity co-evaluation of the advanced video coding standard for cost-effective multimedia communications. *EURASIP Journal on Applied Signal Processing, 2004*(2), 220–235. doi:10.1155/S111086570431019X

Schulzrinne, H., Casner, S., Frederick, R., & Jacobson, V. (2003). *Real-time transport protocol.* RFC1899.

Shimizu, T., Shirai, D., Takahashi, H., Murooka, T., Obana, K., Tonomura, Y., & Ohta, N. et al. (2006). International real-time streaming of 4K digital cinema. *Future Generation Computer Systems, 22*(8), 929–939. doi:10.1016/j.future.2006.04.001

Shirai, D., Kawano, T., & Fujii, T. (2007). *6 Gbit/s uncompressed 4K video stream switching on a 10 Gbit/s network.* Paper presented at the Intelligent Signal Processing and Communication Systems, 2007. ISPACS 2007. International Symposium on.

Shirai, D., Kawano, T., Fujii, T., Kaneko, K., Ohta, N., Ono, S., & Ogoshi, T. et al. (2009). Real time switching and streaming transmission of uncompressed 4K motion pictures. *Future Generation Computer Systems, 25*(2), 192–197. doi:10.1016/j.future.2008.07.003

Shirai, D., Shimizu, K., Sameshima, Y., & Takahashi, H. (2007). 6-Gbit/s Uncompressed 4K Video IP Stream Transmission and OXC Stream Switching Trial Using JGN II. *NTT Technical Review, 5*(1).

Shirai, D., Yamaguchi, T., Shimizu, T., Murooka, T., & Fujii, T. (2006, 2006). *4K SHD real-time video streaming system with JPEG 2000 parallel codec.* Paper presented at the Circuits and Systems, 2006. APCCAS 2006. IEEE Asia Pacific Conference on.

Su, H., Wen, M., Wu, N., Ren, J., & Zhang, C. (2014). Efficient parallel video processing techniques on GPU: From framework to implementation. *The Scientific World Journal*. PMID:24757432

Sühring, K., Heising, G., & Marpe, D. (2009). *H. 264/AVC reference software*. Academic Press.

Sullivan, G. J., & Ohm, J.-R. (2010). *Recent developments in standardization of high efficiency video coding (HEVC).* Paper presented at the SPIE Optical Engineering+ Applications.

Sullivan, G. J., Ohm, J., Han, W.-J., & Wiegand, T. (2012). Overview of the High Efficiency Video Coding (HEVC) Standard. *Circuits and Systems for Video Technology. IEEE Transactions on, 22*(12), 1649–1668.

SYNESIS. (2010). *Megapixel video analytics in difficult surveillance conditions*. Retrieved from http://en.synesis.ru/en/surveillance/contents/megapixel-analytics

Thompson, J., Ge, X.-L., Wu, H.-C., Irmer, R., Jiang, H., Fettweis, G., & Alamouti, S. (2014). 5G wireless communication systems: Prospects and challenges[guest editorial]. *Communications Magazine, IEEE, 52*(2), 62–64. doi:10.1109/MCOM.2014.6736744

Webster, D. (2015). *Cisco Visual Networking Index Predicts IP Traffic to Triple from 2014-2019; Growth Drivers Include Increasing Mobile Access, Demand for Video Services*. Retrieved from http://newsroom.cisco.com/press-release-content?articleId=1644203

Wen-Mei, W. H. (2011). *GPU Computing Gems Emerald Edition*. Elsevier.

Winkler, S., van den Branden Lambrecht, C. J., & Kunt, M. (2001). *Vision and video: models and applications* (Vol. 10). Academic Press.

Wu, N., Wen, M., Su, H., Ren, J., & Zhang, C. (2012). *A parallel H. 264 encoder with CUDA: mapping and evaluation.* Paper presented at the Parallel and Distributed Systems (ICPADS), 2012 IEEE 18th International Conference on. doi:10.1109/ICPADS.2012.46

Zhu, A. D., & Kocak, T. (2011). *Throughput and Coverage Performance for IEEE 802.11ad Millimeter-Wave WPANs*. Academic Press.

Zhu, W., Luo, C., Wang, J., & Li, S. (2011). Multimedia cloud computing. *Signal Processing Magazine, IEEE*, *28*(3), 59–69. doi:10.1109/MSP.2011.940269

Zhuo, Z., & Ping, L. (2006). *A Highly Efficient Parallel Algorithm for H.264 Video Encoder*. Paper presented at the Acoustics, Speech and Signal Processing.

Zrida, H. K., Abid, M., Ammri, A. C., & Jemai, A. (2008). *A YAPI-KPN parallel model of a H264/AVC video encoder*. Paper presented at the Research in Microelectronics and Electronics.

Chapter 7
Quality of Experience (QoE) for Wireless Video Over Critical Communication Systems

Emad Danish
Saudia Airlines, Saudi Arabia

Mazin I. Alshamrani
Ministry of Haj and Umra, Saudi Arabia

ABSTRACT

Video streaming is expected to acquire a massive share of the global internet traffic in the near future. Meanwhile, it is expected that most of the global traffic will be carried over wireless networks. This trend translates into considerable challenges for Service Providers (SP) in terms of maintaining consumers' Quality of Experience (QoE), energy consumption, utilisation of wireless resources, and profitability. However, the majority of Radio Resource Allocation (RRA) algorithms only consider enhancing Quality of Service (QoS) and network parameters. Since this approach may end up with unsatisfied customers in the future, it is essential to develop innovative RRA algorithms that adopt a user-centric approach based on users' QoE. This chapter focus on wireless video over Critical communication systems that are inspired by QoE perceived by end users. This chapter presents a background to introduce the reader to this area, followed by a review of the related up-to-date literature.

INTRODUCTION

The research focus on wireless video over Critical communication systems that are inspired by the Quality of Experience (QoE) perceived by end users. This chapter presents a background to introduce the reader to this area, followed by a review of the related up-to-date literature. In the background, the elements of the triple-Q framework are first introduced. Then communication systems are demonstrated in the context of wireless video communications.

DOI: 10.4018/978-1-5225-2113-6.ch007

THE TRIPLE-Q MODEL

The roots of "quality" have been discussed in (Reeves, and Bednar, 1994) where different but mutually related definitions have been suggested. Quality has been defined as "excellence", "value", "conformance to specifications", and "meeting and/or exceeding customers' expectations". In the context of multimedia communications, quality is attached to the three main elements shown in Figure 1: the network, the end-user, and the Service Provider (SP). Accordingly, a triple-Q model (Moorsel, 2001) has been proposed to combine network's Quality of Service (QoS), customer's Quality of Experience (QoE) and service provider's Quality of Business (QoB). The objective is to increase each quality until an overall optimality is achieved. Recently, the triple-Q model has grabbed even more attention by researchers. For instance, an extended multidimensional approach for this model has been proposed in (Ibarrola et al., 2014).

Vast majority of research efforts have focused on exploring the interaction between QoS and QoE and have overlooked the service provider's interest, until QoB was put into context in the triple-Q model. In a typical multimedia communications scenario (Tektronix, 2008), as illustrated in Figure 2, emphasis is put on the stages where QoS and QoE are considered. In this model, the content passes through several stages over the "data plane" before it is delivered to and consumed by the end-user. Firstly, the audio-visual content is created by the content provider and subsequently passed to the application provider where it is encoded and may be stored. In the third stage, the encoded stream is transported over the network provider's core IP network. This is a controlled network such that QoS parameters can be controlled and guaranteed. Fourthly, the content is transported to the access network, which is another QoS control facility, after which the content is delivered to the Consumer Premise Equipment (CPE) for end-user consumption. This final stage is where user's QoE can be measured. On a reverse path, measured QoE parameters are conveyed back to the application provider over the "control plane". Accordingly, the application provider advises the network provider to set appropriate QoS conditions. This model is said to be an adequate model to help SPs accommodate multimedia services at controllable levels of QoE (Tektronix, 2008). However, the model focuses on the importance of user satisfaction trusting it

Figure 1. The main three players in the multimedia delivery chain

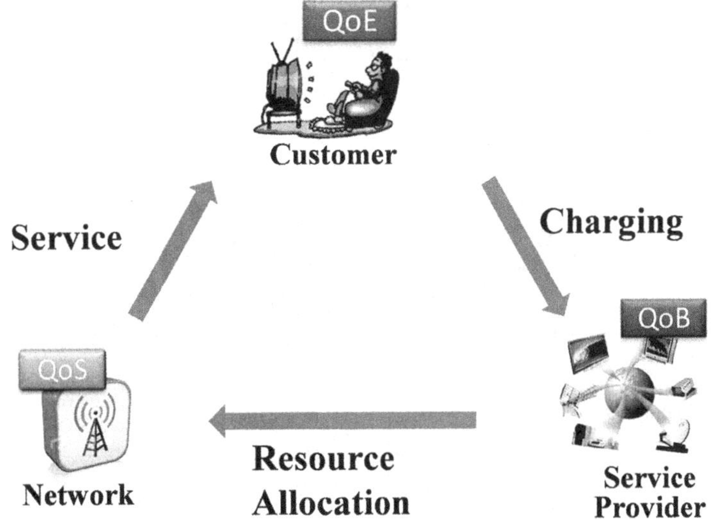

Figure 2. Typical multimedia communications scenario

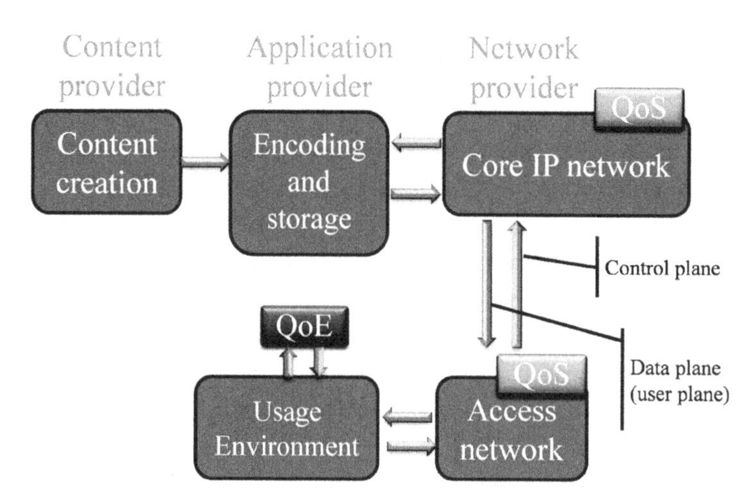

will result in better customer loyalty, which is believed to generate more profit for SPs, but there is no direct link between profit and customer satisfaction. Instead, it is an indirect link emerged as a result of customer loyalty. Consequently, there is a gap between customer satisfaction (QoE) and business revenue (QoB). Therefore, QoB was introduced as a significant element in formulating the triple-Q model.

Figure 3 illustrates the QoS/QoE/QoB interactions within the triple-Q model. This model suggests that QoB is influenced by the user's experience of the service. The service flows from the SP to the customer at an appropriately configured level of QoS. The service is then consumed by the customer at

Figure 3. The Triple-Q model

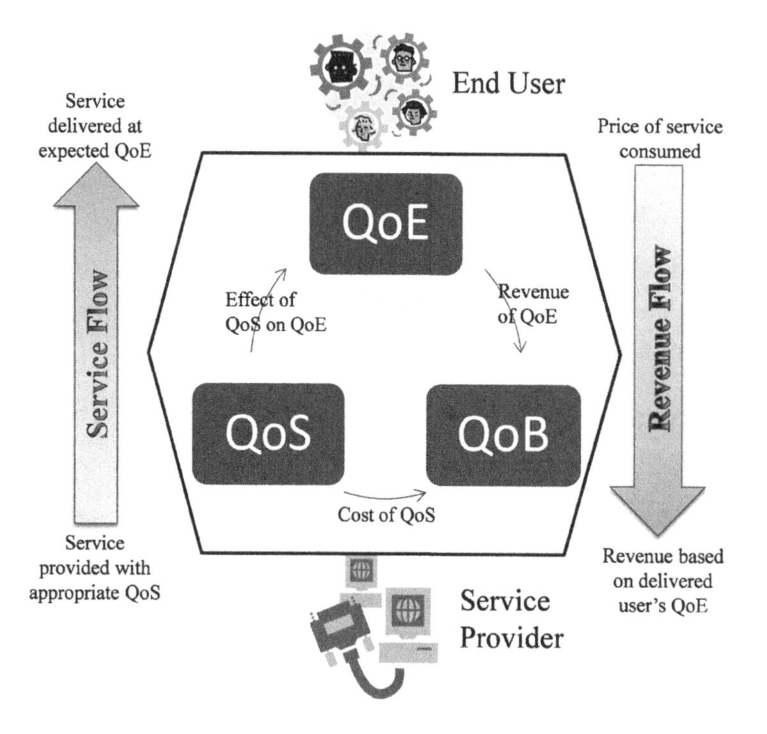

a certain level of QoE. Consequently, revenue flows from the customer to the SP. QoB consists of the revenue along with the cost of delivering the service.

By introducing QoB to the typical scenario Figure 2, a new multimedia communications scenario is produced as shown in Figure 4. In this new model, QoB aims at providing the functionality for new charging and resource allocation decision engine. QoB receives measured QoE indicators from the usage environment, and signals the optimised decisions of resource allocation to the network provider.

A detailed overview on Quality of Service (QoS), Quality of Experience (QoE), and Quality of Business (QoB) is presented in the following sections.

QUALITY OF SERVICE (QoS)

QoS is a dominant network term that describes the service provider's ability to guarantee a certain level of consistent performance to a data flow, by means of monitoring and controlling the network (ITU-T G.1000, 2001). QoS control can be implemented by different techniques, aiming at maintaining a guaranteed transfer of data packets at an expected bit-rate, in the correct sequence and within acceptable delay window. QoS control also includes both the storage and transport of the medium over controlled networks.

The need for QoS comes from the fact that demand on traffic-intensive multimedia applications is increasing at a rate that cannot be met by the limited network bandwidth. This is in addition to applications' sensitivity to delay, jitter and information loss. The study of QoS requirements should consider each multimedia application individually. Hence, multimedia applications were classified in four categories based on their sensitivity (ITU-T G.1010, 2001): interactive, responsive, timely, and non-critical. The interactive category is considered highly sensitive to QoS. Also, the traffic in this category is inelastic and intolerable to delay, and demands data transfer on guaranteed basis. Therefore, it requires a particular lower bound on throughput and upper bounds on delay and jitter for each service. Examples of such

Figure 4. QoB-driven multimedia communications scenario

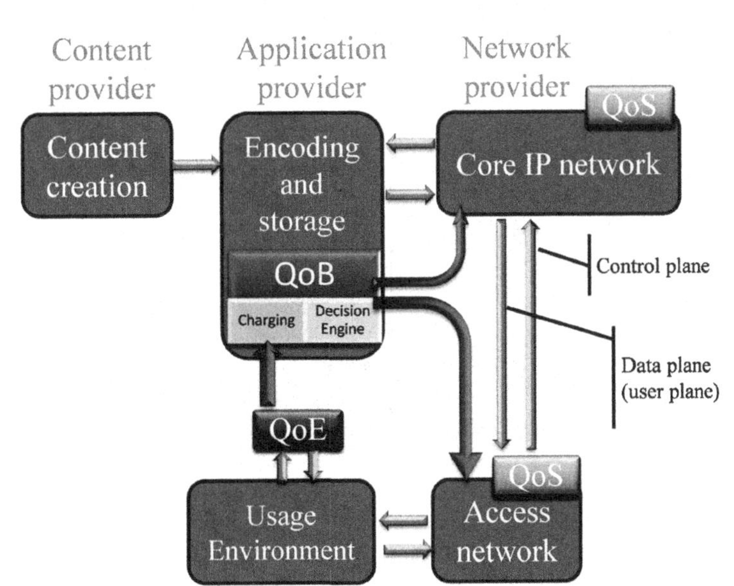

services include real-time voice, interactive gaming, video telephony, and real-time video. The rest of the categories are considered not QoS-sensitive. Also, the traffic in these categories is elastic and tolerates data transfer on best-effort basis. For example, SMS messaging, Web browsing, and audio streaming can accommodate delay to some extent. There are also four inter-related viewpoints by which QoS can be seen (ITU-T G.1000, 2001), as presented in Figure 5. Identification of these viewpoints is needed for differentiated charging where a customer is charged based on the level of quality perceived.

QoS Levels

QoS may be considered at three different levels in a wireless network: the transaction level, the connection level, and the packet level (Abdul-Hameed, 2008).

- **Transaction-Level QoS**: Represented by the probability of transaction completion and the response time.
- **Connection-Level QoS**: Represented by connection establishment probabilities. Two measures are used for this purpose: connection blocking probability and connection dropping probability.
- **Packet-Level QoS (also called application-level QoS)**: Expressed by delay, jitter, packet loss rate, and throughput. These parameters are mostly affected by the network resources in packet-switched networks, and they usually influence the quality perceived by end-users.

Figure 5. The four viewpoints of QoS

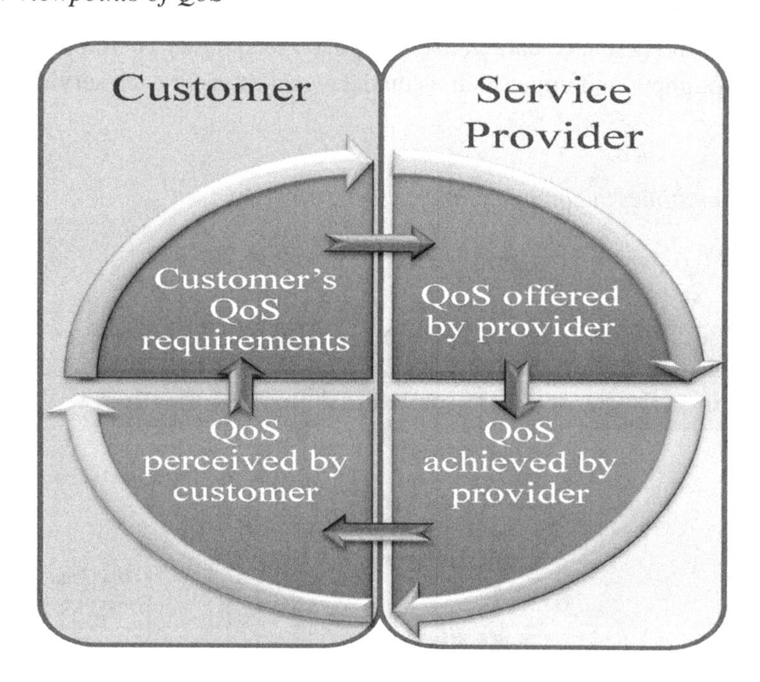

Packet-Level QoS Parameters

Many of the impairments affecting video transmission are IP transmission-related. The transmission QoS parameters causing these impairments are traditionally known as (TeleManagement Forum, 2009):

- **Packet Loss:** the percentage of packets that are not received in the destination. Attenuation in the transmitted radio signal in wireless networks may result in bit errors which lead to the whole packet being dropped. Packet loss is known to be one of the most influencing factors on the received quality for IP-based voice and video services.
- **Delay (Latency):** the time a packet takes from source to destination. Delay can be caused by packets queuing or by the time consumed for processing, propagation, or transmission of a packet. Excessive delay may lead to packet loss, because the receiver may decide to drop a late packet. In video streaming, delay could lead to impairments such as freezed or distorted video frames. A common practice to alleviate the effects of delay is packet buffering, so packets are buffered in the receiving device ahead of the time they are displayed.
- **Jitter:** the variation in delay from packet to packet, i.e., the variation of delay over time. Jitter can be caused by network congestion, but it can also be alleviated by a jitter buffer or by control of bandwidth allocation.
- **Bandwidth:** the data rate supported by the transmission medium. In a congested network, low data rate may lead to delay or jitter as well, ending up with lost packets.

Despite the significance of QoS control, it is a network-centric approach whose main focus is limited to network performance. Hence, excellent QoS is meaningless if the user is unsatisfied. Therefore, QoE is introduced to capitalise on QoS with the aim to assess and improve customer experience.

QUALITY OF EXPERIENCE (QoE)

QoE is a user-centric approach, which explores user's perception and satisfaction about the quality and usability of a particular service (Nokia, 2004). It is also known as the overall acceptability of an application or service, as perceived subjectively by the end-user (ITU-T FG IPTV, 2007; Callet et al., 2013). Recently, QoE has attracted considerable attention from researchers and standardization bodies such as ITU, Broadband Forum (DSL Forum) and TM Forum. This is due to the momentous benefits attached to QoE, from customer loyalty and increased revenue, to maintained competitiveness and reputation. Therefore, researchers' aim has always been to improve QoE through the design and development of appropriate measurement techniques for multimedia services.

Typically QoE is a user aspect that is conceptually measured at the user end. In a communication system, the media service travels across the layers depicted in Figure 6 End-to-end QoE shall consider the impairments introduced across these layers. From the service provider's point of view, only the three higher layers are considered to improve QoE through QoS control.

Figure 6. Layers travelled by the media in communication systems

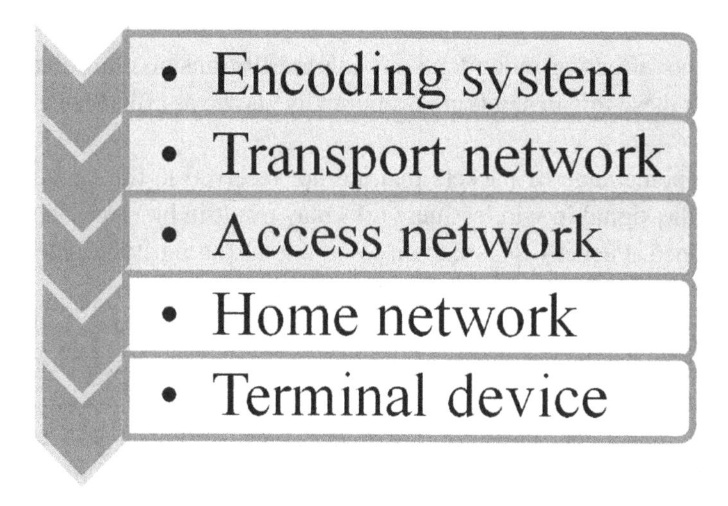

QoE Dimensions

QoE dimensions are those different high level components which altogether comprise the overall QoE. For multimedia services, a summarized high-level collection of the dimensions (Nokia, 2004; Batteram et al., 2010; DSL Forum, 2006) is depicted in *Figure 7*. To measure overall QoE, one metric needs to be employed in each of these dimensions, and the combined suite of metrics constitutes a global QoE measure that is believed to represent the overall value experienced by the customer. However, the majority of QoE assessment techniques in multimedia communications research have adopted the Service Quality dimension with a focus on the Fidelity of the content.

Figure 7. QoE dimensions for multimedia services

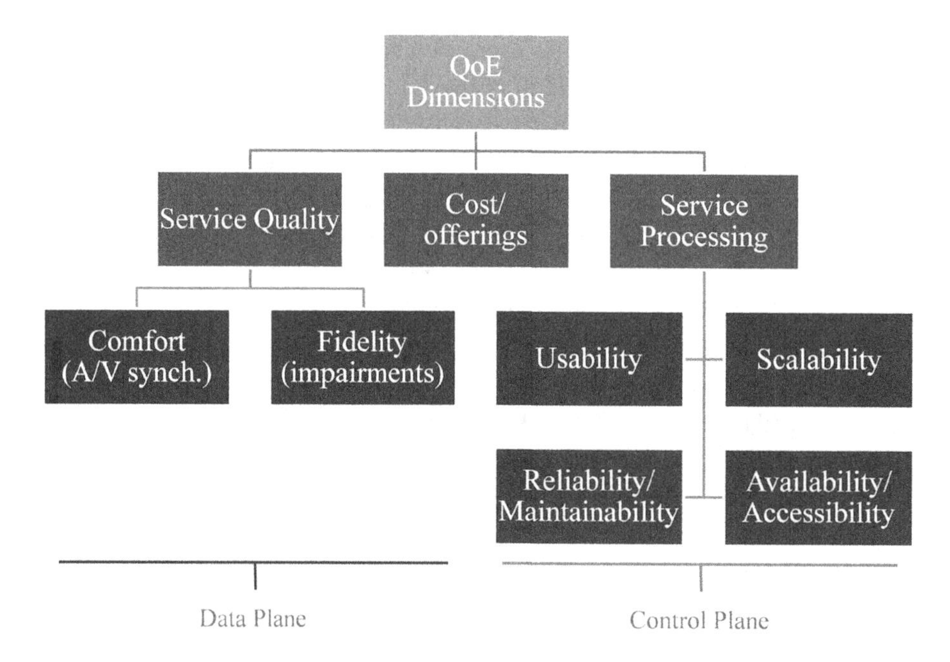

Figure 8. Classification of QoE assessment techniques

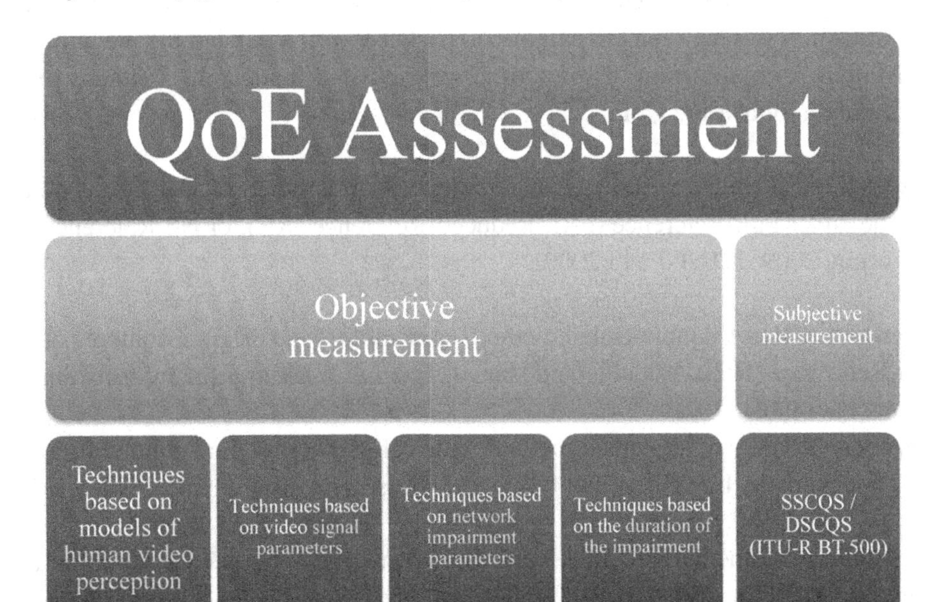

QoE Assessment

To improve and control customer's QoE, researchers have designed and developed various content-specific measurement techniques. These techniques assess QoE and quantify it in a measurable metric. Basically, measurement techniques are either subjective or objective. Subjective techniques are usually conducted in test labs and they are assumed more accurate since they reflect literal human perception. However, researchers have focused on developing objective techniques, which endeavour to emulate human perception, as they are less costly and can be implemented in practical real-time applications. Also, objective metrics can be used to predict QoE. This prediction is essential for the quality estimation of a vast number of scenarios which cannot be fulfilled with subjective techniques.

Two classifications of objective QoE techniques have been introduced in literature. The first is a classification by the DSL Forum (DSL Forum, 2006) which is demonstrated in Figure 8. Objective techniques are either based on human perception, transmitted signal parameters, network impairment parameters or the duration of impairment.

The second classification is the ITU classification of objective QoE assessment model (Takahashi et al., 2008; ITU-T G.1011, 2010; Alreshoodi, and Woods, 2013), which is based on a focus of the assessment model:

- **Parametric Packet-Layer Model:** Uses packet-header information without the content signal (payload information).
- **Parametric Planning Model:** Uses networks' and terminals' quality planning parameters. Therefore, it requires a priori information about the communications system.
- **Media Layer Model**: Analyses the content signal via the Human Vision System (HVS).

- **Bit-Stream Model:** Derives quality by extracting and analysing media characteristics from the coded bit-stream. It lies between the parametric packet-layer model and the media-layer model.
- **Hybrid Model:** A combination of some or all of the listed models. It is effective in exploiting more information to measure QoE.

Depending on the availability of the original content as a reference at assessment time, objective techniques can be either Full-Reference (FR), Reduced-Reference (RR), or No-Reference (NR) (Tele-Management Forum, 2009; ITU-T J.143,2000):

- **Full Reference (FR):** A double-ended method, applicable if the original content is available.
- **Reduced Reference (RR):** A double-ended method, applicable if partial information of the original content is available.
- **No Reference (NR):** A single-ended method, applicable if no original content or information is available. For example, blockiness and frame rate measurements are typical NR metrics.

For video quality assessment, there are several QoE metrics that have been proposed in literature. The most widely accepted three models in literature are briefly introduced (Yasakethu, 2010):

- **Peak-Signal-to-Noise-Ratio (PSNR):** PSNR is a full-reference metric derived by calculating the Mean Squared Error (MSE) in relation to the maximum possible value of the luminance in the video signal. Measured quality is represented by a single number in decibels (dB).
- **Structural Similarity Index (SSIM) (Wang, Lu, & Bovik, 2004):** SSIM uses the structural distortion measurement instead of error measurement. It relies on the fact that the HVS focuses on visualising the structural information of the view rather than the pixel errors. SSIM denotes the quality by a value between 0 and 1 such that the higher the value the better the quality.
- **Video Quality Metric (VQM) (Pinson & Wolf, 2004**): VQM provides an objective assessment of perceptual video quality. Its measurement is highly correlated to the subjective quality scores obtained by human perception in subjective testing. VQM could measure the perceptual effects of video impairments such as blurriness, jerkiness, unnatural motion, noise, blockiness, and colour distortion. These aspects are combined in a single measurement that ranges from 0 to 1 such that the lower the value the better the quality. VQM is a well-established video quality metric standardised by the National Telecommunications & Information Administration (NTIA). The NTIA General Model (Pinson & Wolf, 2004) was independently evaluated by the Video Quality Experts Group (VQEG) and also standardized by ANSI and ITU (ANSI Standard, 2003; ITU-T J.144, 2004; ITU-R BT.1683, 2004).

To date, VQM is said to be the best validated and standardised metric with the highest correlation to video quality as perceived by the human vision system (Brunnström et al., 2012). Indeed, in a performance comparison of objective video quality assessment methods (Chikkerur et al., 2011), VQM is found to be the most accurate in the class of natural visual feature based methods.

Future QoE assessment models are expected to measure QoE at a further level than just the perceptual level (Vriendt et al., 2014). This could offer metrics that are even closer to human subjective scores.

QoE-Based Charging

The concept of charging users based on the level of quality they "actually" receive is a relatively new concept in multimedia communications. The breakthroughs made in objective quality assessment, especially the no-reference models, bring the concept of QoE-based charging to practical grounds. The concept of QoS-based differentiated charging has already been presented in literature. Either based on QoS parameters (Wolter & Moorsel, 2001; Dixit et al., 2004; Butyka et al., 2008; Barachi, Glitho, & Dssouli, 2008; Rodríguez, Rosa, & Bressan, 2013) or based on service level agreements (TeleManagement Forum, 2005). Furthermore, pricing schemes that conjoin supplied bitrate with consumed energy in a cost-benefit analysis approach have been discussed in literature as well (Gizelis & Vergados, 2011). However, QoE-based differentiated charging where QoE is directly associated with user billing have not been addressed, to the author's best of knowledge.

QoE-based differentiated charging could be a motivation to both the user and the service provider. It motivates the service provider to offer the best of quality levels, and the user to be compensated on-the-fly for degraded or poor quality.

QUALITY OF BUSINESS (QoB)

Subject to organisational strategy, objectives or mission statement, QoB could be an immeasurable entity such as reputation, social contribution, political or ideological target. In general business terms QoB is "The effectiveness of business processes in meeting the desired outcome of an organization which fit into the QoE framework through its effectiveness component. (Mahadevan, Chaczko, & Braun, 2004)" However, for a profit-driven organisation such as telecommunication service providers, QoB can be a measurable entity which exhibits the extent to which a targeted or expected revenue or profit can be achieved. So, it inherits the triple-Q vision, which focuses on the revenue side of QoB.

QoB is influenced by two main factors: cost and revenue. Considering the provisioning of multimedia services, cost and revenue analysis would attract the issue of service tariff. However, the service tariff is out of the scope of this Chapter. In this Chapter, QoB is represented by a utility function that considers user charging and cost of service.

Through the triple-Q concept, QoB has not received much attention from researchers. Existing charging models lack a direct link between user charging and user experience. Therefore, QoB needs to bridge this missing link, by considering customer's QoE and the associated cost of service provision.

COMMUNICATION SYSTEMS

In its fundamental form, communication is the transmission of information from one point to another through some form of medium. Hence, in a communication system there are three basic elements as depicted in Figure 9: the transmitter, the receiver, and the channel.

In modern digital communication systems, the transmitted signal passes through a series of processes. These processes are carried out by the functional elements in the transmitter and the receiver as illustrated in Figure 10. Typically, the source encoder removes redundant information from the message signal. The channel encoder adds some information to the data stream to increase its robustness

Figure 9. The communication system

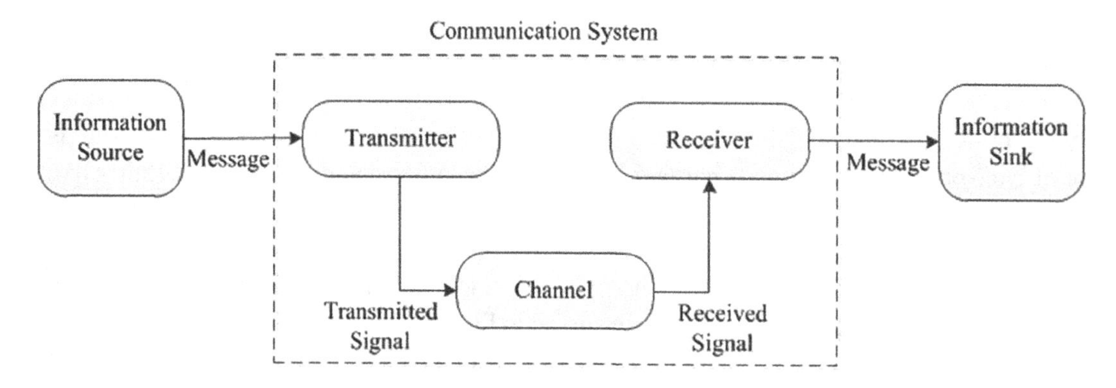

against possible attenuating conditions in the transmission channel. Finally, the modulator represents the data stream by analogue symbols which are suitable for transmission over the channel. The received symbols are then processed in a reverse order by the receiver. Thus, a copy of the original message is reconstructed and delivered.

Figure 10. The digital communication system

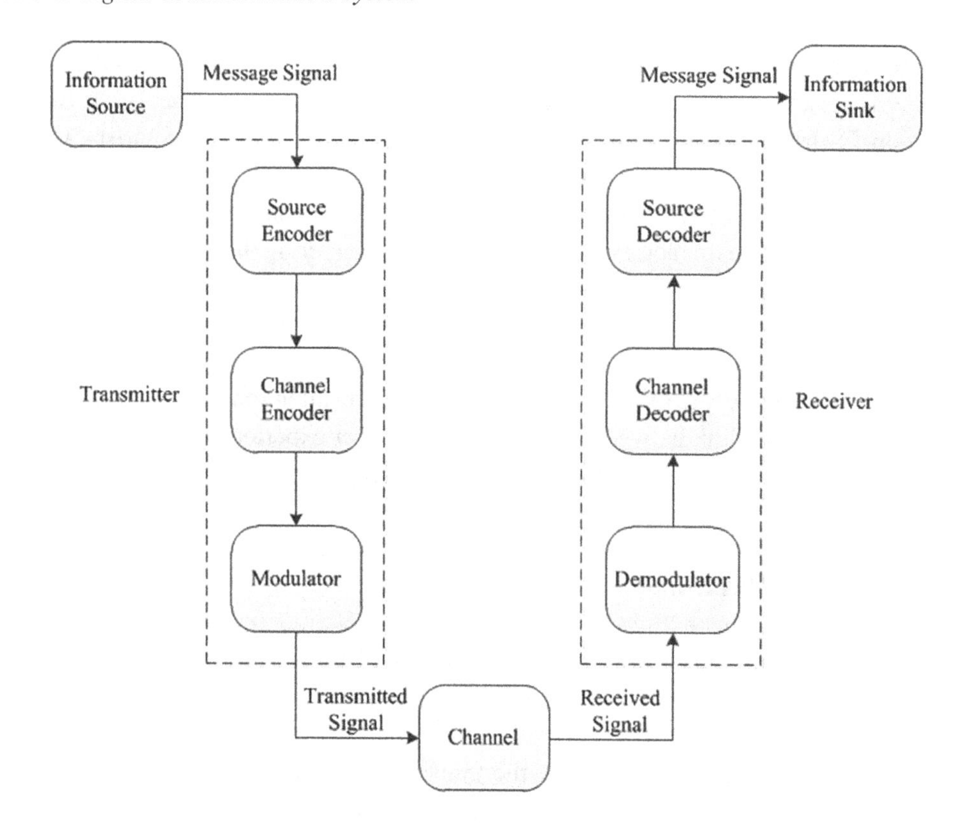

Video Communications System

A typical video communications system takes a design similar to that shown in Figure 10. The elements in Figure 10 are elaborated for video communication systems in the following sections.

Video Sources Coding

Digital video content is normally of large size and telecommunication networks (especially wireless networks) have a limited bandwidth and resources. Therefore, for video to be transmitted over a wireless network (or to be stored), video compression is essentially required to reduce the size of video and hence the data rate to be carried over the network.

Video signal compression is carried out by exploiting the redundant information in the signal in space and time domains. Accordingly, several video compression codecs have been developed and have efficiently reduced the video data rate. One of the well-established and widely used video compression standards is H.264/AVC (ITU-T H.264, 2014). This video coding standard was proposed by ITU Telecommunication Standardisation Sector (ITU-T). It has achieved a considerable enhancement in compression performance over other existing video coding standards. For instance, it could reduce the bitrate of a Standard Definition (SD) video from 3 Mbps to 1.5 Mbps, and a High-Definition (HD) video from 19 Mbps to about 10 Mbps (Abdul-Hameed, 2008). The H. 264/AVC is a block-based video codec, i.e., it identifies and compensates the motion between successive frames then encodes the residual image using a still-image coder. Also, there exists a Scalable Video Coding (SVC) version of H.264, namely, H.264/SVC which exploits the features of SVC video in terms of spatial, temporal, or quality scalability. However, the research conducted in this Chapter has adopted the H.264/AVC standard (ITU-T H.264, 2014).

Generally, the video compression process can be of three types: spatial redundancy reduction, temporal redundancy reduction, and variable length coding. The spatial redundancy reduction reduces the redundancy of pixels within the same video frame. Therefore, if the pixels in the frame are highly similar, the spatial redundancy can be highly reduced. In the temporal redundancy reduction, similarity between a current frame and a reference frame is removed. This relies on the fact that there can be no activity in a scene for some consecutive frames; hence, there is a high similarity between these frames which can be exploited. Temporal redundancy is also performed using-block based motion compensation. This is a technique used in modern video codecs such as H.264. Motion compensation works on blocks of pixels to model the current frame as a translation of some previous frame. Variable length coding contributes a further reduction in bitrate by replacing each data symbol with a variable length unique prefix code word. The length of the code word is identified based on the probability of the data symbol occurrence.

Channel Coding

To protect the transmitted signal from channel-related impairments such as fading and noise, channel coding is used in most communication networks, and especially error-prone wireless channels. Channel coding is also attributed as Forward Error Correction (FEC). Typically, the sender adds redundant bits to the transmitted data; these bits are then extracted by the receiver to correct errors, if any. This saves the sender from the need to retransmit the whole data, and also from waiting for acknowledgement from the receiver through a feedback channel. Therefore, delay-sensitive applications, where retransmissions are not possible, take great advantage of channel coding. However, this protection comes at the expense

Figure 11. Structure of the Code-Word in channel coding

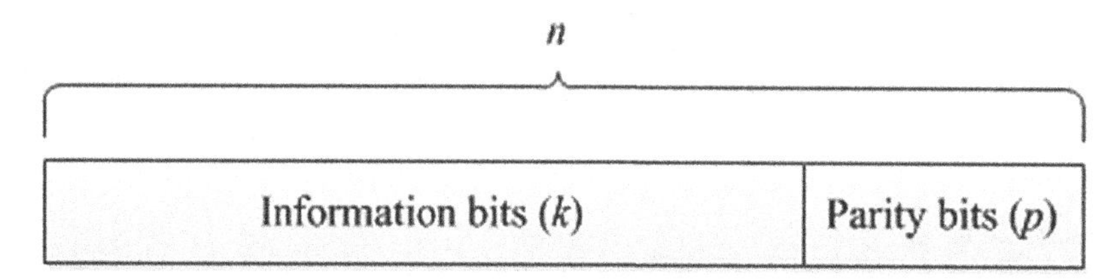

of overhead in terms of the redundant bits added. Typically, the code-word in channel coding consists of information bits and parity bits. This structure is illustrated in Figure 11 where n is the length of the code-word, and the ratio k/n is named the coding rate (Yasakethu, 2010).

Modulation

In order to transmit a bit-stream over an analogue channel, the bit-stream needs to be modulated. In digital communications, modulation involves the formulation of a digital message signal so that it can be physically transmitted inside another signal. Also, modulation enables the transmission of a high frequency signal over a low-frequency channel. That is, it enables high information capacity and high data security. At transmission time, once a carrier is generated at the transmitter, it is modulated with the data to be transmitted. The signal is then demodulated at the receiver. There exist many schemes for digital modulations and they differ in terms of robustness and bandwidth capacity. For example, Amplitude Shift Keying (ASK) modifies the amplitude of the signal; Frequency Shift Keying (FSK) modifies the frequency of the signal; Phase Shift Keying (PSK) modifies the phase of the signal; and Quadrature Amplitude Modulation (QAM) modifies both the phase and the amplitude of the signal (Tektronix, 2009).

Video Transmission Protocols

For multimedia communication applications, the main network protocols are the Internet Protocol (IP), the Transmission Control Protocol (TCP), the User Datagram Protocol (UDP), and the Real-time Transport Protocol (RTP) (Abdul-Hameed, 2008). The IP protocol is the network layer protocol (layer 3) in the OSI model. It facilitates the functions required for transferring blocks of data (IP packets) from a source to a destination over a packet-switched network. IP packets carry the address of the destination system, and hence they are routed between interconnected networks until they reach their designated destination. TCP is the transport layer protocol (layer 4) in the OSI model. It provides a reliable transport service for delay insensitive applications such as email. TCP's reliability is based on robust mechanisms for retransmission and timeout of TCP packets. UDP is also a transport layer (layer 4) protocol in the OSI model. In contrast with TCP, UDP is a connectionless protocol, i.e., it transmits a packet regardless of whether the packet has been received or not. It provides an unreliable but simple transport service without flow control or error recovery functions. Hence, UDP is useful in real-time data transportation such as 2-way conversational applications.

Real-time Transport Protocol (RTP) runs on top of UDP and is typically used by real-time applications such as audio and video over unicast or multicast transmissions. RTP consists of two closely linked parts (Tektronix, 2008):

- **Real Time Protocol:** This part provides time-stamping for timing issues, and sequence-numbering to identify lost or out of order packets, along with other payload-related services. Through these mechanisms, RTP supports end-to-end transport for real-time data transmission.
- **Real Time Control Protocol:** This part is used to for end-to-end monitoring of data delivery.

However, RTP does not offer QoS guarantees or mechanisms to ensure timely delivery for real-time applications. It does not either guarantee the delivery or prevent out-of-order delivery; it only provides the mechanisms for the upper layer to detect missing packets or out-of-order delivery.

Wireless Transmission Modes

In multiuser wireless transmissions, the transmitter can communicate to the receiver in three basic modes: unicast, broadcast, and multicast. As shown in Figure 12, in a unicast transmission each mobile user receives an exclusive service over an exclusively allocated set of subcarriers. Hence, each transmitted service is dedicated to a single receiver solely. In broadcast transmission (Luby, 2012), however, all users receive a single data stream simultaneously over a set of subcarriers allocated to the broadcast group. Thus, a single service is transmitted by the base station and received by all connected users. In the case of multicast, several different services are transmitted by the base station, and each service is received by a multicasting group. Each group consists of subscribers who receive the designated service simultaneously. Radio resources are, therefore, allocated to each multicasting group independently. 3GPP has standardized an architecture of multicasting attributed Multimedia Broadcast Multicast Services (MBMS) (Dixit et al., 2004; Muthusamy, 2011; 3GPP TS, 2014). MBMS introduces point to multipoint communication where data packets are simultaneously transmitted from a single source to multiple destinations. Radio

Figure 12. Subcarrier allocation in Unicast, Broadcast, and Multicast transmissions

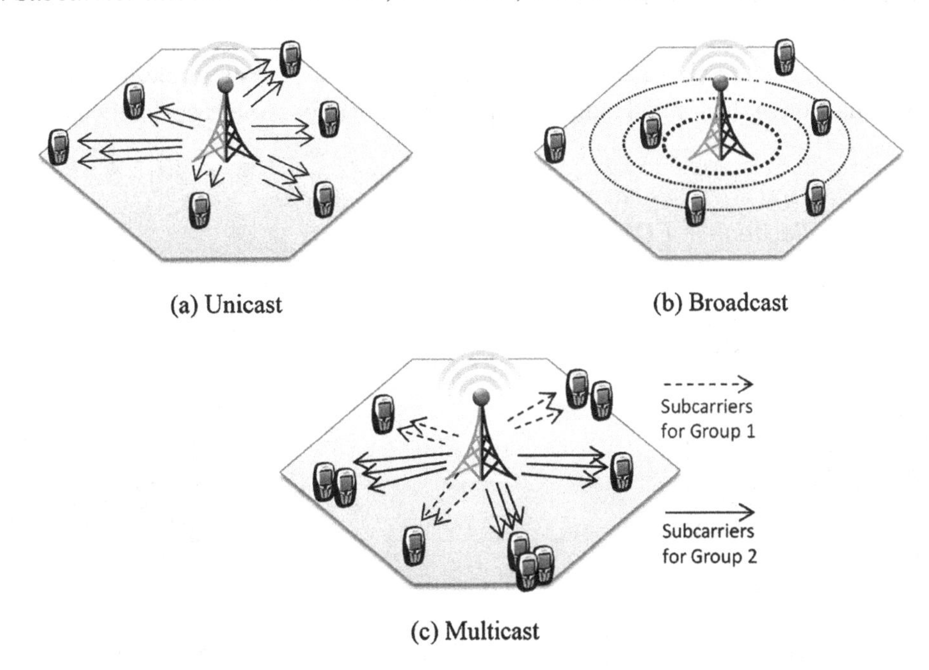

resource efficiency is a prominent aspect of MBMS. For instance, radio transmission cost is independent of the number of subscribers in the cell. Also, MBMS can be set to use only a portion of a cell carrier, leaving the rest for other services such as regular voice and data.

Multiple Access Techniques

Multiple access is the means by which a number of users can share the radio spectrum of a wireless communication channel simultaneously. It is desirable in multiple access systems to perform the sharing of physical resources without significant interference. Therefore, four multiple access techniques can be used to share the wireless spectrum (Rohling, & Gruneid, 1997; Wong et al., 1999; Zander, & Kim, 2001; Wang, Haykin, 2003; Chen, & Chen, 2005; Haykin, & Moher, 2005):

- **Frequency-Division Multiple Access (FDMA):** FDMA assigns each user an exclusive and predetermined band of subcarriers on a continuous-time basis; see Figure 13 (a).
- **Time-Division Multiple Access (TDMA):** TDMA assigns all subcarriers to every user one at a time for a predetermined time slot. That is, the full spectrum is assigned to each user but only for a short duration of time; see Figure 13 (b).
- **Code-Division Multiple Access (CDMA):** CDMA also allows the users to use the available spectrum but their individual signals are coded (encrypted) such that they can be distinguished from each other.
- **Space-Division Multiple Access (SDMA):** In SDMA, the radio spectrum is shared between users by exploiting the spatial distribution of users. That is performed using multi-beam directional antennas, so users can access the channel on the same frequency or in the same time slot simultaneously.

Radio Resource Management

Radio Resource Management (RRM) is a significantly important aspect of wireless communications to allow efficient utilisation of the limited radio resources as much as possible, especially in multi-user and

Figure 13. Spectrum allocation in FDMA and TDMA for four users

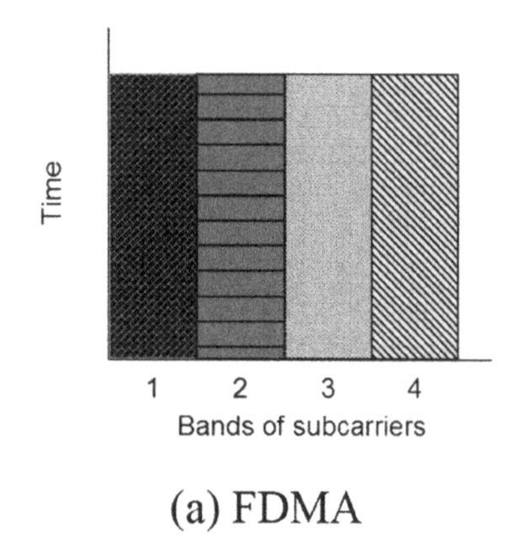

(a) FDMA

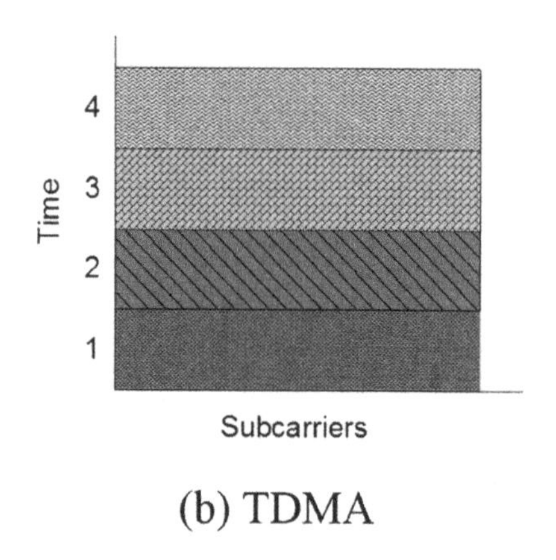

(b) TDMA

multi-cell environments. RRM is also meant to satisfy different QoS requirements for users of different applications. In wireless environments, the channel state changes over time due to a number of phenomena such as channel fading and interference. Hence, RRM's role is to manage and adjust the wireless resources dynamically to meet the desired QoS requirements and improve the spectral efficiency. To achieve this efficiently the development of intelligent Resource Allocation Algorithms (RAA) is required.

Basically, there are two types of RAAs (Xiao, 2010):

- **Static Resource Allocation:** Allocation is based on statistical information; hence it is mainly designed during the planning stage of the wireless system.
- **Dynamic Resource Allocation:** Allocation is based on the measurement of parameters as changes happen in propagation conditions and user traffic requirements. Usually, reassignments of resources happens in a per millisecond intervals.

Radio resources in wireless networks include: transmission power, time-domain resources, frequency-domain resources and spatial-domain resources (Xiao, 2010). RAAs use traffic information and channel information to make a decision about how to efficiently allocate those resources. Finding the optimum resource allocation is believed to be a formidable NP-complete problem (Xiao, 2010). In addition, no general algorithm that is capable of solving such an optimality problem is known yet. However, a number of RAAs have been proposed (some of which are already used in wireless systems), which target specific objectives under certain conditions.

Generally, RRM algorithms can be divided into five areas: power control, handover, admission control, packet scheduling, and load control (Abdul-Hameed, 2008):

- **Power Control**: Algorithms control the level of the transmitted power to keep the interference levels at minimum while the required QoS is accommodated.
- **Handover**: Algorithms handle a mobile terminal's transition from the coverage area of one cell to another.
- **Admission Control**: Algorithms decide whether a user's request to connect can be accepted or not. For instance, this can be based on efficient allocation of the limited resources to maintain users' QoS requirements.
- **Packet Scheduling**: Algorithms determine the bitrates and the sub-channels to allocate to users in addition to the allocation duration. That is, it determines how available bandwidth is shared among users.
- **Load Control**: Algorithms are needed when overload state is encountered. The role of these algorithms is to monitor, detect, and handle congestion.

The RAA contributed in this Chapter exploits both the power control and the packet scheduling algorithms in one joint optimization scheme. The RAAs proposed in the literature usually fall into one of two classes of adaptive bit-loading depending on the approach taken to solve the optimization problem. The two classes are Margin Adaptive (MA) algorithms and Rate Adaptive (RA) algorithms (Sadr, Anpalagan, & Raahemifar, 2009; Tang, 2013):

- **Margin Adaptive (MA)**: Algorithms aim at minimising the transmitted power subject to the constraints of data rate and Bit-Error-Rate (BER). Hence, the required QoS by each user is considered while the optimization problem is formulated.
- **Rate Adaptive (RA)**: Algorithms aim at maximising the data rate (throughput) subject to the constraints of transmit power and BER.

The optimisation problem presented in this Chapter is based on the Margin Adaptive (MA) class; hence power minimisation is targeted while the user assigned bitrate is adjusted.

Orthogonal Frequency-Division Multiplexing (OFDM)

A part and parcel of modern wireless communication systems is Orthogonal Frequency-Division Multiplexing (OFDM) technique. OFDM is a well-established technique in wireless networks due to its support for adaptive multi-user transmission at a high data rate. Therefore, OFDM has been adopted in several digital transmission systems such as IEEE 802.11a/g Wireless Local Area Network (WLAN), IEEE 802.16 Worldwide Interoperability for Microwave Access (WiMAX), Long Term Evolution (LTE), and many others. In OFDM, a high-rate data stream is split into a number of lower rate streams transmitted simultaneously over a number of subcarriers. So, an individual data element normally occupies only a small part of the available bandwidth since the coherent channel bandwidth is divided into many narrow sub-bands. The sub-streams are modulated into symbols and transmitted on the designated orthogonal subcarriers. In particular, a higher order modulation is used to carry more bits/symbol on subcarriers with large gains, whereas subcarriers in deep fade carry one bit/symbol, or are not even used. Hence, OFDM represents a significantly important technology for wireless communications.

Orthogonality is a property that allows a robust transmission of multiple data signals over a common channel, and allows their detection without interference (Wu, 2010). Signals are orthogonal if they are mutually independent of each other. Hence, the principle of an OFDM system is to use narrow, mutually orthogonal subcarriers to carry data (Xiao, 2010). This feature of OFDM is explained by the diagrams in Figure 14. The OFDM technique offers a great deal of spectral efficiency (bandwidth saving) by exploiting subcarrier orthogonality in comparison with the conventional multicarrier technique (Nee, and Prasad, 2000). The subcarriers are overlapping but orthogonal to each other, so that at the sampling instant of a given subcarrier, all other subcarriers are zero-valued, as shown in Figure 14.

OFDM System Model

A typical OFDM system model is illustrated in Figure 15. In a typical OFDM system, after the data bits are modulated, the serial data stream is converted to parallel data sub-streams. Inverse Fast Fourier Transform (IFFT) algorithm is then performed on each sub-stream to convert the modulated symbols (represented by real and imaginary parts) from frequency domain into waveform representations in time domain. Later, the cyclic prefix is added to help avoid inter-symbol interference. Once the signal is received, the cyclic prefix is removed and the signal is recovered using the Fast Fourier Transform (FFT) algorithm. A parallel-to-serial conversion is performed then the data stream is demodulated to obtain an estimated copy of the original information bits.

Figure 14. The advantage of Orthogonal subcarriers in OFDM

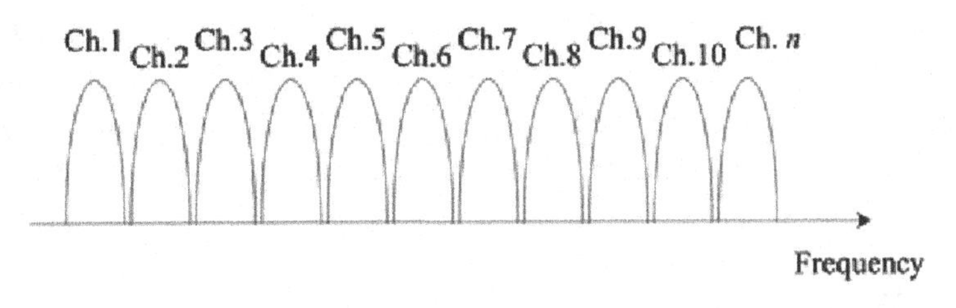

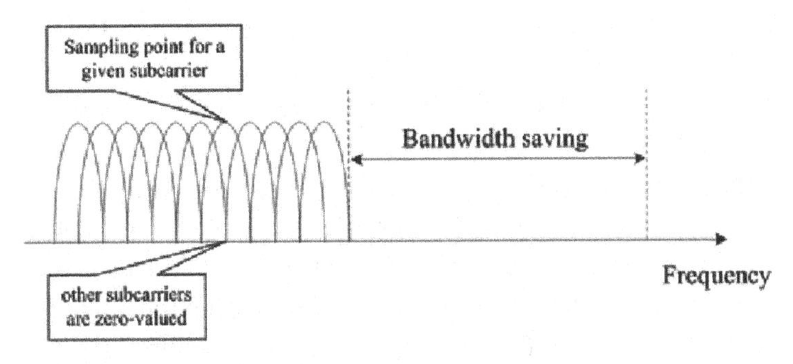

Figure 15. A typical OFDM system model

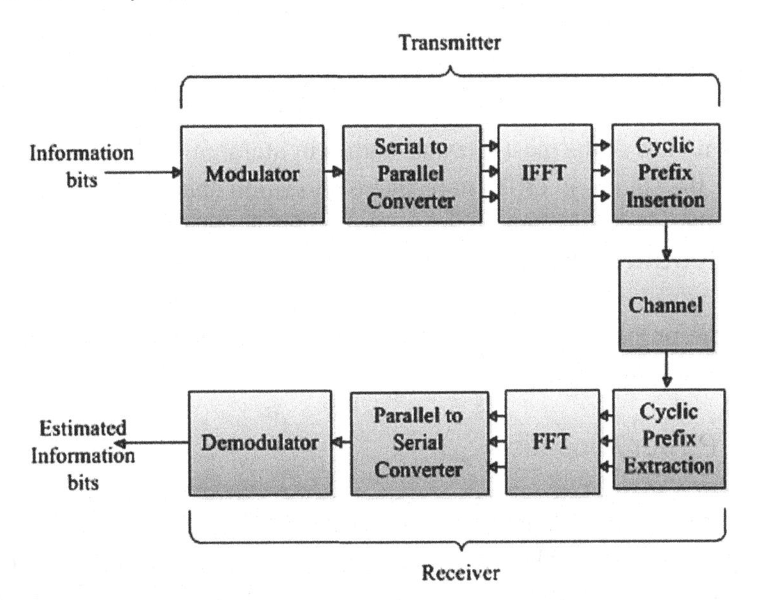

In the early stages of OFDM development, all the subcarriers of OFDM were assigned to a single user at any particular point of time. Later, OFDM has been conjoined with TDMA and FDMA to optimize the sharing of resources between multiple users (Tang, 2013). The OFDM system model studied in this Chapter is a multiuser-enabled OFDM. The multi-user capability is also supported natively by the Orthogonal Frequency-Division Multiple Access (OFDMA) technique, which is an extended version of OFDM.

Advantages of OFDM

The OFDM transmission model offers the following key advantages (Nee & Prasad, 2000; Xiao, 2010):

- **High Spectral Efficiency:** Since subcarriers can partially overlap, as shown in *Figure 14*, the system throughput can be substantially increased.
- **Robustness and Resiliency to Fading and Interference**: Because of the division of a wideband spectrum into orthogonal narrowband sub-channels, the radio frequency interference between users is reduced considerably, and the frequency-selective multipath fading is alleviated. This increases the reliability in complex indoor environments and mobile environments.
- **Flexible Resource Allocation:** With dynamic frequency allocation, OFDM can assign certain subcarriers to users according to users' channel conditions. This allows full use of frequency diversity and multi-user diversity to obtain near optimal system performance.
- **Simple System Implementation:** Since OFDM uses IFFT/FFT instead of Discrete Fourier Transform (IDFT/DFT), the computational complexity in OFDM is remarkably reduced.

LITERATURE REVIEW

This section presents a review of the most relevant studies in literature, to the contributions presented in this Chapter. These are the studies in efficient resource allocation and in QoE prediction.

Related Work in Resource Allocation

The works related to the proposed recourse allocation scheme are reviewed in this section. They are divided based on the main targeted objectives.

Targeting Efficient Power Allocation

Past research investigated several challenges confronting the wireless system, such as dynamic subcarrier allocation, adaptive power efficient allocation, admission control, and capacity planning (Katoozian, Navaie, & Yanikomeroglu, 2009). Power efficient adaptive allocation, however, has inspired considerable research efforts in multicarrier communication systems (Wong et al., 1999; Krongold, Ramchandran, & Jones, 2000; ; Zhang et al., 2002; Kivanc, Li, & Liu, 2003; Shen, Andrews, & Evans, 2003; Song & Li, 2003; Hu et al., 2004; Zhang & Letaief, 2004; Wang, Chen, & Chen, 2005; Reddy, 2007; Kim, Kwon, & Cho, 2007; Shen et al., 2009; Lande et al,, 2011; Papoutsis, & Kotsopoulos, 2011; Haider et al., 2012; Al-Kanj, & Dawy, 2012; Wang et al., 2012; Zhao, Gan, & Zhu, 2012; Ng, Lo, & Schober, 2013; Zhao et al., 2014; Chen et al., 2014; Chen et al., 2015). In OFDM systems, studies have proposed power optimisation algorithms based on adaptive subcarrier and bit allocation (Wong et al., 1999), proportional rate adaptive allocation (Shen, Andrews, & Evans, 2003), Lagrange-multiplier bisection search (Krongold, Ramchandran, & Jones, 2000), or evolutionary genetic algorithms with water-filling technique (Wang, Chen, & Chen, 2005; Reddy, 2007; Lande et al,, 2011) for multiuser OFDM. In another work, subcarrier and power allocation was optimised based on a utility function that quantifies the radio resources a user occupies (Song, & Li, 2003). Similar works in OFDMA systems have proposed fractional programming (Haider et al., 2012; Ng, Lo, & Schober, 2013) and a Lagrangian relaxation-based algorithm (Kivanc, Li, & Liu, 2003) for optimization. Also, minimising receiver energy has been achieved by minimising the time required to receive bits (Al-Kanj, & Dawy, 2012), and by minimising the number of received symbols (Kim, Kwon, & Cho, 2007), in multicasting. In (Wang et al., 2012) only water-filling has been used for power, whereas the objective has been throughput maximisation for unicast transmissions. Therein, for mixed unicast and multicast, throughput has not been maximised for the sake of unicast which is given priority. Also, since content quality is not accounted for, higher bitrate may be offered unnecessarily. In (Shen et al., 2009), the greedy algorithm has been used for power allocation and tested in unicast and multicast. (Papoutsis, and Kotsopoulos, 2011) have proposed resource allocation based on allocating chunks of subcarriers to maximize throughput under limited power constraint. Yet minimal power and content quality have not been addressed. An adaptive subcarrier allocation algorithm based on perfect and partial channel information have been proposed in (Hu et al., 2004) for a Multi-Input-Multi-Output (MIMO-OFDM) system. A power-minimized bit-allocation scheme has been proposed in (Zhang et al., 2002) which considers the processing power as well as the transmission power jointly. This work has not considered subcarrier allocation for multiuser case. Whereas the work in (Zhang & Letaief, 2004) has considered a multiuser environment in a resource allocation problem aiming at minimizing total transmit power and increasing spectral efficiency. However, no bit-loading strategy has been presented (such as greedy algorithms), and content quality has not been considered as well. A basic subcarrier allocation scheme that considered SVC has been presented in (Zhao, Gan, & Zhu, 2012), but without reflection on content quality. In (Zhao et al., 2014), a QoS-Aware resource allocation model has been proposed for multi-view video in multicast networks. The model targets energy efficiency and helps users achieve different data rates according to their QoS requirements. In a related study on resource allocation schemes, typical OFDM subcarrier allocation schemes have been shown to perform better (in terms of SNR) when error control coding techniques are engaged (Lee & Wen, J. 2007). A QoE-driven power allocation scheme has been presented in (Chen et al., 2014; Chen et al., 2015). for SVC video over MIMO systems. QoE has been measured with PSNR, and power is referred to by a utility cost of power. This study has targeted power allocation without referring to resource allocation.

The subcarrier and bit allocation for power efficiency has been also the subject of several research theses (Xiao, 2010; Wu, 2010; Chen, 2010; Tang, 2013; Hu, 2013). The work in (Moorsel, 2001) has presented a resource allocation algorithm for downlink multiuser MIMO-OFDM targeting minimal transmit power. Similarly, energy-efficient scheduling algorithms have been presented in (Chen, 2010) for delay constrained communications for single user and multiuser SISO and MIMO transmission systems. Therein, the total transmission power is minimised while individual user's QoS constraints are satisfied. In another RRM effort, a radio resource allocation scheme has been proposed for relay based cellular networks (Xiao, 2010). This scheme has targeted throughput, power, and user fairness (of allocated resources), but not QoE. Within the context of cognitive radio (spectrum sharing networks), an optimization technique for subcarrier and power allocation has been proposed for the secondary wireless network in the presence of multiple primary users in OFDMA (Tang, 2013). Similarly, power allocation in cognitive radio OFDM has been studied and a sum rate maximisation problem has been presented in (Hu, 2013) to achieve power minimisation. In a quality-driven study, the information about video traffic has been the basis for a scheduling strategy proposing efficient resource allocation in multiuser downlink LTE systems (Chen, 2010). The study has applied the game theory and fuzzy logic concepts; however, it has not considered energy efficiency. In another LTE focused research, an energy efficient resource allocation method has been developed from the view point of interference mitigation rather than power minimization (Ramirez, 2014). Instead of proposing a power allocation scheme, the approach relies on efficient use of interference management techniques and system spare resources such as the bandwidth in order to reduce the energy expenditure in LTE networks.

However, the aforementioned studies have focused on power minimisation only as the main objective in their resource allocation algorithms, regardless of the type or quality of content used.

Targeting Bitrate-Driven Quality and Rate Control

While several research activities have targeted the power allocation problem for efficient resource allocation, a few others have put efforts in considering the content or the quality aspect along with power (Kim et al., 2006; Lee & Wen, 2007; Ngo, Tellambura, & Nguyen, 2009; Chen, 2010; Bakanoğlu et al., 2010; Lee, Koo, & Chung, 2010; Li, Liu, & He, 2010; Li et al., 2011; Huang, & Leung, 2012; Goudarzi, 2013; Hu, 2013; Ramirez, 2014; Gao et al., 2015; Chen et al., 2015;). Based on content quality, rate control has been designed to reduce the bit-stream which would result in reducing energy (Lee, Koo, & Chung, 2010). The quality of the reconstructed received video has been enhanced using packet scheduling algorithms in cross-layer optimization (Li et al., 2011; Gao et al., 2015). In (Ngo, Tellambura, & Nguyen, 2009; Lei et al., 2013), a maximum throughput has been distributed fairly among multicasting groups, with power as a constraint but not guaranteed to a minimum. The system sum rate is maximized to improve quality, but no measures are taken to assess the received quality. User application bits have been used as the basis for a subcarrier and bit allocation algorithm in (Huang & Leung, 2012). Therein, a framework called bitQoS-aware resource allocation, which relies on the QoS requirements of the application at the bit level, has been proposed. In (Kim et al., 2006), the encoding bitrate has been adjusted to the receiving power level in a rate-control method for CDMA. Hence, power minimisation has not been addressed. To fairly maximise the bitrate for multicast groups, the work in (Bakanoğlu et al., 2010) has proposed a mathematical algorithm in two steps. Subcarrier allocation is first done assuming equal powers, then greedy bit-loading is used, but power minimisation has not been presented. Hence, this approach is less efficient compared to the algorithms that provide a global optimisation of objectives. Based on differentiated QoE, the work in (Goudarzi, 2013) has proposed a framework for optimal bitrate assignment in competing SVC video flows. Therein, the rate allocation algorithm targets QoE maximisation, but energy has been out of the scope. In (Gao et al., 2015), an algorithm aiming at fair and balanced QoE distribution among two types of users has been presented. The algorithm is based on equivalent rate scaling to deliver balanced QoE among guaranteed and best effort users, but QoE maximisation has not been targeted. A general framework has been presented in (Katsaggelos et al., 2005) which emphasised on the importance of considering receiver's energy and bitrate-driven quality. However, no algorithms have been proposed to back the presented framework.

Although these studies have relatively considered content quality in one way, in the other way there has been no literal focus on QoE maximisation and power minimisation jointly, let alone QoB. Even in the cases where QoE has been considered, the metrics used were not state-of-the-art QoE metrics.

Recent and Targeting Energy and/or QoE

Since energy-quality optimisation has attracted research efforts recently, a number of studies have considered a joint optimisation of energy and QoE (Chuah, Chen, & Tan, 2010; Sacchi, Granelli, & Schlegel, 2011; Chuah, Chen, & Tan, 2012; Lei et al., 2013; Bisio et al., 2014; Li et al., 2015; Liu et al., 2015; Fei, Xing, & Li, 2015;). For Relay-based OFDMA networks, the energy-quality trade-off has been targeted by an optimisation algorithm (Lei et al., 2013). The focus of this study was on SVC coded video. The work in (Sacchi, Granelli, & Schlegel, 2011) has been based on the game theory by assigning subcarriers randomly and then managing a competition between user pairs to distribute resources among users. In comparison of this method with the GA-based method (proposed in this Chapter), the GA-based method is much more efficient in that it comprehensively brings all inputs together and makes the decision of allocation. Also, the GA-based exploits more solutions than the game theory could exploit in each iteration. For SVC coded video in multiuser MIMO-OFDM, a QoE-aware resource allocation algorithm has been proposed (Li et al., 2015). Therein, quality modelling has been designed in a similar way to that in this Chapter. The algorithm in (Li et al., 2015) has optimised a set of network parameters including power allocation, however, it has not considered power minimisation. In the context of cloud computing, but not in the context of resource allocation, an optimisation algorithm has been introduced in (Liu et al., 2015) for the trade-off between energy and QoE. Another study based on the energy-QoE trade-off (Bisio et al., 2014) has proposed a rate allocation algorithm for web navigation over satellite networks. In (Fei, Xing, & Li, 2015), subcarrier allocation and power allocation have been addressed separately, not as one problem in one algorithm. Nash bargaining method has been used for the subcarrier allocation algorithm, whereas the power allocation algorithm has been based on a fuzzy decision model. The objective has been to maximise QoE through maximising the bitrate for three services: file download, IPTV, and VoIP. Chuah *et al.* (Chuah, Chen, & Tan, 2012) have proposed a resource allocation algorithm for multicast SVC video based on dynamic programming. The algorithm targets QoE (using PSNR) maximisation under power and access constraints. The resources presented have been power, multicast access, and SVC frame-rate, spatial resolution, and quality. The authors have extended their work in (He & Liu, 2014) where they have added packet loss rate with PSNR to estimate video quality. However, this QoE metric lacks the perceptual aspect offered by the Video Quality Model (VQM). Also, the optimisation algorithm has been extended to join MCS and transmission power allocation.

In (He & Liu, 2014), Lagrange dual decomposition has been used to maximise QoE by addressing three problems separately: packet scheduling, subcarrier allocation, and power allocation. Later, the author has joined scheduling and subcarrier allocation in a sub-optimal solution which excluded power. Hence, resource allocation has been presented but not in a joint optimisation algorithm. The work presented in (Khan et al., 2014) has advanced an energy efficient approach for HTTP streaming over LTE. The approach targets mobile device energy consumption (not transmission power) by making use of inactive states during HTTP streaming. No QoE-based resource allocation has been presented, but rather an extension to an existing algorithm such that device energy is considered. In (Liu, Ho, & Chou, 2014), the authors have proposed a spectrum (subcarrier) and power allocation algorithm considering the video content for multicast over femtocells. The optimisation algorithm is content-aware but does not enhance quality which is measured with PSNR. Also for video over LTE, (Perera et al., 2014) have presented a QoE-aware resource allocation approach to improve QoE. Therein, important (critical) video frames are selected and allocated to the most robust OFDM Resource Blocks (RB). PSNR is then used to measure QoE. The study has not considered energy. In the context of cognitive radio, a subcarrier and power allocation scheme has been proposed for video over OFDMA (Saki & Shikh-Bahaei, 2015). The authors have developed an algorithm to improve QoE (measured with PSNR) for the secondary receiver. (Wang et al., 2014) have presented a green allocation scheduler solution that adopts a *free-lunch* approach in multicast multimedia delivery. Targeting fairness, the allocation scheduler shifts the interference power from capacity-saturated users to unsaturated users. Hence, throughput is improved for unsaturated users, which leads to an overall more efficient power and enhanced QoE. In (Xu et al., 2014), QoE (measured in PSNR) has been employed as the basis to optimise two problems separately: energy efficiency and bandwidth efficiency. The proposed resource allocation algorithm has been developed using a nonlinear fractional programming approach and dual decomposition to address the formulated mixed-integer nonlinear optimisation problem. For multi-radio access in heterogeneous wireless networks, a QoE-based resource allocation scheme has been presented (Yang et al., 2014). Therein, a multimedia QoE metric has been developed using the so called multiplicative exponent weighting method. The algorithm, which has been developed using Lagrangian analysis, targets QoE maximisation by considering a rate factor and an energy factor subject to the constraints of transmit bandwidth, rate, and power.

The foregoing studies in this section have more or less targeted two objectives: QoE and energy in varying contexts. However, none of them have considered a joint optimisation of QoE, power, QoB, CSI, user fairness, and rate control for unicast, broadcast, and multicast. Also, the QoE metrics used were not state-of-the-art perceptually motivated QoE metrics (Yang et al., 2014).

Recent and Most Relevant Works

In this section, the recent studies that are believed to be closely relevant to the main contribution in this Chapter are presented and discussed. Li *et al.* (Li, Chen, & Tan, 2012) have proposed a QoE-aware resource allocation algorithm for SVC video over multiuser MIMO-OFDM. A QoE-rate relation has been developed to model video quality in a similar manner to that in this Chapter; however, it was not clear which QoE metric has been used. With the objective to maximise QoE, a simple bit-loading algorithm has been proposed. The algorithm increments the loaded bits according to channel CSIs. Consequently, total transmit power is incremented, modulation is selected, and QoE is incremented. This process continues until a total power constraint is reached with the maximum limit for modulation bits. The authors consider this level of power is the optimum for high QoE. This algorithm exposes several drawbacks. With the power constraint as a stopping criterion, high QoE may be achieved unnecessarily at the expense of high power. That is, users may be assigned a high bitrate they do not require. Therefore, the algorithm does not guarantee a fair and balanced distribution of quality and power. Also, it does not show any rate control mechanism, nor does it consider service provider's QoB.

QoE-based resource allocation has been presented also for multiuser-multiservice OFDM transmission targeting energy efficiency and quality guarantee (Li et al., 2012). Therein, three multimedia services have been considered: IPTV, file download, and VoIP. The quality metric used for video (IPTV) has been PSNR (mapped to MOS) which lacks the perceptual aspect discussed in the problem statement of this Chapter. Lagrange-based optimisation has been presented for the resource allocation algorithm. The algorithm follows a greedy-like technique by assigning the subcarrier with best CSI to each user, then gradually increasing the number of subcarriers allocated to each user, until a minimum QoE threshold is reached. Hence, the algorithm seeks to minimise power by utilising as much subcarriers as possible. The main drawback in this work is that it guarantees a minimum QoE but does not maximise QoE. By relying on a fixed bitrate approach, QoE is fixed but not enhanced. Since fixed bitrate has been assumed for each user, fairness and rate control has not been considered. Also, the proposed algorithm has been compared only with the exhaustive search algorithm. No other comparable algorithms are shown, and service provider's QoB is not considered.

For femtocell networks, a QoE-oriented resource allocation has been presented in (Li et al., 2014). This study has considered three multimedia services: audio stream, data stream, and video stream. To maximise QoE and minimise power, the algorithm operates on two stages iteratively. First, subject to the power and RB constraints, either power or Resource Blocks (RB) are allocated to users until QoE reaches 4.5 on the MOS scale. Second, power is decreased, and QoE is consequently decreased, until a *cost-effective threshold* defined by the algorithm is observed. This method does not guarantee user fairness and limits QoE at a MOS value below 4.5. Also, QoE has been measured with SSIM and then mapped to MOS. Furthermore, QoB has not been in the scope.

The work in (Ma et al., 2012) has proposed an energy efficient power allocation algorithm based on constrained Particle Swarm Optimisation (PSO). This study does not improve QoE but rather use it in the power allocation algorithm with the objective to increase power efficiency. Hence, only power is optimised, but the balance between quality and power is not considered. Power and quality have been unified in one objective function, but the authors have not normalised them on one scale for this function. Also, the PSO-based algorithm allocates subcarriers before power is allocated. This is less efficient than genetic algorithms where subcarrier and power allocation is addressed in one shot. QoE is represented by PSNR mapped to MOS, and QoB has not been considered as well.

The majority of the reviewed algorithms seek an optimum solution in the quality-energy trade-off using different approaches. The solution they achieve may not be an optimal one because their algorithms lack a decision maker. Ideally, it is the role of the decision maker to identify an optimum solution in multi-objective optimisations. Also, none of these works have been "perceptually motivated" in the selection of the QoE metric. Moreover, differentiated charging by incorporating service provider's QoB has not been in the focus of these studies.

Related Work in QoE Prediction

Estimation of the perceptual quality of multimedia content in mobile environments is a significant issue for communication systems (Reiter, 2009). Hence, it has been an area of significant interest to researchers in video quality. In this section, varying research efforts in this field are summarised. The summaries are divided based on the QoS parameters employed in the proposed systems.

QoS Parameters from the Application Layer

A number of studies have considered QoS parameters from the application layer only, such as video codec and bitrate (Eden, 2007; Lee, Jung, & Sim, 2010; Joskowicz, & Ardao, 2011; Lin et al., 2012; Malekmohamadi et al., 2014). No-Reference (NR) algorithms have been used to estimate PSNR (Eden, 2007), or to measure quality using video quality assessment in the compressed domain (C-VQA) (Lin et al., 2012). The Reduced-Reference (RR) and Full-Reference (FR) methods proposed, on the one hand they have been based on parametric non-machine learning algorithms using a standard video quality metric (VQM) (Joskowicz, & Ardao, 2011), or a derived VQM (Lee, Jung, & Sim, 2010). But on the other hand, they have been based on machine learning in the 3D video domain (Malekmohamadi et al., 2014).

QoS Parameters from the Physical Layers

Another group of research studies have focused on QoS parameters solely from the physical layer, such as packet loss and delay (Suzuki, Kutsuna, & Tasaka, 2008; Argyropoulos et al., 2011; Mushtaq, Augustin, & Mellouk, 2012; Pokhrel et al., 2013; Alreshoodi, & Woods, 2013). Either machine learning has been used to assess the QoS/QoE correlation (Mushtaq, Augustin, & Mellouk, 2012), or a fuzzy expert system has been used for QoE estimation (Pokhrel et al., 2013; Alreshoodi, & Woods, 2013). The majority have employed the Mean Opinion Score (MOS) as a quality measure. Also, multiple regression analysis has been the method used by Suzuki *et al.* (Suzuki, Kutsuna, & Tasaka, 2008) to estimate QoE from MAC–level QoS over an IEEE 802.11e wireless LAN. In a different approach Argyropoulos *et al.* (Argyropoulos et al., 2011) have proposed a bit-stream type of model which uses the visibility of packet losses to predict objective quality scores.

QoS Parameters from the Application and the Physical Layers

For an end-to-end prediction of video quality, hybrid models have come to light and have consolidated both application layer and physical layer parameters (Bellini et al., 1996; Kelly, Maulloo, & Tan, 1998; Hamam et al., 2008; Khan, Sun, and Ifeachor, 2010; Hewage, & Martini, 2011; Joskowicz, & Ardao, 2011; Brunnström et al., 2012; Khan, Sun, & Ifeachor, 2012; Aguiar et al., 2012; Alreshoodi, & Woods, 2013; Joskowicz, Sotelo, & Ardao, 2013; Aguiar et al., 2014; Joskowicz, & Sotelo, 2014; Nightingale et al., 2014; Paudel et al., 2014; Ghalut, & Larijani, 2014; Tychogiorgos, & Leung, 2014; Khan et al., 2015;). Khan *et al.* (Khan, Sun, and Ifeachor, 2010; Khan et al., 2015) have proposed two learning models to estimate video quality in PSNR normalized to MOS. The proposed models have been Adaptive Neural Fuzzy Inference System (ANFIS) (Khan, Sun, & Ifeachor, 2010; Khan et al., 2015) and non-linear regression analysis (Khan, Sun, & Ifeachor, 2010; Khan et al., 2015). They have also studied the effect of individual QoS parameters on end-to-end video quality for H.264 encoded videos. The authors have validated both models with subjective testing in (Khan, Sun, & Ifeachor, 2010; Khan, Sun, & Ifeachor, 2010). However, in all studies they have not considered spatial video resolution as a QoS factor.

Fuzzy Inference Systems (FIS) have been used in a similar study (Alreshoodi & Woods, 2013) that has also considered one spatial resolution (QCIF). This study has concluded that the FIS-based method outperformed the non-linear regression-based method (Khan, Sun, & Ifeachor, 2010) with regard to prediction accuracy. FIS has also been used to model a taxonomy of a QoE evaluation system for haptic virtual reality applications (Hamam et al., 2008). Therein, both QoS and QoE parameters have been used as input parameters to estimate the overall QoE. Fuzzy logic control has also been used in an application to an H.261 encoder to maximize QoE of video (Bellini et al., 1996).

Random Neural Networks (RNN) have been adopted by several research efforts in quality prediction. A real-time estimator has been proposed by Aguiar *et al.* utilizing RNN (Aguiar et al., 2012), and Multiple Artificial Neural Network (MANN) (Aguiar et al., 2014) as a prediction engine for abstract Wireless Mesh Networks (WMN). However, few parameters from the application and the physical layers have been investigated in both studies. Also, Paudel *et al.* (Paudel et al., 2014) have employed RNN in a quality estimation engine in their study of Media Access Control (MAC) parameters' effect on QoE. A non-intrusive RNN-based method has also been proposed by (Ghalut & Larijani, 2014) for video quality prediction over LTE networks. This work has addressed parameters from both the application and physical layers, and relied on PSNR as an objective QoE metric. Further content-focused studies have presented a RR metric for 3D video (Hewage & Martini, 2011) based on PSNR, or just an investigation of QoS impact on QoE for videos encoded with High Efficiency Video Coding (HEVC) (Nightingale et al., 2014). Joskowicz *et al.* have extended their parametric model (Joskowicz & Ardao, 2011) to accommodate random packet loss (Joskowicz & Sotelo, 2013), and later, have presented an application of their model (Joskowicz & Sotelo, 2014) on broadcast digital television. A simple lightweight non-linear model has also been proposed by Brunnström *et al.* (Brunnström et al., 2012) for no-reference video prediction.

From this literature review, it can be observed that learning-based techniques, in the Artificial Intelligence (AI) domain, have been the prime focus for developing objective QoE prediction models. It is also noted that existing proposals of video quality prediction tend to consider either the encoder's compression artefacts, or network impairments, or the features of video content, but rarely all three.

SUMMARY

This chapter introduced the area of research in this Chapter, and reviewed the state-of the-art related works in this area. First, the general framework of the contributions in the Chapter was presented, which is the triple-Q model consisting of QoS, QoE, and QoB. Later, the communication system was introduced with focus on video communication systems in the wireless environment. In this system, the stages where QoS and QoE are affected were explained in detail. These stages are video coding, channel coding and modulation, and video transmission protocols. The wireless transmission modes studied in this Chapter: unicast, broadcast, and multicast, were also described. This was followed by an introduction to the multiple access technologies in wireless systems; TDMA, FDMA, CDMA, and SDMA. Since the main focus of this Chapter is the allocation of wireless resources, RRM was presented and the RRM areas studied in the Chapter were highlighted. Finally, OFDM was explained for it constitutes the core technology upon which the RRA in this Chapter is based.

The latter part of the chapter reviewed the related literature in two areas. First, the related work in radio resource allocation. This was categorised based on the elements targeted by the algorithms. These elements were power, bitrate, quality, and QoE. Second, the related work in QoE prediction. Therein, three types of algorithms were presented based on the selection of QoS parameters. These were: physical layer parameters, application layer parameters, and joint physical and application layers parameters.

REFERENCES

3GPP TS. (2014). *Multimedia Broadcast/Multicast Service (MBMS); Architecture and functional description. 3GPP TS 23.246 V12.1.0.* Author.

Abdul-Hameed, O. (2008). *Quality of Service for Multimedia Applications over Wireless Networks* (Doctoral Thesis). University of Surrey, Guildford, UK.

Aguiar, E., Riker, A., Mu, M., Zeadally, S., Cerqueira, E., & Abelem, A. (2012). Real-time QoE prediction for multimedia applications in wireless mesh networks. *Proc.IEEE Consumer Communications and Networking Conference*, 592–596.

Aguiar, E., Riker, A., Cerqueira, E., Abelém, A., Mu, M., Braun, T., & Zeadally, S. et al. (2014). A real-time video quality estimator for emerging wireless multimedia systems. *Wireless Networks*, *20*(7), 1759–1776. doi:10.1007/s11276-014-0709-y

Al-Kanj, L., & Dawy, Z. (2012). Energy-aware resource allocation in OFDMA wireless multicasting networks in*Proc. IEEE 19th International Conference on Telecommunications*, 1–5. doi:10.1109/ICTEL.2012.6221318

Alreshoodi, M., & Woods, J. (2013). An empirical study based on a fuzzy logic system to assess the QoS/QoE correlation for layered video streaming. *Proc. IEEE International Conference on Computational Intelligence and Virtual Environments for Measurement Systems and Applications,* 180–184. doi:10.1109/CIVEMSA.2013.6617417

Alreshoodi, M., & Woods, J. (2013). QoE prediction model based on fuzzy logic system for different video contents. *Proc. IEEE European Modelling Symposium*, 635–639. doi:10.1109/EMS.2013.106

Alreshoodi, M., & Woods, J. (2013). Survey on QoE\QoS correlation models for multimedia services. *International Journal of Distributed and Parallel Systems*, *4*(3), 53–72. doi:10.5121/ijdps.2013.4305

ANSI Standard. (2003). *Digital transport of one-way video signals - parameters for objective performance assessment,* ANSI Standard ATIS-0100801.03.2003(R2013).

Argyropoulos, S., Raake, A., Garcia, M.-N., & List, P. (2011). No-reference bit stream model for video quality assessment of H.264/AVC video based on packet loss visibility. *Proc. IEEE International Conference on Acoustics, Speech and Signal Processing,*1169–1172. doi:10.1109/ICASSP.2011.5946617

Bakanoğlu, K., Mingquan, W., Hang, L., & Saurabh, M. (2010). Adaptive resource allocation in multicast OFDMA systems. *Proc. IEEE Wireless Communications and Networking Conference*, 1–6. doi:10.1109/WCNC.2010.5506213

Barachi, M. E., Glitho, R., & Dssouli, R. (2008). Charging for multi-grade services in the IP Multimedia Subsystem. *Proc.International Conference on Next Generation Mobile Applications, Services and Technologies*, 10–17.

Batteram, H., Damm, G., Mukhopadhyay, A., Philippart, L., Odysseos, R., & Urrutia-Valdes, C. (2010). Delivering quality of experience in multimedia networks. *Bell Labs Technical Journal*, *15*(1), 75–193.

Bellini, A., Leone, A., Rovatti, R., Franchi, E., & Manaresi, N. (1996). Analog fuzzy implementation of a perceptual classifier for videophone sequences. *IEEE Transactions on Consumer Electronics*, *42*(3), 787–794. doi:10.1109/30.536186

Bisio, I., Delucchi, S., Lavagetto, F., & Marchese, M. (2014). *Transmission rate allocation over satellite networks with quality of experience - based performance metrics in*. Livorno: Advanced Satellite Multimedia Systems Conference and Signal Processing for Space Communications Workshop. doi:10.1109/ASMS-SPSC.2014.6934576

Brunnström, K., Sedano, I., Wang, K., Barkowsky, M., Kihl, M., Andrén, B., & Aurelius, A. et al. (2012). 2D no-reference video quality model development and 3D video transmission quality. *Proc. International Workshop on Video Processing and Quality Metrics for Consumer Electronics*, 1–6.

Butyka, Z., Jursonovics, T., & Imre, S. (2008). New fair QoS-based charging solution for mobile multimedia streams. *International Journal of Virtual Technology and Multimedia*, *1*(1), 3–22. doi:10.1504/IJVTM.2008.017107

Le Callet, P., Möller, S., & Perkis, A. (2013). Qualinet White Paper on Definitions of Quality of Experience. *European Network on Quality of Experience in Multimedia Systems and Services (COST Action IC 1003).*

Chen, J. (2010). *Resource Allocation for Delay Constrained Wireless Communications* (Doctoral Thesis). University College London, London, UK.

Chen, X., Hwang, J., Wang, C., & Lee, C. (2014). A near optimal QoE-driven power allocation scheme for SVC-based video transmissions over MIMO systems. *Proc. IEEE International Conference on Communications*, 1675–1680. doi:10.1109/ICC.2014.6883563

Chen, X., Hwang, J., Lee, C., & Chen, S. (2015). A near optimal QoE-driven power allocation scheme for scalable video transmissions over MIMO systems. *IEEE Journal of Selected Topics in Signal Processing*, *9*(1), 76–88. doi:10.1109/JSTSP.2014.2336603

Chikkerur, S., Sundaram, V., Reisslein, M., & Karam, L. J. (2011). Objective video quality assessment methods: A classification, review, and performance comparison. *IEEE Transactions on Broadcasting*, *57*(2), 165–182. doi:10.1109/TBC.2011.2104671

Chuah, S., Chen, Z., & Tan, Y. (2010). An optimized resource allocation algorithm for scalable video delivery over wireless multicast links. *International Packet Video Workshop, Hong Kong*, 41–47. doi:10.1109/PV.2010.5706818

Chuah, S., Chen, Z., & Tan, Y. (2012). Energy-efficient resource allocation and scheduling for multicast of scalable video over wireless networks. *IEEE Transactions on Multimedia*, *14*(4), 1324–1336. doi:10.1109/TMM.2012.2193560

Dixit, S., & Wu, T. (Eds.). (2004). *Content networking in the mobile Internet*. Hoboken, NJ: John Wiley & Sons. doi:10.1002/047147827X

Forum, D. S. L. (2006). *Triple-play services quality of experience (QoE) requirements.* DSL Forum, Technical Report TR-126.

Eden, A. (2007). No-reference estimation of the coding PSNR for H.264-coded sequences. *IEEE Transactions on Consumer Electronics*, *53*(2), 667–674. doi:10.1109/TCE.2007.381744

Fei, Z., Xing, C., & Li, N. (2015). QoE-driven resource allocation for mobile IP services in wireless network *Science China. Information Sciences*, *58*(1), 1–10.

Gao, Q., Fei, N., Xing, C., & Kuang, J. (2015). Balancing QoE and fairness of heterogeneous traffics based on Equivalent Rate Scaling. *China Communications*, *12*(1), 136–144.

Ghalut, T., & Larijani, H. (2014). Non-intrusive method for video quality prediction over LTE using random neural networks (RNN). *International Symposium on Communication Systems, Networks Digital Signal Processing*, 519–524. doi:10.1109/CSNDSP.2014.6923884

Gizelis, C., & Vergados, D. (2011). A survey of pricing schemes in wireless networks. *IEEE Communications Surveys and Tutorials*, *13*(1), 126–145. doi:10.1109/SURV.2011.060710.00028

Goudarzi, P. (2013). On the differentiated QoE enforcement between competing scalable video flows over wireless networks. *Wireless Communications and Mobile Computing*, *13*(7), 633–649.

Haider, F., Wang, C., Haas, H., Hepsaydir, E., & Ge, X. (2012). Energy-efficient subcarrier-and-bit allocation in multi-user OFDMA systems. *Proc. IEEE 75th Vehicular Technology Conference,*1–5. doi:10.1109/VETECS.2012.6240331

Hamam, A., Eid, M., El Saddik, A., & Georganas, N. (2008). A fuzzy logic system for evaluating quality of experience of haptic-based applications. In M. Ferre (Ed.), *Haptics: Perception, Devices and Scenarios* (Vol. 5024, pp. 129–138). Springer Berlin Heidelberg. doi:10.1007/978-3-540-69057-3_14

Haykin, S. (Ed.). (2003). *Communication Systems* (4th ed.). Wiley.

Haykin, S., & Moher, M. (2005). *Modern wireless communications.* Pearson/Prentice Hall.

He, L., & Liu, G. (2014). Quality-driven cross-layer design for H.264/AVC video transmission over OFDMA system. *IEEE Transactions on Wireless Communications*, *13*(12), 6768–6782. doi:10.1109/TWC.2014.2364603

Hewage, C., & Martini, M. (2011). Reduced-reference quality assessment for 3D video compression and transmission. *IEEE Transactions on Consumer Electronics*, *57*(3), 1185–1193. doi:10.1109/TCE.2011.6018873

Hu, J. (2013). *Resource Allocation and Optimization Techniques in Wireless Relay Networks (*Doctoral Thesis). Loughborough University, Loughborough, UK.

Hu, Z., Zhu, G., Xia, Y., & Liu, G. (2004). Multiuser subcarrier and bit allocation for MIMO-OFDM systems with perfect and partial channel information (Vol. 2). Academic Press.

Huang, C., & Leung, C. (2012). BitQoS-aware resource allocation for multi-user mixed-traffic OFDM systems. *IEEE Transactions on Vehicular Technology*, *61*(5), 2067–2082. doi:10.1109/TVT.2012.2189030

Ibarrola, E., Saiz, E., Zabala, L., Cristobo, L., & Xiao, J. (2014). A new global quality of service model: QoXphere. *IEEE Communications Magazine*, *52*(1), 193–199. doi:10.1109/MCOM.2014.6710083

ITU-R BT.1683. (2004). *Objective perceptual video quality measurement techniques for standard definition digital broadcast television in the presence of a full reference.* ITU-R Recommendation BT.1683.

ITU-T FG IPTV. (2007). *Definition of Quality of Experience (QoE).* International Telecommunication Union, Geneva, Switzerland, Technical Report ITU-T FG IPTV.

ITU-T G. 1000. (2001). *Communications quality of service: a framework and definitions (2001).* International Telecommunication Union, Technical Report ITU-T G.1000.

ITU-T G. 1010. (2001). *End-user multimedia QoS categories.* International Telecommunication Union, Technical Report ITU-T G.1010.

ITU-T G.1011. (2010). *Multimedia Quality of Service and performance – Generic and user-related aspects - Reference guide to quality of experience assessment methodologies.* ITU, Switzerland, Recommendation ITU-T G.1011.

ITU-T H.264. (2014). *H.264: Advanced video coding for generic audiovisual services.* ITU, Switzerland, Recommendation ITU-T H.264.

ITU-T J. 143. (2000). *User requirements for objective perceptual video quality measurements in digital cable television.* International Telecommunication Union, Geneva, Switzerland, Technical Report ITU-T J.143.

ITU-T J.144. (2004). *Objective perceptual video quality measurement techniques for digital cable television in the presence of a full reference.* ITU-T Recommendation J.144.

Joskowicz, J., & Ardao, J. C. L. (2011). Combining the effects of frame rate, bit rate, display size and video content in a parametric video quality model. *Proceedings of the 6th Latin America Networking Conference*, 4–11. doi:10.1145/2078216.2078218

Joskowicz, J., Sotelo, R., & Ardao, J. (2013). Towards a general parametric model for perceptual video quality estimation. *IEEE Transactions on Broadcasting*, *59*(4), 569–579. doi:10.1109/TBC.2013.2277951

Joskowicz, J., & Sotelo, R. (2014). A model for video quality assessment considering packet loss for broadcast digital television coded in H.264. *International Journal of Digital Multimedia Broadcasting*, *2014*, 11. doi:10.1155/2014/242531

Katoozian, M., Navaie, K., & Yanikomeroglu, H. (2009). Utility-based adaptive radio resource allocation in OFDM wireless networks with traffic prioritization. *IEEE Transactions on Wireless Communications*, *8*(1), 66–71. doi:10.1109/T-WC.2009.080033

Katsaggelos, A., Zhai, F., Eisenberg, Y., & Berry, R. (2005). Energy-efficient wireless video coding and delivery. IEEE Wireless Communications, 12(4), 24–30.

Kelly, F., Maulloo, A., & Tan, D. (1998). Rate control for communication networks: Shadow prices, proportional fairness and stability. *The Journal of the Operational Research Society*, *49*(3), 237–252. doi:10.1057/palgrave.jors.2600523

Khan, A., Sun, L., & Ifeachor, E. (2010). Learning models for video quality prediction over wireless local area network and universal mobile telecommunication system networks. *IET Communications*, *4*(12), 1389–1403. doi:10.1049/iet-com.2009.0649

Khan, A., Sun, L., & Ifeachor, E. (2012). QoE prediction model and its application in video quality adaptation over UMTS networks. *IEEE Transactions on Multimedia*, *14*(2), 431–442. doi:10.1109/TMM.2011.2176324

Khan, A., Sun, L., Ifeachor, E., Fajardo, J., Liberal, F., & Koumaras, H. (2015). Video quality prediction models based on video content dynamics for H.264 video over UMTS networks. *International Journal of Digital Multimedia Broadcasting*, *2010*, 1–17. doi:10.1155/2010/608138

Khan, S., Schroeder, D., El Essaili, A., & Steinbach, E. (2014). Energy-efficient and QoE-driven adaptive HTTP streaming over LTE. *Proc. IEEE Wireless Communications and Networking Conference*, 2354–2359.

Kim, H., Duong, D., Yun, J., Jang, Y., & Ko, S. (2006). Power-aware rate-control for QoS provisioning over CDMA networks. *Digest of Technical Papers International Conference on Consumer Electronics*, 139–140.

Kim, J., Kwon, T., & Cho, D.-H. (2007). Resource allocation scheme for minimizing power consumption in OFDM multicast systems. *Communications Letters, IEEE*, *11*(6), 486–488.

Kivanc, D., Li, G., & Liu, H. (2003). Computationally efficient bandwidth allocation and power control for OFDMA. *IEEE Transactions on Wireless Communications*, *2*(6), 1150–1158. doi:10.1109/TWC.2003.819016

Krongold, B., Ramchandran, K., & Jones, D. (2000). Computationally efficient optimal power allocation algorithms for multicarrier communication systems. *IEEE Transactions on Communications*, *48*(1), 23–27. doi:10.1109/26.818869

Lande, S., Helonde, S., Pande, R., & Pathak, H. (2011). *Adaptive Subcarrier and Bit Allocation for Downlink OFDMA System with Proportional Fairness. International Journal of Wireless & Mobile Networks*, *3*(5), 125–140.

Lee, G. R., & Wen, J. (2007). The performance of subcarrier allocation schemes combined with error control codings in OFDM systems. *IEEE Transactions on Consumer Electronics*, *53*(3), 852–856. doi:10.1109/TCE.2007.4341556

Lee, S., Jung, K., & Sim, D. (2010). Real-time objective quality assessment based on coding parameters extracted from H.264/AVC bitstream. *IEEE Transactions on Consumer Electronics*, *56*(2), 1071–1078. doi:10.1109/TCE.2010.5506041

Lee, S., Koo, J., & Chung, K. (2010). Content-aware rate control scheme to improve the energy efficiency for mobile IPTV. *Proc. IEEE International Conference on Consumer Electronics*, 445–446.

Lei, Y., Xiang, W., Chunping, H., Jianjun, L., & Yonghong, H. (2013). Trade-off optimization for scalable video coding streaming in relay-based OFDMA networks. *China Communications*, *10*(5), 99–113.

Li, B., Li, S., Xing, C., Fei, Z., & Kuang, J. (2012). A QoE-based OFDM resource allocation scheme for energy efficiency and quality guarantee in multiuser-multiservice system. *Proc. IEEE Globecom Workshops*, 1293–1297. doi:10.1109/GLOCOMW.2012.6477768

Li, F., Liu, G., & He, L. (2010). A low complexity algorithm of packet scheduling and resource allocation for wireless VoD systems. *IEEE Transactions on Consumer Electronics*, *56*(2), 1057–1062. doi:10.1109/TCE.2010.5506039

Li, M., Chen, Z., & Tan, Y. (2012). QoE-aware resource allocation for scalable video transmission over multiuser MIMO-OFDM systems. *Proc. IEEE Visual Communications and Image Processing*, 1–6. doi:10.1109/VCIP.2012.6410733

Li, M., Chen, Z., Tan, P., Sun, S., & Tan, Y. (2015). QoE-aware video streaming for SVC over multiuser MIMO–OFDM systems. *Journal of Visual Communication and Image Representation*, *26*, 24–36. doi:10.1016/j.jvcir.2014.10.011

Li, P., Chang, Y., Feng, N., & Yang, F. (2011). A cross-layer algorithm of packet scheduling and resource allocation for multi-user wireless video transmission. *IEEE Transactions on Consumer Electronics*, *57*(3), 1128–1134. doi:10.1109/TCE.2011.6018865

Li, P., Wang, Y., Zhang, W., & Huang, Y. (2014). QoE-oriented two-stage resource allocation in femtocell networks. *Proc. IEEE Vehicular Technology Conference*, 1–5. doi:10.1109/VTCFall.2014.6966143

Lin, X., Ma, H., Luo, L., & Chen, Y. (2012). No-reference video quality assessment in the compressed domain. *IEEE Transactions on Consumer Electronics*, *58*(2), 505–512. doi:10.1109/TCE.2012.6227454

Liu, H., Ho, H., & Chou, C. (2014). Content-aware spectrum and power allocation for video multicast in two-tier femtocell networks. *Proc. IEEE Wireless Communications and Networking Conference*, 3213–3217. doi:10.1109/WCNC.2014.6953056

Liu, Y., Li, C., & Yang, Z. (2015). Tradeoff between energy and user experience for multimedia cloud computing. *Computers & Electrical Engineering*, *47*, 161–172. doi:10.1016/j.compeleceng.2015.04.016

Luby, M. (2012). *Broadcast Delivery of Multimedia Content to Mobile Users*. Qualcomm, Technical Report.

Ma, W., Zhang, H., Zheng, W., Lu, Z., & Wen, X. (2012). MOS-driven energy efficient power allocation for wireless video communications. *Proc. IEEE Globecom Workshops*, 52–56. doi:10.1109/GLOCOMW.2012.6477543

Mahadevan, V., Chaczko, Z., & Braun, R. (2004). Mastering the mystery through "SAIQ" metrics of user experience in telecollaboration business systems. *Proc. IADIS International Conference on WWW/Internet,* 1029–1034.

Malekmohamadi, H., Fernando, A., Danish, E., & Kondoz, A. (2014). Subjective quality estimation based on neural networks for stereoscopic videos. *Proc. IEEE International Conference on Consumer Electronics*, 107–108. doi:10.1109/ICCE.2014.6775929

Moorsel, A. (2001). Metrics for the internet age: quality of experience and quality of business. *Proc. International Workshop on Performability Modeling of Computer and Communication Systems*, 34, 26–31.

Moorsel, A. (2001). *Metrics for the internet age: quality of experience and quality of business.* Hewlett Packard (HP), HP Labs Technical Report HPL-2001-179.

Mushtaq, M., Augustin, B., & Mellouk, A. (2012). Empirical study based on machine learning approach to assess the QoS/QoE correlation. *Proc. European Conference on Networks and Optical Communications,* 1–7. doi:10.1109/NOC.2012.6249939

Muthusamy, S. (2011). *Increasing broadcast and multicast service capacity and quality using LTE and MBMS.* Teleca, White paper.

Nee, R., & Prasad, R. (2000). OFDM for wireless multimedia communications. Boston: Artech House.

Ng, D., Lo, E., & Schober, R. (2013). Energy-efficient resource allocation in OFDMA systems with hybrid energy harvesting base station. *IEEE Transactions on Wireless Communications*, *12*(7), 3412–3427. doi:10.1109/TWC.2013.052813.121589

Ngo, D., Tellambura, C., & Nguyen, H. (2009). Efficient resource allocation for OFDMA multicast systems with fairness consideration. *Radio and Wireless Symposium,* 392–395.

Ngo, D., Tellambura, C., & Nguyen, H. (2009). Efficient resource allocation for OFDMA multicast systems with spectrum-sharing control *Vehicular Technology. IEEE Transactions on*, *58*(9), 4878–4889.

Nightingale, J., Wang, Q., Grecos, C., & Goma, S. (2014). The impact of network impairment on quality of experience (QoE) in H.265/HEVC video streaming. *IEEE Transactions on Consumer Electronics*, *60*(2), 242–250. doi:10.1109/TCE.2014.6852000

Nokia. (2004). *Quality of Experience (QoE) of mobile services: Can it be measured and improved?* Nokia, White paper.

Papoutsis, V., & Kotsopoulos, S. (2011). Chunk-based resource allocation in multicast OFDMA systems with average BER constraint. *Communications Letters, IEEE*, *15*(5), 551–553.

Paudel, I., Pokhrel, J., Wehbi, B., Cavalli, A., & Jouaber, B. (2014). Estimation of video QoE from MAC parameters in wireless network: a random neural network approach. *Proc. International Symposium on Communications and Information Technologies*, 51–55. doi:10.1109/ISCIT.2014.7011868

Perera, R., Fernando, A., Mallikarachchi, T., Arachchi, H., & Pourazad, M. (2014). QoE aware resource allocation for video communications over LTE based mobile networks. *Proc. International Conference on Heterogeneous Networking for Quality, Reliability, Security and Robustness*, 63–69. doi:10.1109/QSHINE.2014.6928661

Pinson, M. H., & Wolf, S. (2004). A new standardized method for objectively measuring video quality. *IEEE Transactions on Broadcasting*, *50*(3), 312–322. doi:10.1109/TBC.2004.834028

Pokhrel, J., Wehbi, B., Morais, A., Cavalli, A., & Allilaire, E. (2013). Estimation of QoE of video traffic using a fuzzy expert system.*Proc. IEEE Consumer Communications and Networking Conference*, 224–229. doi:10.1109/CCNC.2013.6488450

Ramirez, R. (2014). *Low Complexity Radio Resource Management for Energy Efficient Wireless Networks (*Doctoral Thesis). The University of Edinburgh, Edinburgh, UK.

Reddy, Y. (2007). Genetic algorithm approach for adaptive subcarrier, bit, and power allocation. *Proc.IEEE International Conference on Networking, Sensing and Control*, 14–19. doi:10.1109/ICNSC.2007.372925

Reeves, C., & Bednar, D. (1994). Defining quality: Alternatives and implications. *Academy of Management Review*, *19*(3), 419–445.

Reiter, U. (2009). Perceived quality in consumer electronics - from quality of service to quality of experience. *Proc. IEEE International Symposium on Consumer Electronics,*958–961. doi:10.1109/ISCE.2009.5156963

Rodríguez, D., Rosa, R., & Bressan, G. (2013). A billing system model for voice call service in cellular networks based on voice quality. *Proc. IEEE International Symposium on Consumer Electronics*, 89–90. doi:10.1109/ISCE.2013.6570267

Rohling, H., & Gruneid, R. (1997). Performance comparison of different multiple access schemes for the downlink of an OFDM communication system. *Proc. IEEE 47th Vehicular Technology Conference*, 3, 1365–1369. doi:10.1109/VETEC.1997.605406

Sacchi, C., Granelli, F., & Schlegel, C. (2011). A QoE-oriented strategy for OFDMA radio resource allocation based on min-MOS maximization. *IEEE Communications Letters*, *15*(5), 494–496. doi:10.1109/LCOMM.2011.031411.101672

Sadr, S., Anpalagan, A., & Raahemifar, K. (2009). Radio resource allocation algorithms for the downlink of multiuser OFDM communication systems. *IEEE Communications Surveys Tutorials*, *11*(3), 92–106.

Saki, H., & Shikh-Bahaei, M. (2015). Cross-layer resource allocation for video streaming over OFDMA cognitive radio networks. *IEEE Transactions on Multimedia*, *17*(3), 333–345. doi:10.1109/TMM.2015.2389032

Shen, J., Yi, N., Wu, B., Jiang, W., & Xiang, H. (2009). A greedy-based resource allocation algorithm for multicast and unicast services in OFDM system. *InternationalConference on Wireless Communications Signal Processing,* 1–5.

Shen, Z., Andrews, J., & Evans, B. (2003). Optimal power allocation in multiuser OFDM systems. *Proc. IEEE Global Telecommunications Conference*, 1, 337–341.

Song, G., & Li, Y. (2003). Adaptive subcarrier and power allocation in OFDM based on maximizing utility. *Proc.IEEE Semiannual Vehicular Technology Conference*, 2, 905–909.

Suzuki, T., Kutsuna, T., & Tasaka, S. (2008). QoE estimation from MAC-level QoS in audio-video transmission with IEEE 802.11e EDCA. *Proc. IEEE International Symposium on Personal, Indoor and Mobile Radio Communications,* 1–6. doi:10.1109/PIMRC.2008.4699471

Takahashi, A., Hands, D., & Barriac, V. (2008). Standardization activities in the ITU for a QoE assessment of IPTV. *Communications Magazine, IEEE*, *46*(2), 78–84.

Tang, J. (2013). *Mathematical optimization techniques for resource allocation and spatial multiplexing in spectrum sharing networks* (Doctoral Thesis). Loughborough University, Loughborough, UK.

Tektronix. (2008). *A Guide to IPTV The Technologies, the Challenges and How to Test IPTV.* Tektronix, Inc.

Tektronix. (2009). *A Guide to MPEG Fundamentals and Protocol Analysis.* Tektronix, Inc., Technical Report.

TeleManagement Forum. (2005). SLA Management Handbook, Concepts and Principles. TeleManagement Forum.

TeleManagement Forum. (2009). Best Practice: Video over IP SLA Management. *TeleManagement Forum.*

Tychogiorgos, G., & Leung, K. (2014). Optimization-based resource allocation in communication networks. *Computer Networks*, *66*, 32–45. doi:10.1016/j.comnet.2014.03.013

Vriendt, J. De, Vleeschauwer, D., & Robinson, D. C. (2014). QoE model for video delivered over an LTE network using HTTP adaptive streaming. *Bell Labs Technical Journal*, *18*(4), 45–62.

Wang, S., Guo, W., Khirallah, C., Vukobratović, D., & Thompson, J. (2014). Interference allocation scheduler for green multimedia delivery. *IEEE Transactions on Vehicular Technology*, *63*(5), 2059–2070. doi:10.1109/TVT.2014.2312373

Wang, X., Dai, H., Zhang, H., & Li, F. (2012). Channel-aware adaptive resource allocation for multicast and unicast services in orthogonal frequency division multiplexing systems. *IET Communications*, *6*(17), 3006–3014. doi:10.1049/iet-com.2012.0111

Wang, Y., Chen, F., & Chen, G. (2005). Adaptive subcarrier and bit allocation for multiuser OFDM system based on genetic algorithm. *Proc. IEEE International Conference on Communications, Circuits and Systems*, 1, 242–246.

Wang, Z., Lu, L., & Bovik, A. C. (2004). Video quality assessment based on structural distortion measurement. *Signal Processing: Image Communication, 19*(2), 121–132.

Wolter, K., & Moorsel, A. (2001). *The Relationship between Quality of Service and Business Metrics: Monitoring, Notification and Optimization.* Hewlett Packard (HP), HP Labs Technical Report HPL-2001-96.

Wong, C., Cheng, R., Lataief, K., & Murch, R. (1999). Multiuser OFDM with adaptive subcarrier, bit, and power allocation. *IEEE Journal on Selected Areas in Communications*, *17*(10), 1747–1758. doi:10.1109/49.793310

Wu, F. (2010). *Synchronization and Resource Allocation in Downlink OFDM Systems* (Doctoral Thesis). University of Plymouth, Plymouth, UK.

Xiao, L. (2010). *Radio Resource Allocation in Relay Based OFDMA Cellular Networks* (Doctoral Thesis). Queen Mary, University of London, London, UK.

Xu, Y., Hu, R. Q., Wei, L., & Wu, G. (2014). QoE-aware mobile association and resource allocation over wireless heterogeneous networks. *Proc. IEEE Global Communications Conference*, 4695–4701. doi:10.1109/GLOCOM.2014.7037549

Yang, F., Yang, Q., Fu, F., & Kwak, K. S. (2014). A QoE-based resource allocation scheme for multi-radio access in heterogeneous wireless network. *Proc. International Symposium on Communications and Information Technologies*, 264–268. doi:10.1109/ISCIT.2014.7011913

Yasakethu, S. (2010). *Perceptual Quality Driven 3D Video Communications* (Doctoral Thesis). University of Surrey, Guildford, UK.

Zander, J., & Kim, S. (2001). *Radio resource management for wireless networks*. Boston: Artech House.

Zhang, Q., Ji, Z., Zhu, W., & Zhang, Y.-Q. (2002). Power-minimized bit allocation for video communication over wireless channels. *IEEE Transactions on Circuits and Systems for Video Technology*, *12*(6), 398–410. doi:10.1109/TCSVT.2002.800322

Zhang, Y., & Letaief, K. (2004). Multiuser adaptive subcarrier-and-bit allocation with adaptive cell selection for OFDM systems. *IEEE Transactions on Wireless Communications*, *3*(5), 1566–1575. doi:10.1109/TWC.2004.833501

Zhao, Q. L., Gan, Z., & Zhu, H. (2012). Adaptive resource allocation method over OFDMA system for H.264 SVC transmission. *IEEE International Conference on Signal Processing*,2, 1435–1438. doi:10.1109/ICoSP.2012.6491845

Zhao, Q., Mao, Y., Leng, S., & Jiang, Y. (2014). QoS-aware energy-efficient multicast for multi-view video in indoor small cell networks. *Proc. IEEE Global Communications Conference*, 4478–4483. doi:10.1109/GLOCOM.2014.7037513

Chapter 8
QoE-Driven Efficient Resource Utilisation for Video Over Critical Communication Systems

Emad Abdullah Danish
Saudia Airlines, Saudi Arabia

Mazin I. Alshamrani
Ministry of Haj and Umra, Saudi Arabia

ABSTRACT

Research in network resource utilisation introduced several techniques for more efficient power and bandwidth consumption. The majority of these techniques, however, were based on Quality of Service (QoS) and network parameters. Therefore, in this study a different approach is taken to investigate the possibility of a more efficient resource utilisation if resources are distributed based on users' Quality of Experience (QoE), in the context of 3D video transmission over WiMAX access networks. In particular, this study suggests a QoE-driven technique to identify the operational regions (bounds) for Modulation and Coding Schemes (MCS). A mobile 3D video transmission is simulated, through which the correlation between receiver's Signal-to-Noise Ratio (SNR) and perceived video quality is identified. The main conclusions drawn from the study demonstrate that a considerable saving in signal power and bandwidth can be achieved in comparison to QoS-based techniques.

INTRODUCTION

In the recent decade, 3D video has attracted a larger portion of the consumer market, due to the added dimension of entertainment. However, since it contains additional information to be transmitted over the network, an extra burden is added to the already congested network's capacity. Therefore, the issue of resource utilisation becomes more than ever a pressing issue for networks in general, and wireless access networks in particular.

Accordingly, research in network resource utilisation introduced several techniques for more efficient power and bandwidth consumption. The majority of these techniques, however, were based on Quality of Service (QoS) and network parameters. Therefore, in this study a different approach is taken to inves-

DOI: 10.4018/978-1-5225-2113-6.ch008

tigate the possibility of a more efficient resource utilisation if resources are distributed based on users' Quality of Experience (QoE), in the context of 3D video transmission over WiMAX access networks. In particular, this study suggests a QoE-driven technique to identify the operational regions (bounds) for Modulation and Coding Schemes (MCS). A mobile 3D video transmission is simulated, through which the correlation between receiver's Signal-to-Noise Ratio (SNR) and perceived video quality is identified. The main conclusions drawn from the study demonstrate that a considerable saving in signal power and bandwidth can be achieved in comparison to QoS-based techniques.

Recently, 3D video has gained an accelerated momentum in the consumer entertainment sector. This is attributed to the added dimension of sensation it offers to the viewer, represented by the ability to sense the depth added to conventional 2D video. Hence, one of the developed techniques for 3D video representation is a 2D colour image accompanied by a grayscale depth map, known as colour-plus-depth representation. Consequently, the stereoscopic 3D view is generated (rendered) at the display terminal following a technique known as Depth Image Based Rendering (DIBR) (Fehn, 2004). This added dimension of depth introduces extra demand on network resources and exhibits further user aspects such as binocular disparity and visual discomfort contributing to the overall 3D perception. In addition, the transmission of 3D video to mobile devices is confronted by the limited bandwidth, bit-errors and finite device power. Therefore, more efficient resource utilisation techniques are needed; techniques that reduce the power requirement and avail extra bandwidth.

Efficient resource utilisation has been an area of interest to several research studies. Hence, different approaches were taken with the aim to save on scarce resources such as bandwidth and power. Many of these approaches often considered Quality of Service (QoS) and network parameters as the basis for their findings. Unlike QoS-based techniques, in this study, efficiency in resource utilisation is suggested through exploiting Adaptive Modulation and Coding (AMC) at the physical layer together with users' QoE (Martini, 2013). Hence, from a user-centric approach, this chapter presents a study proposing a more efficient technique for AMC in terms of power and bandwidth utilisation, on the basis of consumer's perception of 3D video.

Section 5 of this chapter presents the methodology followed to propose the QoE-driven resource utilisation. This includes the simulation model and the proposed approach for band AMC transition thresholds. The results are presented in section 6. Then the contribution of this chapter is summarised in the last section.

CONTENT-AWARE VIDEO QUALITY MODELLING

Content-Aware Approach

As the required transmit power is a function of bitrate, so is the quality of the transmitted video. From this perspective, identifying the proper bitrate in a multiuser unicast video transmission becomes a crucial element of the problem since a high bitrate would result in excess power for unnoticeable quality in the video. In contrast, a low bitrate would result in tremendous power saving for a poor user experience.

Drawing on the importance of user's quality of experience (QoE), this work employs a perceptual video quality metric to identify an appropriate bitrate allocation to users. The NTIA general Video Quality Model (VQM) (Pinson & Wolf, 2004) is selected based on its performance and accreditation by both ANSI (ANSI Standard, 2003) and ITU (ITU-T J.144, 2004) as an objective video quality assessment standard. In the conducted experiments, the test video sequences are encoded at different quality levels and exposed to information loss, and then the average video quality is measured with VQM. The measured VQM (QoE) is mapped to the corresponding bitrate. Deriving the required bitrate on this basis offers the advantage of incorporating several quality aspects and video content characteristics into a representative bitrate. Hence, this approach is called the content-aware approach.

Quality Modelling

The aim of quality modelling is to establish a mapping between video bitrate and video quality so that the required data bits by a user can be identified based on the targeted video quality for that user. In a real-time scenario, such a mapping would not yet exist, as it is content dependent and fluctuates according to bit errors at transmission time. Therefore, in this work, quality modelling is developed on an off-line basis, and it is assumed that such a mapping will be made possible on practical grounds. The simulation environment of quality modelling is presented in Appendix A.

Figure 1 depicts the process followed to develop the required mapping of video bitrate to video quality. Each video sequence is source coded based on the H.264/AVC standard (ITU-T H.264, 2014), utilizing the H.264/AVC JM Reference Software (H.264/AVC, 2014). Varying levels of the encoder Quantization Parameter (QP) are used to generate different quality levels of each sequence. The average video bitrate of each sequence is also recorded for later mapping with the quality metric. To simulate transmission errors on the physical layer, the encoded .264 bit-streams are exposed to packet loss patterns, which interpret disparate levels of bit-error-rates, Pe (10^{-5}, 10^{-4}, 10^{-3}). The packet loss traces are generated based on the Gilbert-Elliot Model (Elliott, 1963; Hasslinger & Hohlfeld, 2008), considering the assumed video coding parameters. In the final stage (QoE metric), average video quality is measured with VQM (Pinson and Wolf, 2004; NTIA - US Department of Commerce, 2013). The recorded quality on the VQM scale is mapped to the corresponding recorded average bitrate. This mapping is represented by rate–distortion (R–D) diagrams, which reveal the bitrate requirement of each video in order to achieve a particular level of quality perception. This method caters for any type of video content carrying any characteristics, since end-to-end video transmission conditions are incorporated altogether in the identified representative bitrate.

For increased confidence in the collected data, the process of generating packet errors is repeated 10 times per BER per encoded video, and the average VQM is computed later.

Figure 1. Quality modelling for the test video sequences

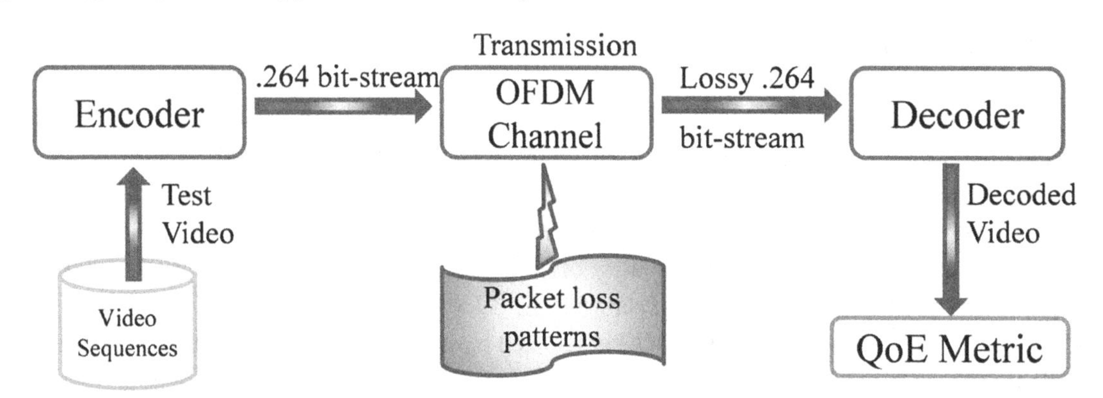

Packet Loss Model

Transmission bit errors are simulated by introducing burst packet losses to the transmitted packet stream. Bit losses are distributed as bursts with a Mean Burst Length (MBL) along the bit loss trace. Thereafter, the bit loss trace is mapped into a packet loss trace such that a lost bit in any given packet is considered a loss of the packet.

Packet loss traces considered for this simulation are generated using the simple Gilbert-Elliot model (Elliott, 1963; Hasslinger & Hohlfeld, 2008) which is a widely used model to simulate burst packet losses in transmission channels. Moreover, Gilbert-Elliot packet loss model is well known to approximate the packet loss of the Rayleigh fading channel with high accuracy (Han et al., 2011; Sun, 2002).

To abstract the layers from the application layer to the physical layer, the Network Abstraction Layer (NAL) units (encapsulated in RTP packets) are used to represent the packets received or dropped. This layers abstraction concept assumes that channel coding is present in order to identify which packets to drop as packets are received. However, to simplify the problem, no channel coding schemes were employed in the simulation because channel coding would only increase the required bitrate, which will not affect the purpose of the simulation. The purpose of the quality modelling simulation is to map the bitrate to perceptual video quality at different levels of bit error rates. Hence, performance of the resource allocation algorithm will not be affected whether a channel coding scheme is literally used or not.

The Gilbert-Elliot packet erasure model is based on a Markov chain with two states, GOOD and BAD, as illustrated in Figure 2.

Accordingly, the probability of switching from state GOOD→BAD is denoted as PGOOD→BAD, while the probability of switching from state BAD→GOOD is denoted by PBAD→GOOD. The probability of a packet loss in state GOOD is denoted as PGOOD and the probability of a packet loss occurring in state BAD is denoted as PBAD. Then, the steady state packet loss rate is defined as (Hasslinger and Hohlfeld, 2008):

$$PLR = \frac{P_{GOOD} \bullet P_{BAD \to GOOD}}{\left(P_{GOOD \to BAD} + P_{BAD \to GOOD}\right)} + \frac{P_{BAD} \bullet P_{GOOD \to BAD}}{\left(P_{GOOD \to BAD} + P_{BAD \to GOOD}\right)} \quad (1)$$

Figure 2. Two-state Markov Model

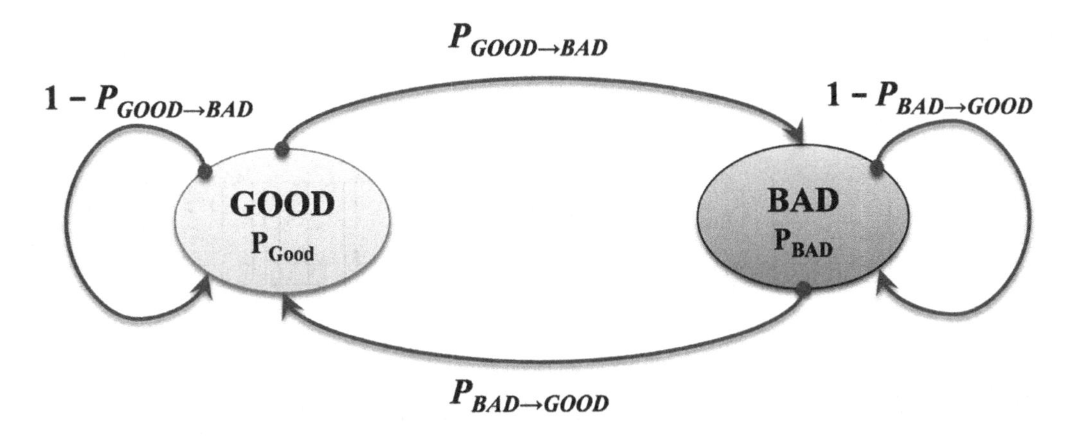

It is commonly assumed that no packet losses occur in state GOOD and all packets are lost in state BAD (Alwis et al., 2012). Accordingly, the PLR is calculated by:

$$PLR = \frac{P_{GOOD \to BAD}}{\left(P_{GOOD \to BAD} + P_{BAD \to GOOD}\right)} \quad (2)$$

In similar vein to (Shi et al., 2010), the MBL of the packet loss trace is:

$$MBL = \frac{1}{P_{BAD \to GOOD}} \quad (3)$$

In Gilbert-Elliot model, the probability of a packet being lost or received depends on the previous packet, and this is how bursts of lost packets are produced. Therefore, in the generation process once the state in the model is moved to BAD, the role of the MBL is to keep the state in BAD for as long as it satisfies the required burst size (MBL).

Consequently, packet loss traces of sufficient length for the encoded video packets are generated. An example is illustrated in Figure 3 to show the distribution pattern of an error trace when bit-error-rate is 10^{-3}. Indeed, 10 different versions of each trace are produced to simulate 10 different instances of transmission errors. This is an analogy to real-life transmission where errors start at random instances of time.

CONTENT-AWARE AND ENERGY-EFFICIENT RESOURCE ALLOCATION SCHEME

In a multi-user multi-carrier OFDM transmission of video content, considering the perceptual video quality and energy consumption, the allocation of wireless resources represents a challenging non-deterministic polynomial (NP-hard) problem. On the one hand, video quality relies on the assigned bitrate to each receiver. On the other hand, the transmit energy required by the same bitrate depends on instantaneous channel state of each subcarrier allocated to each user.

Figure 3. Pattern of a packet loss trace when $Pe=10^{-3}$

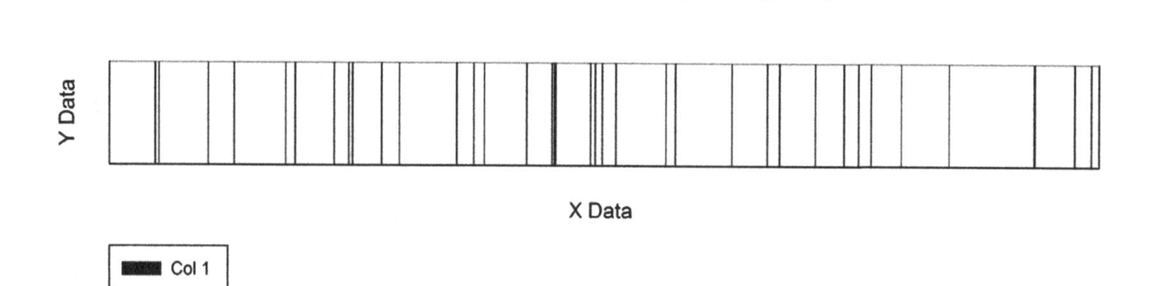

Also, it becomes evident that the less the data rate the less the power required. Hence, power can always be decreased by providing lower data rates; however, this will result in poor video quality at the user's end. Therefore, video quality is introduced as a counteractive objective to limit the lower extent at which data rates could reach. In fact, increasing the quality will cause the data rate to rise. Apparently, a trade-off between total required energy (P_T) and average achievable quality (Q_{avg}) is observed with both being functions of bitrate. This conflicting relation develops the multi-objective problem, with the aim to minimize required energy and maximize received content quality by identifying the proper data rate to allocate to each receiver simultaneously.

Accordingly, the resource allocation problem is twofold: allocation of subcarriers to users, and allocation of bitrate to users based on the requirement of received video. Considering the instantaneous channel state on each user's allocated subcarrier, a Multi-Objective Optimization Problem (MOOP) is formulated. The key question is: What is the optimum bitrate and subcarrier allocation among users that will attain minimum P_T at high Q_{avg} for a maximum utility value, f ? Therefore, the evolutionary genetic algorithm (GA) is nominated as a suitable technique to achieve a sub-optimal solution for the problem at hand.

The development of the proposed resource allocation scheme has taken two stages. In the first stage, two nested genetic algorithms (GA) were used. The second stage developed a significantly enhanced version of the first stage, where the nested two genetic algorithms (GA) were merged into a single GA for global optimization of the problem. The optimization frequency of the proposed resource allocation scheme is assumed to be equal to the frequency of the radio channel. That is, for every OFDM transmission frame, the scheme seeks the optimal allocation of resources. The practicality concern of this assumption is left for future work; however, it is assumed that the computational requirements of the scheme can be fulfilled by operating on machines of a very high processing power that exceeds the radio channel frequency.

Resource Allocation Scheme with Two Nested GAs

In the first stage of this research, the proposed scheme uses the genetic algorithm (GA) twice in a nested manner: once in bitrate searching and once in subcarrier allocation. Hence it searches for subcarrier allocation within the bitrate allocation.

The block diagram in Figure 4 shows the processing flow of the resource allocation scheme in its basic form, which is based on using GA twice. First, a generation of chromosomes is generated that suggests bitrate assignment to users. Each chromosome (solution) is taken as a candidate solution and average quality is calculated based on the proposed quality model. To calculate the power budget of the solution, the joint power allocation algorithm is utilised. This algorithm is another round of GA; i.e., another population of suggested subcarrier-to-user allocations is generated and iterated until a sub-optimal power allocation is reached. Thereafter, the second chromosome of the first GA is taken into a similar process.

This scheme achieved good results; however, it was inefficient in terms of processing time. Therefore, to overcome this inefficiency, the scheme is developed in its second stage to a lower complexity version that utilises one GA for global optimisation of the MOOP.

Figure 4. Resource allocation scheme with nested two-levels of genetic algorithms

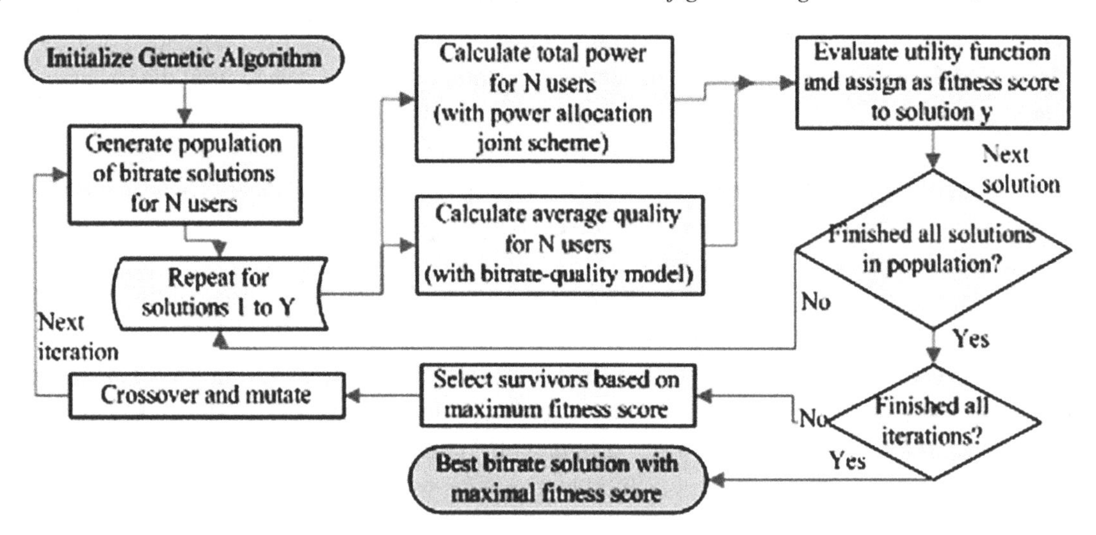

Resource Allocation Scheme with Two Nested GAs

This is the second stage of this research task, where a significant enhancement is carried out to the optimization scheme. The nested two genetic algorithms (GA) are merged into a single GA for global optimization of both algorithms. This has introduced substantial performance efficiency with lower complexity in the proposed algorithms.

Process merging is performed on the chromosome structure and decoding level. Hence, the GA's binary chromosome is formed, as shown in Figure 5. Thanks to the binary representation of data, any number of users, subcarriers, or bitrate allocations can be accommodated by this chromosome. Accordingly, during the optimization process, a single GA is used to search for a sub-optimal solution of both: users' required bitrates and subcarrier and bit allocation. The acquired sub-optimal solution shall satisfy the aforementioned objectives of P_T, Q_{avg}, and f.

The proposed scheme to address the MOOP is the Content-aware and Energy-efficient Resource Allocation Scheme (CaERAS). Figure 6 illustrates a high-level block diagram of the CaERAS scheme, identifying input/output elements. Once the input parameters are known to the OFDM channel, the optimization algorithms will propose the most sub-optimal allocation of bitrates and subcarriers among users, which satisfy the targeted objectives of minimum energy and maximum quality for a designated utility function.

Figure 5. Structure of GA binary chromosome in the proposed scheme

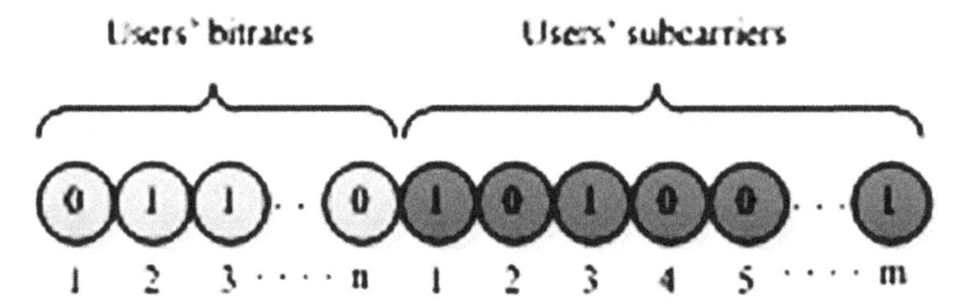

Figure 6. Main components and I/Os of the CaERAS scheme

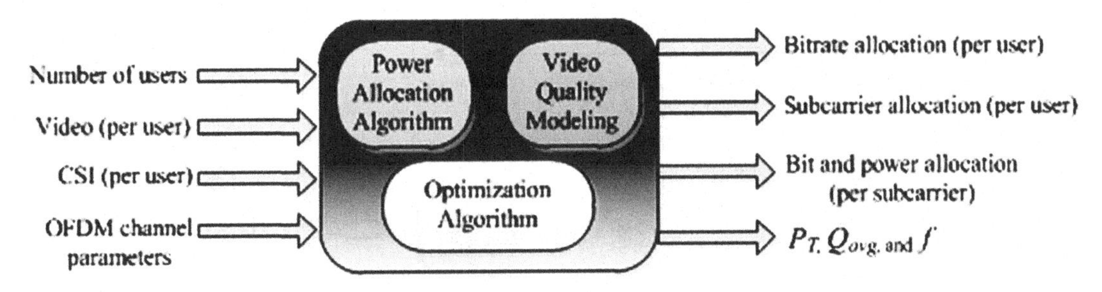

CaERAS initiates the search for a solution through four main stages, as presented by the algorithmic flow in Figure 7. First, a search for a solution is initiated by generating and decoding a random GA population of chromosomes that take the structure in Figure 5. Second, the power allocation scheme extracts the decoded bitrate and subcarrier data of each chromosome and employs a greedy algorithm to calculate minimum total power of the chromosome. Third, the associated average video quality is acquired through R–D relations. Fourth, a utility score is evaluated to value the candidate chromosome. This process is repeated as a GA problem until a number of predefined iterations are exhausted. The solution with maximum f is considered the surviving sub-optimal solution.

To further clarify the subcarrier, bit, modulation, and user allocation as generated by a population of chromosomes, Figure 8 illustrates an example of a detailed snapshot of a sample allocation (subcarrier-to-user and bitrate-to-user and bit-to-subcarrier allocation). This snapshot is taken just after chromosomes are decoded in Figure 7, and indicated by the oval highlighted shape at the end of chromosome decoding in Figure 7.

Figure 7. Schematic diagram of the proposed CaERAS scheme

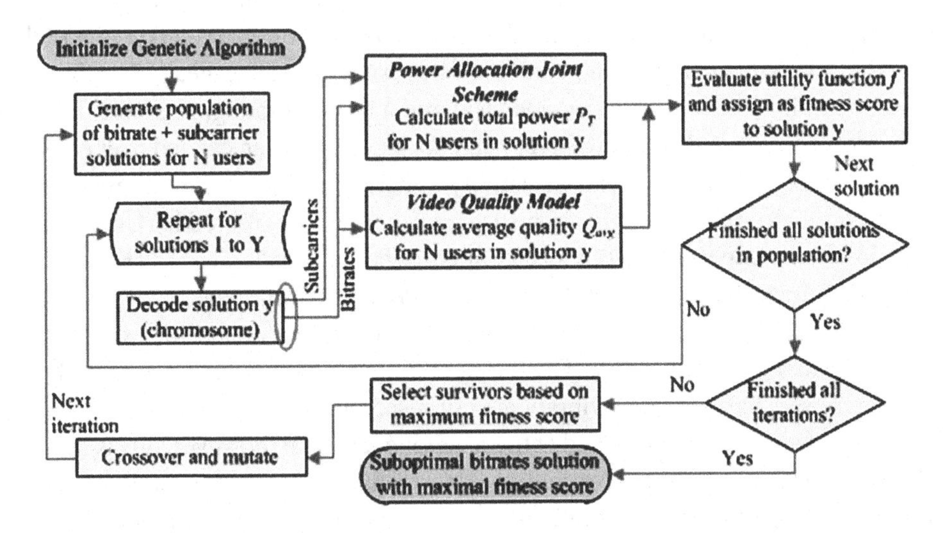

Figure 8. Example of resources as allocated by a population of decoded chromosomes

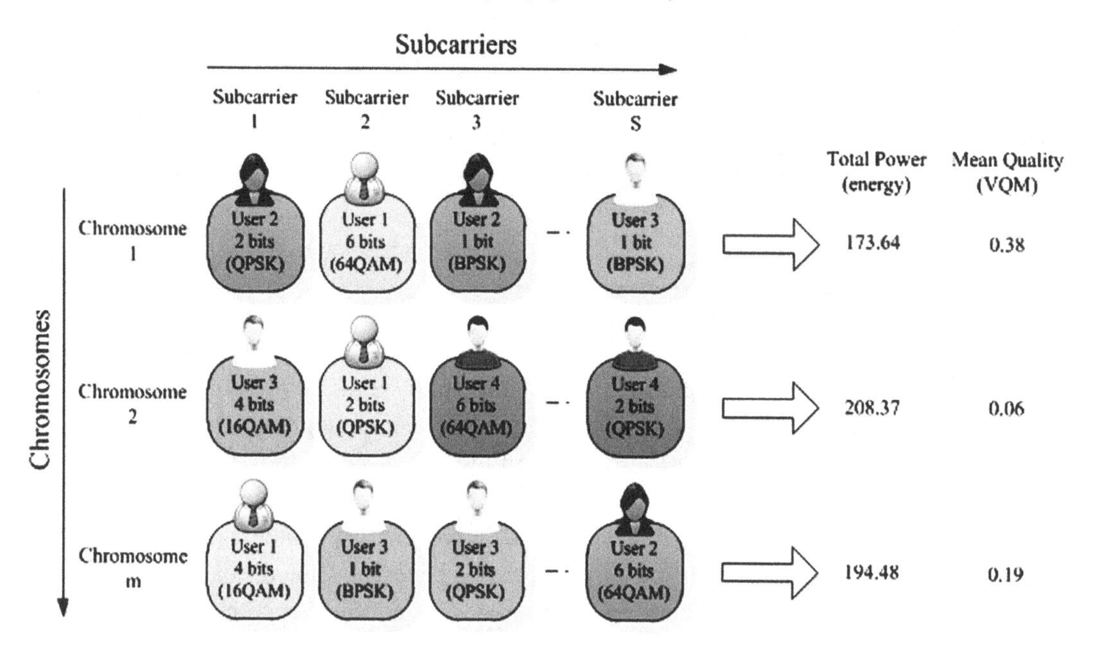

QoE PREDICTION FOR WIRELESS VIDEO

In Ever since modern portable consumer devices (tablets, smartphones, etc.) were brought into existence, user consumption of video content has been gaining an accelerated momentum. According to a recent research study by Cisco (Cisco White Paper FLGD, 2014), 61% of the global internet traffic will be carried over wireless and mobile devices. The same study stated that video is expected to dominate up to 90% of the global internet traffic by 2018. Moreover, in the mobile domain, video was found to prevail by 66% of the mobile data traffic in 2014 (Khan, 2014). In line with this trend, 3D video services are expected to capture a significant portion of the Consumer Electronics (CE) market due to the advancement of immersive video technologies and enabling infrastructures (Hewage, & Martini, 2011). Since video in general, and 3D video in particular, is considered a resource-hungry service, it becomes more sensitive to any problems along the video delivery chain, especially when carried over error-prone wireless networks to portable CE devices. This sensitivity is translated into different levels of degradation in user experience. Add to this the highly sophisticated functionality offered by CE devices nowadays which elevate consumers' expectations of video quality to that of broadcast level (Khan et al., 2012). A level of quality that fully utilises the capabilities of their devices (Nightingale et al., 2014). As a result, a change in paradigm from a quality of service (QoS) approach to a quality of experience (QoE) approach becomes inevitable (Reiter, 2009). Therefore, adopting a broad understanding of video quality as perceived by end-users has become a vibrant area of research.

Perceptual video quality can be measured at the receiving terminal; however, this is impractical since the reference video is absent and subjective quality assessment is impractical as well. Therefore, in order to meet QoE requirements of users, it is vitally necessary to predict, monitor and possibly control video quality (Khan et al., 2012). QoE prediction, however, requires an overarching understanding of those QoS factors that are the most influential on QoE (Reiter, 2009; DSL Forum, 2006). Hence, it becomes equally important to model the relationship between QoS and QoE such that QoE can be predicted without the reference video (no-reference). Video quality can be measured in an intrusive, Full-Reference (FR), or non-intrusive, No-Reference (NR) mode. Intrusive models require availability of the original video, unlike non-intrusive (NR) models which can fill this gap for prediction by providing less accurate measurement but are sufficiently reliable for real-time video streaming.

In multimedia communications, the QoS/QoE correlation attracted considerable attention by the research community. Researchers in this area were unanimous in concluding that a tight relationship exists between QoS and QoE (Alreshoodi & Woods, 2013; Kim & Choi, 2010; Khirman, & Henriksen, 2002; Kim et al., 2008; Orosz et al., 2014; Moller, 2009). Once the QoS/QoE relationship is identified and modelled, it can be used in either one of two approaches (DSL Forum, 2006; Kim & Choi, 2010):

- For known QoS parameter measurements, the anticipated user's QoE can be predicted.
- For a targeted QoE level for a user, the required application layer and physical layer performance could be deduced.

A model that can estimate end-to-end 3D video quality in the context of wireless video streaming is yet to be thoroughly explored. This motivates the investigation of prediction models that adopt key factors of QoS which have a substantial effect on perceived QoE. There are several QoS factors that influence end-to-end video quality, but their joint effect is obscure and their interactions are believed to be non-linear (Khan et al., 2012; DSL Forum, 2006). For this, learning models represent a feasible approach to model the QoS/QoE correlation since they have the ability to learn then predict in a manner similar to human reasoning. Different techniques (Aroussi & Mellouk, 2014) have been used by researchers to realize predictive QoE models through QoS/QoE correlation analysis. Regression-based models, Artificial Neural Networks (ANN), and Fuzzy Inference Systems (FIS) are few to mention. However, most of the research in this area discussed partial solutions and overlooked some influential QoS parameters across the video delivery layers. Therefore, to develop a more generic prediction model, the work in this chapter adopts a hybrid model that considers key QoS parameters from both the application layer and the physical layer. The selected QoS parameters have not been addressed so far in the context of 3D video. Hence, the proposed model allows an optimal estimation of 3D quality given QoS constraints. 3D video is considered because there are still many aspects related to 3D quality evaluation caused by the complexity of 3D perception as opposed to 2D perception (Goldmann & Ebrahimi, 2011). Nevertheless, it should be highlighted that the proposed prediction model can be applied to both 2D and 3D video. There is no significant difference except that the same QoS factors could cause disparate distortions to 3D quality perception.

The presented NR prediction model of perceived 3D video quality (QoE) represents a significant addition to video communications since it provides several advantages to both the consumer and the service provider:

- Through automated real-time QoE monitoring, service providers can control and maintain desired quality levels to the user through the management of controllable QoS parameters, such as video codec, bitrate, signal power, modulation, etc. Such parameters can be fine-tuned on-the-fly as channel conditions can be sent to the service provider over auxiliary feedback channels.
- For service charging, achieved quality could be used as a criterion for quality-based billing that employs state-of-the-art charging schemes in real-time.

- More efficient QoE-based resource allocation in multi-user mobile environments can be achieved and is made possible. This results in more efficient bandwidth utilization and power consumption.

Hence, the work presented in this chapter proposes a cross-layer non-intrusive prediction model in the context of wireless 3D video streaming. The model is designed on the basis of the fuzzy logic inference systems (FIS). Accordingly, the main contributions in this chapter are twofold:

- An evaluation of QoS/QoE correlation for 3D video streaming employing cross-layer key QoS parameters.
- A FIS-based non-intrusive prediction model utilizing the datasets constructed from the QoS/QoE correlation.

QoE-DRIVEN OPERATIONAL REGIONS IN ADAPTIVE MODULATION AND CODING

Simulation Model

The simulation system presented in Figure 9 is designed and implemented to study the effect of mobile wireless transmission errors on users' perception of 3D video. 3D video sequences, in the form of 2D-colour plus depth-map, are tested. As illustrated in Figure 9 the selected 3D video sequences are encoded into H.264 bit-streams, which are projected to the wireless channel error traces, and then decoded. Later, an objective quality metric of 3D video (Yasakethu et al., 2011), which emulates human perception of the received video, is employed to measure the 3D quality perception.

Figure 9. The simulation system

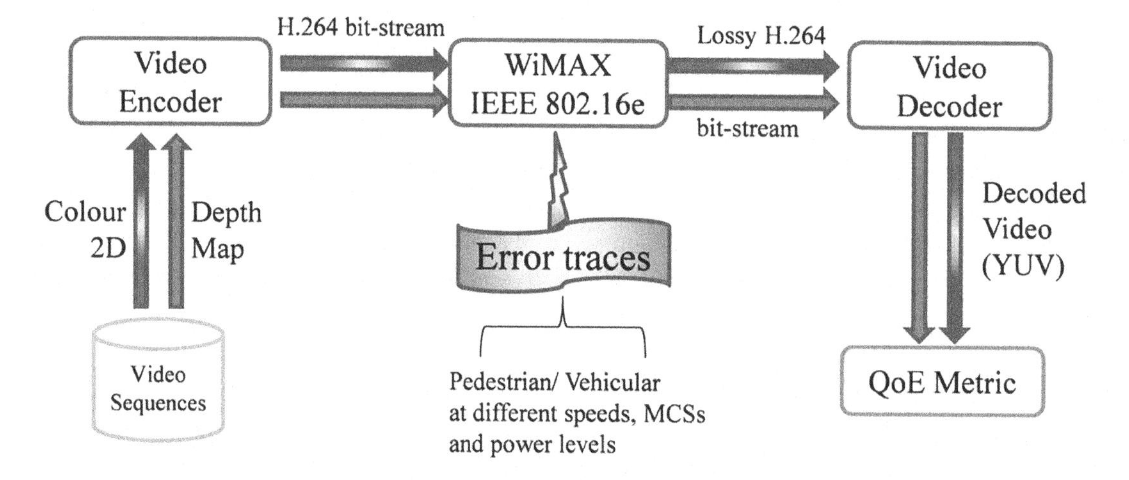

Figure 10. The test video sequences

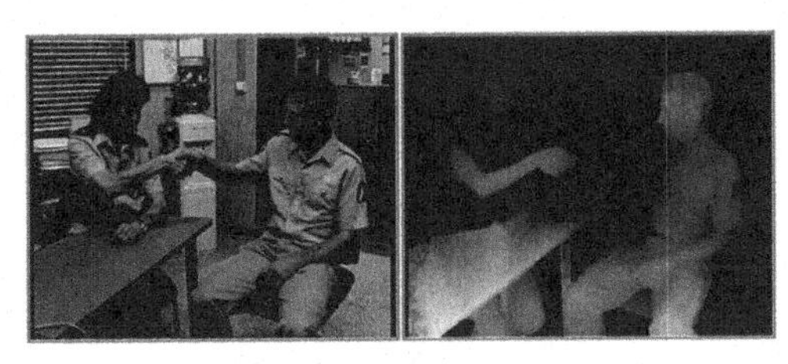

(a) *Interview* video

(b) *GT Fly* video

Video Content and Encoding

The test video sequences, in the form of colour plus depth, are shown in Figure 10. Both videos are 720x576 Standard Definition (SD) with 25 frames/second (fps), and 10 seconds long, making 250 frames.

Based on the H.264 standard (ITU-T H.264, 2014), H.264/AVC Reference JM Software encoder (H.264/AVC, 2014) is used for source video coding. For both test videos, rate control is assumed with a constant bitrate distribution of 80% and 20% to colour and depth, respectively. Based on a recommended bitrate for SD video (Issa et al., 2010). Furthermore, the bitrate allocated to the Interview video is 1Mbps, 200Kbps for colour and depth channels, respectively. Whereas the GT Fly video is allocated 2Mbps, 400Kbps similarly (Issa et al., 2010).

The Wireless Channel

The experiment simulates the transmission of stereoscopic video over Worldwide Interoperability for Microwave Access (WiMAX IEEE 802.16e) networks (So-In, 2010). The simulations consider various Signal-to-Noise Ratio (SNR) levels using different Modulation and Coding Schemes (MCSs). Two channel model environments at two different mobility speeds are tested: a pedestrian walking at the speed of 3 km/hour, and a vehicular running at the speed of 60 km/hour. The WiMAX simulation model presented in (Liew, 2006) is considered for this testing. The SNR-to-PER relationship in this model is shown in *Figure 11* for the vehicular scenario. This figure describes the Signal-to-Noise Ratio (SNR) at the receiver for different Packet Error Rate (PER) levels. For each MCS, this information tells the expected signal quality at the receiver given a PER.

Transmission error patterns are generated in this WiMAX simulation model (Liew, 2006). The error traces are designed such that a radio frame is transmitted every 5ms, which means 200 frame bursts are transmitted per second. The error trace is formed of 3000 frames for a period of 15 seconds. A number of error bits are introduced in the 15 seconds error trace depending on the MCS used.

This structure of the error trace is described in Figure 12 for the case of MCS QPSK 1/2. In this example, a bit value of 1 means a successfully received bit, and a bit value of 0 means the bit has changed, i.e., lost. If a single bit of a slot's 48 bits is changed then the whole data slot is considered lost. Therefore, MCSs that carry more bits per data slot are more sensitive to channel errors.

Later, based on the target video packet size, the bit-error traces are used to produce packet-error traces. 10 different versions of each error trace are generated, that is 10 versions per MCS per BER. Hence, simulations are performed for all the versions of each case and quality is measured for every

Figure 11. Channel SNR vs. Packet error rate for the vehicular scenario

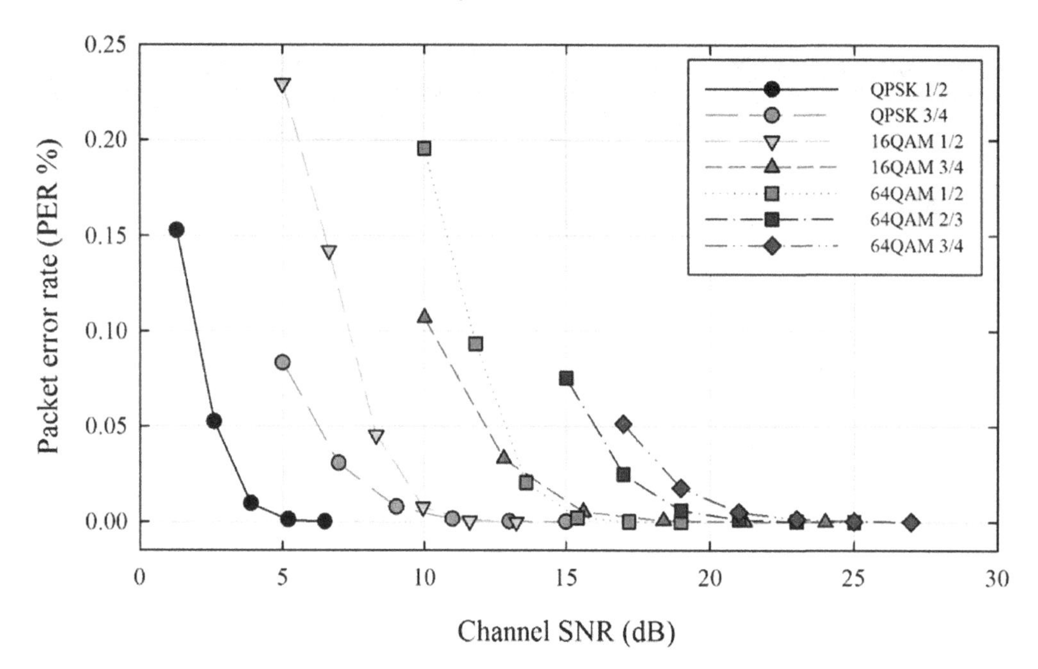

Figure 12. How error patterns simulate for MCS QPSK 1/2

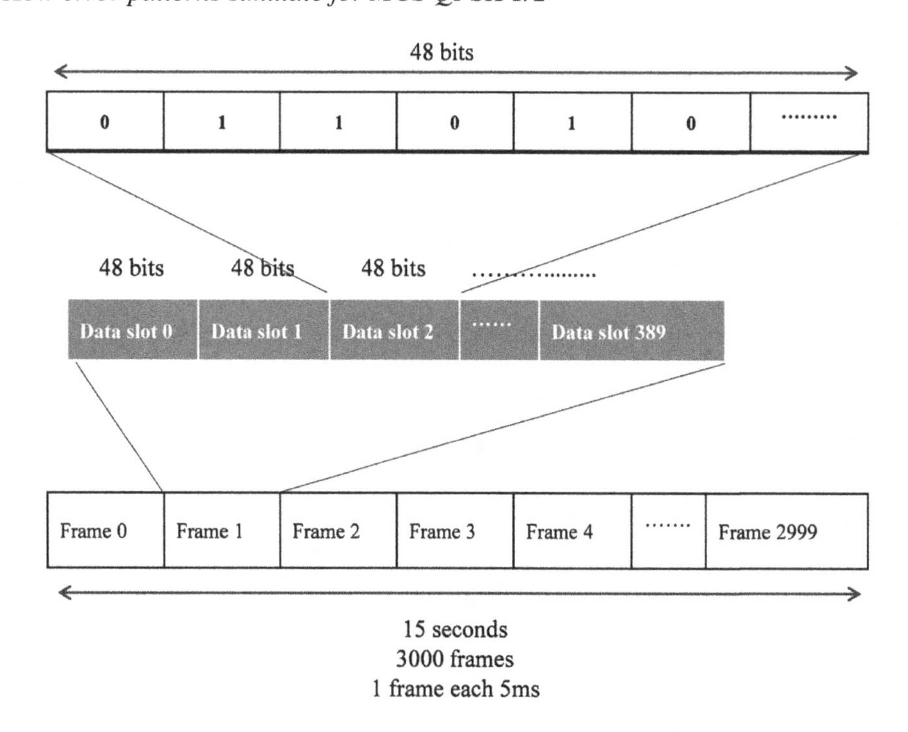

instance then the average quality is computed. This is because in real-life scenarios bit errors could occur at any time during transmission. For comprehensive experimenting, all supported MCSs by the WiMAX model where tested.

The WiMAX model (Liew, 2006) used in these simulations offers a downlink capacity of 390 data slots (30 subcarriers x 13 time symbols) in a Time Division Duplex (TDD) frame every 5ms. Since the number of bits allocated to the data slot depends on the MCS used, the maximum channel transmission rate ranges from 3.744 Mbps for QPSK 1/2 (48 bits per data slot), to 16.848 Mbps for 64QAM 3/4 (216 bits per data slot). The data bits capacity of each MCS in the tested WiMAX model is depicted in Table 1.

Both video components, the colour and the depth, are simulated using the aforementioned wireless transmission channel model. Once the distorted video bit-streams are received, they are decoded and the 3D quality perception is measured using the 3D quality metric (Yasakethu et al., 2011).

QoE Metric

The objective 3D video quality metric (Yasakethu et al., 2011) is employed to measure the perceptual quality of the received distorted colour image and depth map.

SNR Thresholds for Band AMC Transition

Based on the aforementioned simulations, a QoE-based method is proposed to identify SNR thresholds of operational MCSs where transitions can occur in adaptive modulation. In Adaptive Modulation and Coding (AMC), once AMC sub-channel allocation is triggered, the MCS can be negotiated between the Base Station (BS) and the Mobile Station (MS) and changed every 5ms. Hence, the transition of the MS between MCS regions is permitted. The downlink operational MCS region is bounded by receiver's SNR threshold levels, as illustrated in Figure 13. If the SNR steps out of the designated operating region, the receiver (MS) requests a change to a new operational MCS region.

The IEEE standard (IEEE Standard 802.16, 2012) suggests that the SNR thresholds bounding each operational region shall be identified at bit-error-rate (BER) of 10^{-6}. In contrast, the proposal in this chapter suggests that SNR thresholds are identified based on viewer's perceptual QoE.

RESULTS

To visualise the simulation results, Figure 14 and Figure 15 shows the quality achieved over different levels of SNR for each MCS. As depicted on Figure 14 and Figure 15, the two tested video sequences were simulated for different environments and at different bitrates. This is so the desired conclusion for these simulations is shown to have covered different test conditions. In Figure 14 and Figure 15, GT Fly

Table 1. Uplink capacity of information bits in the WiMAX model (Liew, 2006)

Modulation	QPSK		16QAM		64QAM		
Code Rate	1/2	3/4	1/2	3/4	1/2	2/3	3/4
Data TX capacity (bits/data slot)	48	72	96	144	144	192	216

Figure 13. SNR Thresholds for Operational MCS Regions in a Cell

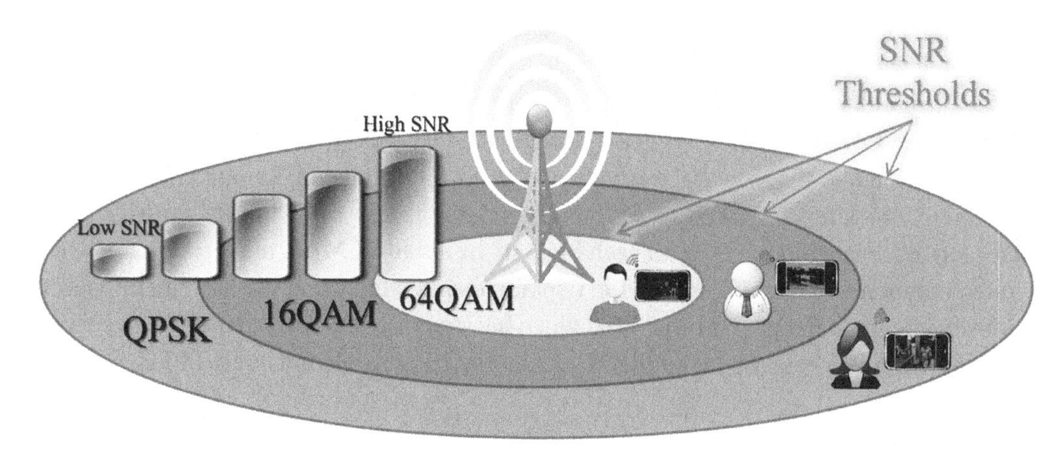

scored a higher maximum performance (QoE) than Interview since it is allocated double the bitrate. This, however, was at the expense of extra power required, which is signified by the higher SNR.

The data collected in Figure 14 and Figure 15 is used to infer the SNR thresholds suggested at QoE=0.8, in order to compare it with those thresholds based on BER.

The two methods for identifying the operational MCS regions (SNR bounds) are compared in Figure 16 and Figure 17. The first method is based on BER= 10^{-6} (IEEE Standard 802.16, 2012), and the second is based on QoE=0.8 which corresponds to 80% on the scale of the quality metric used (Yasakethu et al., 2011). This QoE level is designated as the reference in this proposal since it is regarded by the research community a sufficiently high and acceptable quality perception level. QoE=0.8 is said to be the threshold of minimum quality level for user acceptability of a "pleasing viewing experience" (Song, 2011).

Figure 14. Average quality performance of 3D videos with respect to channel SNR: 'Interview' video for a pedestrian at 3km/h (bitrate 1.2 Mbps)

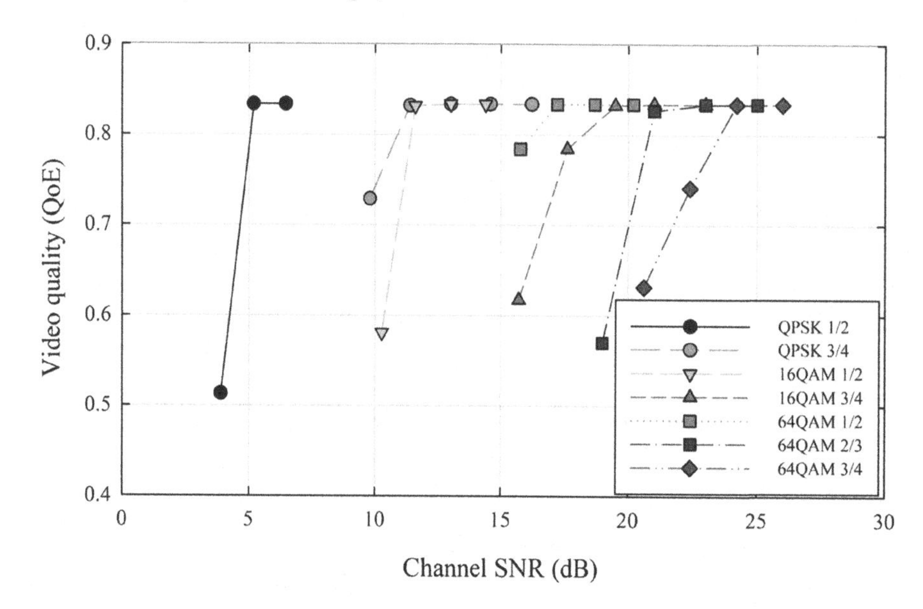

Figure 15. Average quality performance of 3D videos with respect to channel SNR: 'GT Fly' video for a vehicle at 60km/h (bitrate 2.4 Mbps)

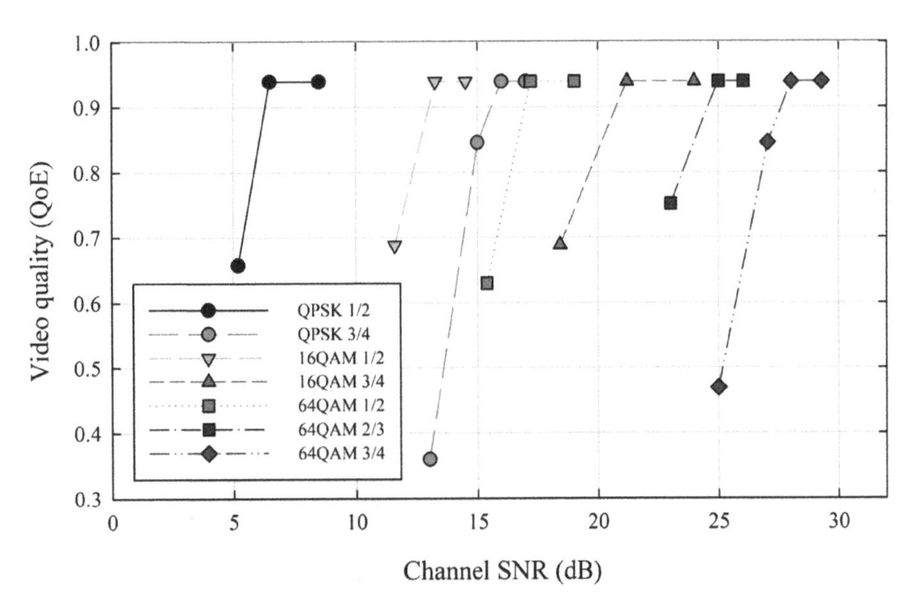

Figure 16. SNR bounds of the MCS operational regions: 'Interview' video

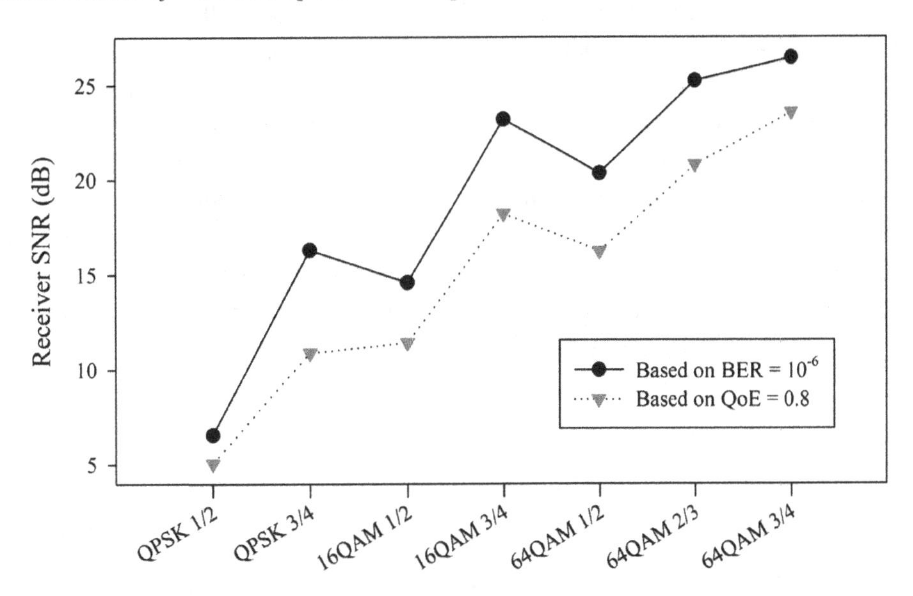

The comparison in Figure 16 and Figure 17 implies that a significant bandwidth and power efficiency can be achieved by considering quality perception (QoE) as the basis for MCS selection (through AMC), compared to the traditional bit-error-rate based technique. The proposed QoE based selection for AMC bands not only is power-efficient, but is also bandwidth-efficient since it keeps a MS allocated to a higher-order band for longer than it used to be (with the BER based method).

Figure 17. SNR bounds of the MCS operational regions: 'GT Fly' video

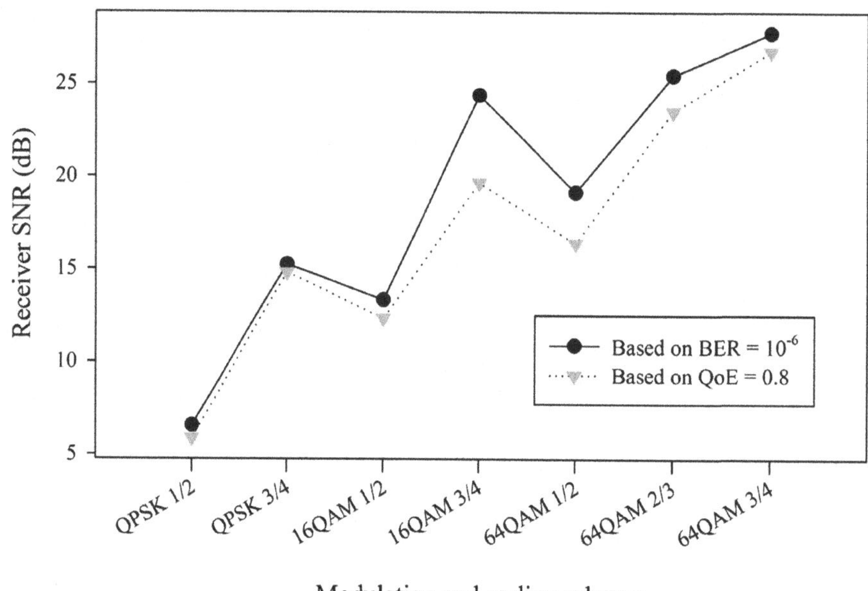

This conclusion becomes evident by comparing the receiver's SNR (in dB) when $BER= 10^{-6}$ and when QoE=0.8. The lower value SNR (when QoE is considered) implies that transition to a higher-order MCS can occur earlier. In terms of bandwidth efficiency, this early transition to a higher AMC band means more data bits per slot can be carried. Table 2 summarizes the percentage of additional data bits that can be used as a result of such early transition. For instance, a transition from 16QAM 1/2 to 16QAM 3/4 allows for 50% additional bandwidth to the user at a lower SNR. In terms of power efficiency, the lower value SNR shows that the receiver can operate on the same MCS at a lower power level, which is controlled by the transmitting base station. For example, for the Interview sequence, the operating SNR threshold for QPSK 3/4 based on QoE requires 5.4 dB less power than the BER ased method. The percentage of power that could be saved in each MCS is shown in Table 2.

Table 2. Percentage of bandwidth and SNR savings

Modulation	Code Rate	Percentage of additional data bits/slots to carry	Percentage of SNR Saved	
			Interview	*GT Fly*
QPSK	1/2		22.87%	10.53%
	3/4	50%	33.13%	2.82%
16QAM	1/2	33.33%	21.66%	7.43%
	3/4	50%	21.40%	19.37%
64QAM	1/2	0%	20.26%	14.37%
	2/3	33.33%	17.57%	7.58%
	3/4	12.5%	10.97%	3.57%

SUMMARY

This work exhibits a reflective study of 3D video transmission over WiMAX mobile broadband networks. The study considered QoE as the basis for identifying the SNR thresholds at which a mobile receiver is transitioned from one MCS to another. Simulations were performed using a WiMAX model that exploited several levels of SNR, bit-error-rates, and modulation and coding schemes in two radio transmission environments. The momentous conclusion drawn from this study denotes that considering QoE as the basis for modulation scheme selection in AMC can be advantageous with respect to power and bandwidth efficiency, as opposed to the techniques based on network parameters solely. This efficiency climbs up to 33% less power and up to 50% more bandwidth.

This conclusion is believed to emphasize a promising approach for an AMC that considers viewer's perception for radio resource efficiency in the resource allocation process. The proposed approach could lead to a better utilization of energy, and a higher data transfer rate simultaneously. This chapter broadly outlined the third contribution in this thesis. Therein, a study of video transmission over WiMAX networks has been conducted. Motivated by the QoE-based resource allocation achievements, the aim of the study has been to investigate the usefulness of relying on QoE rather than bit-error-rate (BER) in determining the bounding thresholds of Modulation and Coding Schemes (MCS) in Adaptive Modulation and Coding (AMC). The simulations have exploited varying levels of received Signal-to-Noise Ratios (SNR), BER, and MCS for a pedestrian and a vehicular transmission environment. SNR thresholds at which receivers are transitioned have been identified based on two methods: QoE of 0.8 and BER of 10^{-6}. The QoE-based method has recorded lower levels of SNR. This means receivers' transition between MCS regions can be delayed. This delay results in making use of the higher bitrate in higher order MCSs for a longer period of time, and avoiding the need for extra power to transit a user to a higher order MCS. Hence, bandwidth and power utilisation is improved. This improvement has reached up to 33% less power and up to 50% more bandwidth. For wireless video communications, this conclusion affirms the claims that QoE-based service provisioning could reduce energy consumption and enhance user experience.

REFERENCES

Alreshoodi, M., & Woods, J. (2013). Survey on QoE\QoS correlation models for multimedia services. *International Journal of Distributed and Parallel Systems*, *4*(3), 53–72. doi:10.5121/ijdps.2013.4305

Alwis, C., Arachchi, H., Silva, V., Fernando, A., & Kondoz, A. (2012). Robust video communication using random linear network coding with pre-coding and interleaving. *Proc.IEEE International Conference on Image Processing*, 2269–2272.

Aroussi, S., & Mellouk, A. (2014). Survey on machine learning-based QoE-QoS correlation models. *Computing, Management and Telecommunications (ComManTel), 2014 International Conference on*, 200–204.

Cisco White Paper, F. L. G. D. (2014). *Cisco Visual Networking Index: Forecast and Methodology.* 2013–2018 Cisco, USA, White Paper FLGD 11684.

Elliott, E. (1963). Estimates of error rates for codes on burst-noise channels. *The Bell System Technical Journal*, *42*(5), 1977–1997. doi:10.1002/j.1538-7305.1963.tb00955.x

Fehn, C. (2004). *Depth-image-based rendering (DIBR), compression, and transmission for a new approach on 3D-TV*. Academic Press.

Forum, D. S. L. (2006). *Triple-play services quality of experience (QoE) requirements*. DSL Forum, Technical Report TR-126.

Goldmann, L., & Ebrahimi, T. (2011). Towards reliable and reproducible 3D video quality assessment. *Proc. SPIE*, 8043, 302–307. doi:10.1117/12.887037

H.264/AVC JM. (2014). *H.264/AVC JM Reference Software*. Available: http://iphome.hhi.de/suehring/tml/

Han, S., Joo, H., Lee, D., & Song, H. (2011, May). An end-to-end virtual path construction system for stable live video streaming over heterogeneous wireless networks. *IEEE Journal on Selected Areas in Communications*, *29*(5), 1032–1041. doi:10.1109/JSAC.2011.110513

Hasslinger, G., & Hohlfeld, O. (2008). The Gilbert-Elliott model for packet loss in real time services on the internet. *Proc. GI/ITG Conference - Measuring, Modelling and Evaluation of Computer and Communication Systems,*1–15.

Hewage, C., & Martini, M. (2011). Reduced-reference quality assessment for 3D video compression and transmission. *IEEE Transactions on Consumer Electronics*, *57*(3), 1185–1193. doi:10.1109/TCE.2011.6018873

IEEE Standard 802.16-2012. (2012). *IEEE Standard for Air Interface for Broadband Wireless Access Systems*. IEEE, New York, Standard 802.16-2012.

Issa, O., Li, W., & Liu, H. (2010). Performance evaluation of TV over broadband wireless access networks. *IEEE Transactions on Broadcasting*, *56*(2), 201–210. doi:10.1109/TBC.2010.2046979

ITU-T H.264. (2014). *H.264: Advanced video coding for generic audiovisual services*. ITU, Switzerland, Recommendation ITU-T H.264.

ITU-T J.144. (2004). *Objective perceptual video quality measurement techniques for digital cable television in the presence of a full reference*. ITU-T Recommendation J.144.

Khan, A., Sun, L., & Ifeachor, E. (2012). QoE prediction model and its application in video quality adaptation over UMTS networks. *IEEE Transactions on Multimedia*, *14*(2), 431–442. doi:10.1109/TMM.2011.2176324

Khan, N. (2014). *Quality-Driven Multi-User Resource Allocation and Scheduling Over LTE for Delay Sensitive Multimedia Applications* (Doctoral Thesis). Kingston University, Kingston, UK.

Khirman, S., & Henriksen, P. (2002). Relationship between quality-of-service and quality-of-experience for public internet service. *Proc. Passive and Active Network Measurement Workshop*.

Kim, H., & Choi, S. (2010). A study on a QoS/QoE correlation model for QoE evaluation on IPTV service. *Proc. International Conference on Advanced Communication Technology,*2, 1377–1382.

Kim, H.-J., Lee, D. H., Lee, J. M., Lee, K.-H., Lyu, W., & Choi, S.-G. (2008). The QoE evaluation method through the QoS-QoE correlation model. *Proc.International Conference on Networked Computing and Advanced Information Management*, 2, 719–725.

Liew, C., Worrall, S., Goldshtein, M., Navarro, A., & Mota, M. (2006). *WIMAX modelling*. Project Deliverable SUIT_208 D2.3.

Martini, M. (2013). Cross-layer design for quality-driven multi-user multimedia transmission in mobile networks. *IEEE COMSOC MMTC E-Letter*, *8*(2), 18–20.

Moller, S., Engelbrecht, K.-P., Kühnel, C., Wechsung, I., & Weiss, B. (2009). A taxonomy of quality of service and quality of experience of multimodal human-machine interaction. *Proc. International Workshop on Quality of Multimedia Experience*, 7–12. doi:10.1109/QOMEX.2009.5246986

Nightingale, J., Wang, Q., Grecos, C., & Goma, S. (2014). The impact of network impairment on quality of experience (QoE) in H.265/HEVC video streaming. *IEEE Transactions on Consumer Electronics*, *60*(2), 242–250. doi:10.1109/TCE.2014.6852000

NTIA - US Department of Commerce. (2013). *Video Quality Metric (CVQM) Software*. Available: http://www.its.bldrdoc.gov/resources/video-quality-research/guides-and-tutorials/cvqm-overview.aspx

Orosz, P., Skopkó, T., Nagy, Z., Varga, P., & Gyimóthi, L. (2014). A case study on correlating video QoS and QoE. *Proc. IEEE Network Operations and Management Symposium*, 1–5. doi:10.1109/NOMS.2014.6838399

Pinson, M., & Wolf, S. (2004). A new standardized method for objectively measuring video quality. *IEEE Transactions on Broadcasting*, *50*(3), 312–322. doi:10.1109/TBC.2004.834028

Reiter, U. (2009). Perceived quality in consumer electronics - from quality of service to quality of experience. *Proc. IEEE International Symposium on Consumer Electronics*,958–961. doi:10.1109/ISCE.2009.5156963

Shi, Z., Zou, H., Rank, M., Chen, L., Hirche, S., & Muller, H. J. (2010). Effects of packet loss and latency on the temporal discrimination of visual-haptic events. *IEEE Transactions on Haptics*, *3*(1), 28–36.

So-In, C., Jain, R., and Tamimi, A.-K. (2010). Capacity evaluation for IEEE 802.16e mobile WiMAX. *Journal of Computer Systems, Networks, and Communications*, 1:1–1:12.

Song, W., Tjondronegoro, D., & Docherty, M. (2011). Saving bitrate vs. pleasing users: where is the break-even point in mobile video quality?*Proc. ACM International Conference on Multimedia*, 403–412. doi:10.1145/2072298.2072351

ANSI Standard. (2003). *Digital transport of one-way video signals - parameters for objective performance assessment.* ANSI Standard ATIS-0100801.03.2003(R2013).

Sun, P.-T. (2002). *Similarity of Discrete Gilbert-Elliot and Polya Channel Models to Continuous Rayleigh Fading Channel Model* (Doctoral Thesis). National Chiao Tung University, Taiwan.

Yasakethu, S., Worrall, S., Silva, D., Fernando, W., & Kondoz, A. (2011). A compound depth and image quality metric for measuring the effects of packet loss on 3D video. *Proc. IEEE International Conference on Digital Signal Processing*,1–7. doi:10.1109/ICDSP.2011.6004998

Chapter 9
Streaming Coded Video in P2P Networks

Muhammad Salman Raheel
University of Wollongong, Australia

Raad Raad
University of Wollongong, Australia

ABSTRACT

This chapter discusses the state of the art in dealing with the resource optimization problem for smooth delivery of video across a peer to peer (P2P) network. It further discusses the properties of using different video coding techniques such as Scalable Video Coding (SVC) and Multiple Descriptive Coding (MDC) to overcome the playback latency in multimedia streaming and maintains an adequate quality of service (QoS) among the users. The problem can be summarized as follows; Given that a video is requested by a peer in the network, what properties of SVC and MDC can be exploited to deliver the video with the highest quality, least upload bandwidth and least delay from all participating peers. However, the solution to these problems is known to be NP hard. Hence, this chapter presents the state of the art in approximation algorithms or techniques that have been proposed to overcome these issues.

INTRODUCTION

Mission Critical Communication Systems and their related Multimedia Services plays a vital role in different sectors of life which includes; Disaster Recovery, Intelligence control and various utility sectors. Hence, to provide a reliable communication, it is important to consider strict set of requirements as compared to wired or wireless communication systems. A mission critical system may be used to support low bandwidth data communication such as voice or text services or it may be used to support more bandwidth killing applications such as video streaming services. Each service comprises of different set of challenges and requirements such as voice services, requires an acceptable delay and packet loss for reliable quality of voice. Whereas, the text service is more delay tolerant, hence it has less bandwidth issues as compared to other real time applications. Similarly, video or multimedia streaming over the internet encounters delay, latency and bandwidth issues.

DOI: 10.4018/978-1-5225-2113-6.ch009

This chapter discusses the state of the art in dealing with the resource optimization problem for smooth delivery of video across a peer to peer (P2P) network comprising of different coding techniques to support the basic requirements of a mission critical system to support multimedia. A multimedia stream comprises of a combination of a speech, audio, text, animation or video content that is transmitted on request to the destination node. There are a number of different media streaming architectures that have been proposed such as; the traditional client server based architecture (Hareesh K & Manjaiah DH, 2011); the media is streamed through a central server to clients upon request. However, a large number of users may not be accommodated due to high bandwidth bottleneck at the server. To overcome the bandwidth issues, another well-known solution being the content distribution network (CDN) architecture was proposed (Zhijie S, Jun L, Roger Z & Athanasios VV, 2011). In CDNs, dedicated servers are deployed at different geographical locations to accommodate a large number of requests from the users. However, the disadvantage of CDN's architecture is that it produces large signalling over-head. Therefore, the authors provide a distributed architecture known as a peer to peer (P2P) network that depends on the network user's resources to share the multimedia content especially video delivery over the internet (Anh TN, Baochun L & Frank E, 2010). Each node in the network behaves as a sender, a receiver or a relay node to forward the content from one peer to another peer. Each peer joins the network to form an overlay architecture. The advantage of using P2P networks is its self-adaptive and self-configuration properties which reduces the overall load at the server and increases the network bandwidth.

Furthermore, a P2P network is a mixture of several heterogeneous peers (nodes with variable upload and download bandwidths) connected with each other to form an overlay. Hence, it requires a sender peer to store and send multiple versions of a similar content. But P2P networks can be further utilised to deliver a variety of content and in conjunction with other application level techniques such as video coding and more specifically coding that encodes video into different layers. The multiple versions of the video content are generated using different video coding techniques such as multiple descriptive coding (MDC) (Vivek KG, 2001) or scalable video coding (SVC) (Zhengye L, Yanming S, Keith WR, Shivendra SP & Yao W, 2009). In MDC, the video is encoded into several different descriptors whereas the quality of the video depends on the number of descriptors received. The descriptors are then forwarded over multiple paths to the destination node such that each descriptor can be decoded independently. However, the authors discuss that by applying MDC over video content incurs considerable bit rate overhead and is computationally complex. On the hand, in SVC, the video is encoded into a base layer and several enhancement layers where the quality of the video depends upon the number of layers received. The base layer carries the basic information of the video whereas the enhancement layers are used to further improve the quality of a base layered video. Hence, each higher layer is dependent over the lower layer. The overall advantage of using SVC over MDC is that it quickly adapts to current network conditions with considerably less overhead complexity.

The chapter is divided into following sub sections. Section II comprises of a brief overview of P2P networks, its applications, challenges and types. Section III gives an overview of video coding techniques such as SVC or MDC. Section IV extends the study towards video streaming across P2P streaming architectures that consider streaming the video using a single source. Section V and Section VI studies the usage of SVC and MDC techniques for streaming video across P2P networks. Moreover, the usage of seed servers, helper nodes and cloud servers is discussed that helps to improve the quality. Section VII provides a comparison of the all the proposed models discussed in the previous sections. Finally, Section VIII concludes the chapter with a discussion about the limitations in the state of the arts and some possible solution.

P2P NETWORKS

A P2P network is a kind of virtual network that is built over the physical network. It is made up of various heterogeneous peers connected with each other that have different upload and download capacities, storages and processing powers as shown in Figure 1. These networks have the following features as discussed below;

- As compared to Client Server (Hareesh K & Manjaiah DH, 2011) and CDN (Zhijie S, Jun L, Roger Z & Athanasios VV, 2011), P2P networks don't rely on any centralised entity. The maintenance and monitoring across the network is distributed among peers.
- P2P technology is considered to be an appropriate approach to be used for resource sharing services such as Bittorrent (2015, August) Retrieved from http://www.bittorrent.com, SopCast (2015, August) Retrieved from http://www.sopcast.com, Napster (2015, August) Retrieved from http://www.napster.com.
- Nodes in the P2P network organize themselves using a discovery process. Hence, no particular indexing is required.
- Each peer in the network behaves as a client or a server. So, at any time, peers can not only download data but it can upload the downloaded data for other users which helps to reduce the overall load at the server.
- P2P networks are considered to be robust as each peer shares its resources among other peers; hence a single point of failure has little impact on the overall system performance.

Figure 1. P2P network architecture

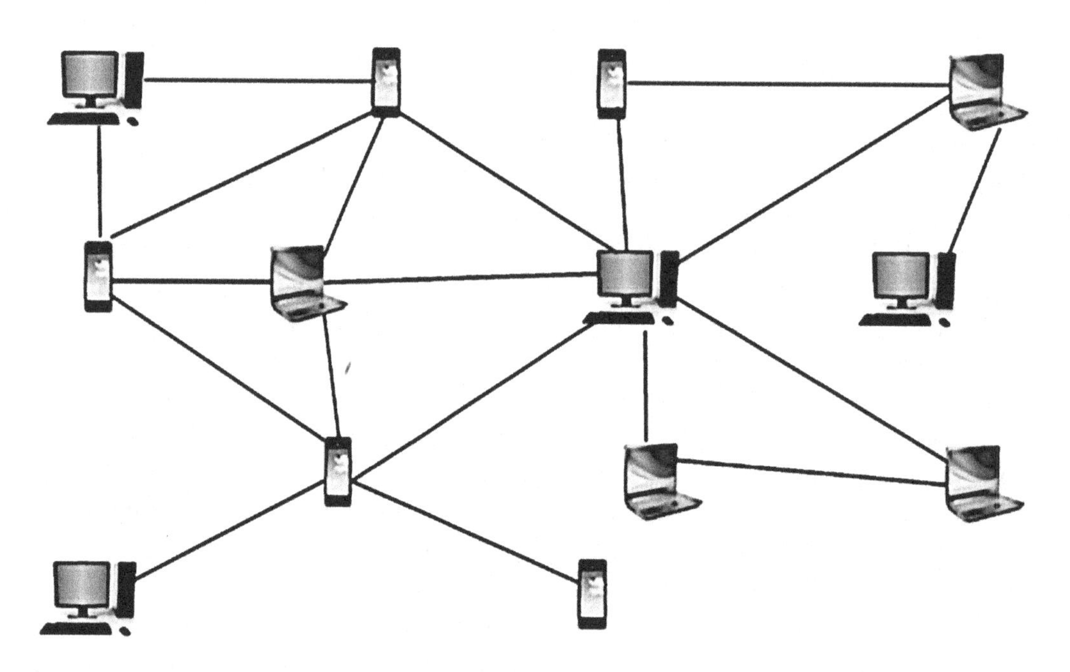

Applications of P2P Networks

P2P networks are used in a number of applications due to their advantageous features. The detailed description of different applications of P2P networks are discussed below;

- **Content Sharing:** Each peer forwards the request for the requested content which travels through different peers to reach the source which then forwards the content. Examples of such networks include; Freenet (August, 2015) Resources from http://freenet.sourceforge.net, Free Haven (R. Dingledine, M. Freedman & D. Molnar, 2001), Gnutella (G. Kan, 2001) or Publis (M. Waldman, L. Cranor & A. Rubin, 2001).
- **Extensive Computation:** A P2P network reduces the need of super computers for distributing the computation load to solve complex problems. This reduces the system cost.
- **Entertainment: A** P2P network is considered to be a promising application that provides an interactive gaming over the internet.
- **Instant Messages:** People use instant messaging to communicate and share the information over the internet. It is a type of application that uses P2P technology to identify the route and provides information of the peer's availability such as Skype (S. Guha, N. Daswani & R. Jainet, 2006).
- **Global Work Environment:** P2P networks allow users to work and cooperate with each other while located at different geographical locations. Magi (August, 2015) Resources from http://www.endeavors.com are one of the examples that provide this collaboration.
- **Collaborative Caching:** P2P networks help the enterprise to share the most common content among users using their local caches (Wang, X, 2002). This reduces the overall system cost.
- **Data Sharing:** P2P manages to share the local databases available at each peer to be shared with the centralised servers. This helps to reduce the cost of sharing the information.

Challenges in P2P Networks

There are certain problems that still exist in P2P networks as discussed below:

- **Heterogeneous Peers:** Peers with diverse characteristics such as variable upload/download capacities, transmission powers or energies join the network to form an overlay architecture. Therefore, an incentive or credit based techniques could be required to handle the heterogeneous nature of nodes.
- **Peers Availability:** In P2P, peers join or leave the network randomly which makes the network unpredictable. Due to this, data or information may not be available for all times and the request for such data is not completed. In order to overcome such an issue a replication strategy can be used that can duplicate the data available at different peers.
- **Network Performance:** The performance of the network largely depends on the peers' connectivity and the network topology at the time a request is made. Because, if a same request is made at different intervals of time, it may have a different impact over the network performance. Hence, content replication and caching techniques can be used to improve the overall network performance.
- **Reliability**: In order to improve the network performance and handle heterogeneous peers, replication strategies are used. However, it becomes hard to maintain the reliability of the content as it

gets outdated after a certain time. Hence, different approaches are required that may validate the copies because the data can be modified by anyone.

- **Resource Discovery:** The most important requirement of a P2P network is to discover the resources (i.e. videos). There are flooding based approaches available that broadcast the request of the requesting peer until the sources are identified such as in Gnutella (G. Kan, 2001). However, they are not considered to be an appropriate approach as huge traffic is introduced in the network.
- **Security Threats**: P2P networks encounter several security threats; by allowing other nodes to access the content of a node, the node is more susceptible to attack where it acts only as a client.
- **Incentives and Fairness:** It is important to provide incentives to the peers in the network that contribute more to the community.

Types of P2P Networks

In order to provide reliable data delivery, the P2P networks are classified into three different network types; centralised P2P networks, decentralised P2P structured networks and decentralised P2P unstructured networks as discussed in (S. German, W. Markus & U. Herwig, 2006).

Centralised P2P Networks

In a centralised P2P network, a centralised peer or a server is available that maintains the information about the content available across nodes in the network using a global indexing approach. Whenever a peer joins the network, it notifies the central peer about the content it has to share. Figure 2 shows an example of a centralised network in which a peer requests a query for a video A to the server. The Server checks which peer has this video; it notifies the requesting peer by sending a response message that this particular peer has video A. The peer can then directly download the video through that peer. These networks are easy to build and consume less bandwidth while discovering the content. NAPSTER (August, 2015) Retrieved from http://www.napster.com is one such network type that shares files among peers.

Figure 2. Centralised P2P network

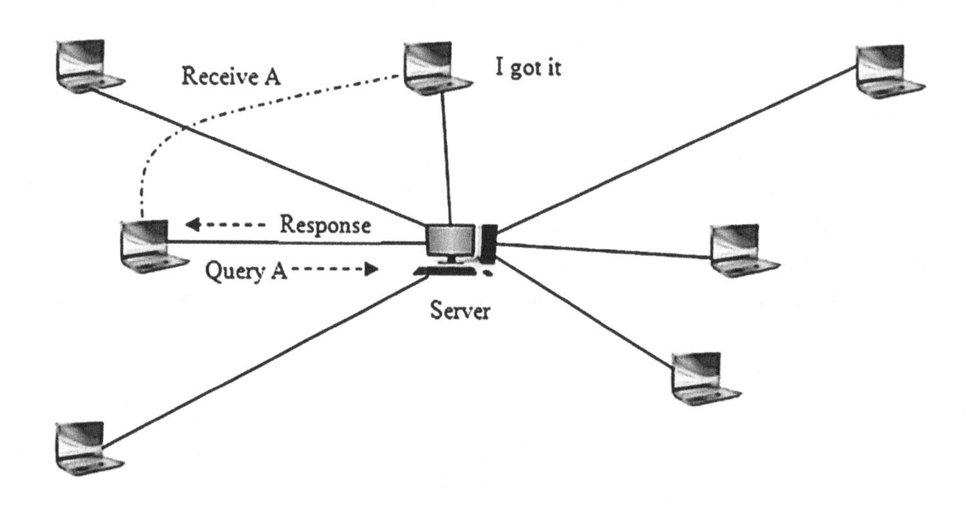

Decentralised Structure P2P Network

In a structured P2P network, the peers are organized into a specific topology and the content location is determined using the deployed P2P protocol. A Distributed Hash Table (DHT) is used to provide a lookup service in the network. DHT stores the key value pairs such that any participating peer is able to retrieve the value associated to a particular key. These keys are then mapped over different peers in the network in order to provide an efficient way of content discovery as shown in Figure 3. Examples of decentralised P2P networks include CAN (S. Ratnasamy, P. Francis, M. Handley, R. Karp & S. Shenker, 2001), Tapestry (Zhao. B, Huang, L, Stribling, J, Rhea. S.C, Joseph, A.D & Kubiatowicz. JD, 2004) and Chord (Stoica. I, Morris. R, Liben. ND, Karger. D.R, Kaashoek. M.F, Dabek. F & Balakrishnan. H, 2004).

Decentralised Unstructured P2P Overlays

In a decentralised unstructured overlay, whenever a peer joins the network and makes a request, it doesn't have any information about the network topology as shown in Figure 4. Hence, the flooding based approach is used to discover the content. However, this approach is not an appropriate solution specially for identifying rare content as this burdens the network with an additional load of requests such as Gnutella (Klingberg. T & Manfredi. R, 2002).

VIDEO CODING TECHNIQUES

One critical application that is ideal for P2P networks is video streaming. The problem becomes more difficult when the peers are heterogeneous and hence have different capabilities. Hence, P2P it becomes hard for such peers to meet the stringent bandwidth requirements of a particular video request. Hence, one way to solve such issues is to use an appropriate codec that can stream the video at different rates. Hence, when the channel condition changes, video can be sent at higher or lower rates based on the link bandwidths. There are a number of different audio/video coding standards that are available which are

Figure 3. Decentralised structured P2P network

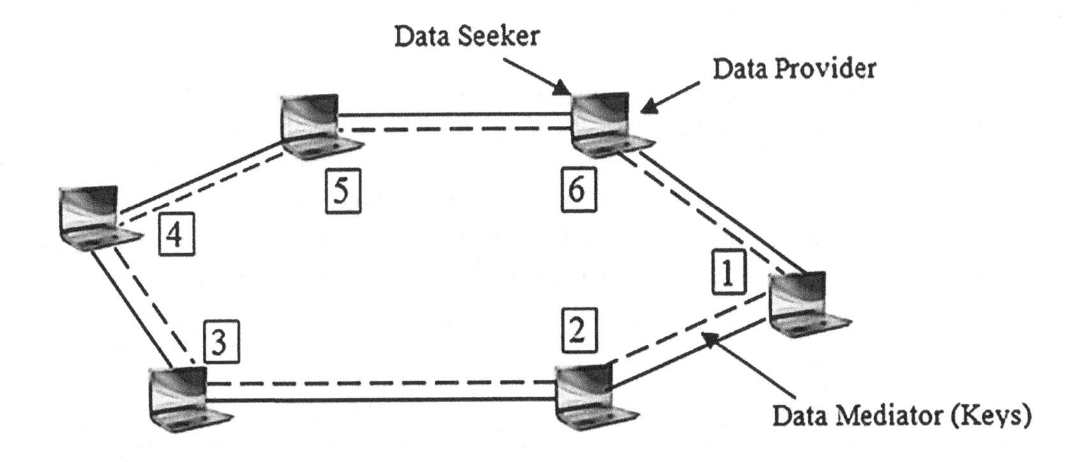

Figure 4. Decentralised unstructured P2P network

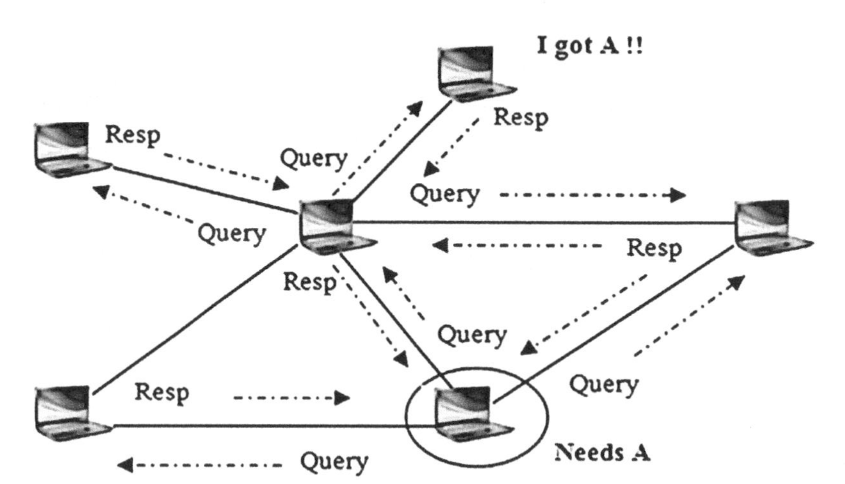

used for efficient video delivery over IP. ITU-T and ISO/IEC are the two most known organizations that provide different coding standards. ITU-T coding standards are denoted by H.26X (such as H.263 or H.264). On the other end ISO/IEC video coding standards are denoted by MPEG-x (e.g. MPEG-1, MPEG-2 etc.).

The ITU-T coding standards are particularly designed for real time application e.g. video conferencing whereas, the ISO/IEC standards are particularly designed to handle storage videos, video broadcast and streaming applications (Kofler. I, Seidl. J, Timmerer. C, Hellwagner. H, Djama. I & Ahmed. T, 2008) (T. Ahmed, A. Nafaa & A. Mehaoua, 2003). In most of the cases, both organizations have worked independently to provide different standards of videos but in some cases they produced joint video coding standards from which the best known is H.264 (also known as MPEG-4 AVC) (T. Ahmed, G. Buridant & A. Mehaoua, 2001) (T. Ahmed, A. Mehaoua & G. Buridant, 2001) was developed with the further extension to produce H.264/SVC.

In the next section, the most widely known video coding techniques for streaming multimedia over the networks are discussed. These are categorized as: Scalable video coding (SVC) (Zhengye L, Yanming S, Keith WR, Shivendra SP & Yao W, 2009) and multiple descriptive coding (MDC) (Vivek KG, 2001).

Scalable Video Coding (SVC)

SVC is a well-known type of layer coding technique, which is an extension of the H.264 standard that is known to be the most promising approach for streaming media over heterogeneous nodes (Leszek. C, 2006). In SVC, each video stream is coded into multiple layers comprising of a base layer and several enhancement layers. The base layer carries the basic information of the video whereas the enhancement layers can further improve the quality of the base layer. Hence, it is important to receive the base layer, if a base layer gets corrupted or not received then it is useless to transmit an enhancement layer.

Types of SVC Coding

SVC is performed on a video stream to provide sub streams based on three different categories: Temporal, Spatial and Quality scalabilities (M. Wien, H. Schwarz & T. Oelbaum, 2007). These scalabilities are discussed in detail as follows;

Temporal Scalability

Temporal scalability is used when the video is partitioned into a temporal base layer and one or more temporal enhancement layers. The term temporal represents the ability to represent the video with different frame rates. It is well illustrated in Figure 5. Each encoded video is composed of three kinds of frames; I (intra), P (predictive) and B (bi-predictive). In the past, video coding standards such as MPEG-2 or H.263, the temporal scalability is performed by encoding the video into different layers based on different frame rates. For example, if a video is comprised of I, B and P frames, then I frames can represent the base layer whereas P and B frames can be decoded as enhancement layers. However, in H.264/SVC the temporal scalability is performed on the structure of group of pictures (GOPs). Hence, each frame is divided into different layers with I, P and B frames in each layer. It is important to remember that it is not necessary that the base layer is only encoded using I frames; however, the first frame should be coded as an I frame.

Spatial Scalability

The spatial scalability is performed to encode the video into different resolutions i.e. each higher layer improves the resolution of the lower layer in order to provide better quality of the video as shown in Figure 6. In order to improve the image quality received, a H.264 encoder uses an ILP (Inter layer prediction) module. The main idea for using this module is to increase the prediction of reused data from the previous layers. ILP is used to provide three different types of motion predictions:

Figure 5. Temporal scalability

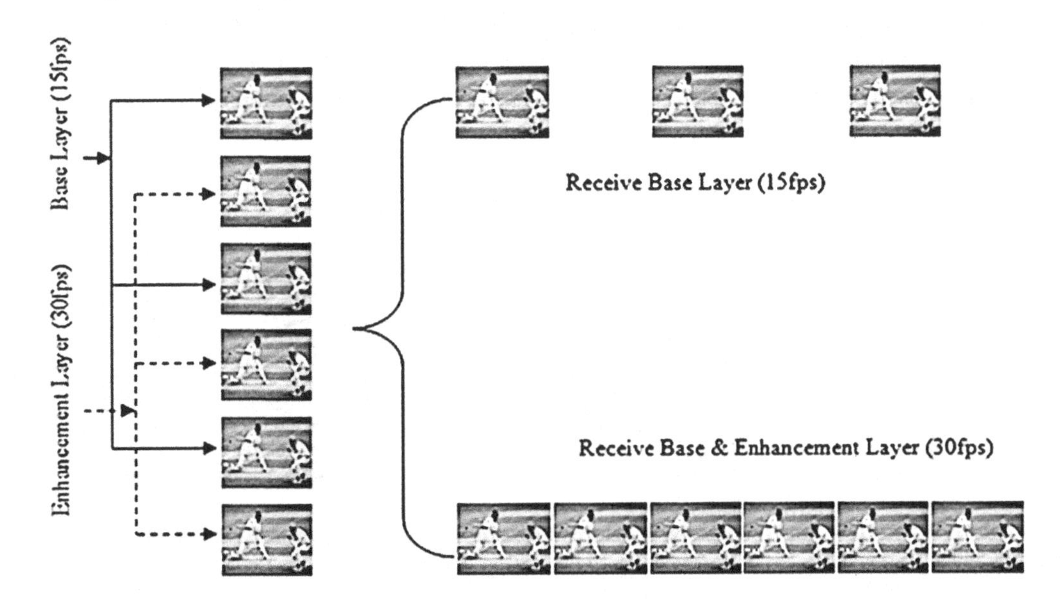

- **Inter Layer Motion Prediction:** In this type, the motion vectors used in lower layers are used in higher layer.
- **Inter Layer Intra Texture Prediction:** SVC can support the texture prediction for the same reference layer in internal blocks. This block prediction can be used by higher layers for prediction of other blocks. So, the advantage of this module is that it improves the resolution of the lower layers by calculating the difference amongst them.
- **Inter Layer Residual Prediction:** It has been investigated that if two consecutive layers have the same motion information then the inter layer register highly correlates with each other. Therefore, in SVC this inter layer residual prediction is used just after the motion compensation in order to check for redundancies.

SNR Scalability

SNR scalability is used to provide a video with different quality levels. Each layer is assigned a different quantization parameter. Three different types of SNR scalabilities are available: Coarse Grain Scalability, Medium Grain Scalability and Fine Grain Scalability.

- **Coarse Grain Scalability:** In this, each layer has a different prediction procedure whereas the references have the same quality level as for the SNR scalability in MPEG 2 standard. It is also considered to be a special case of SNR scalability in which the consecutive layers have the same resolution. This scalar granularity mode is explained in Figure 7(a).
- **Medium Grain Scalability:** This type of scalability uses the base layer and enhancement layers as a reference for the prediction module to improve the efficiency. But, the disadvantage of this approach is that in the case where only the base layer is received, it introduces drift effect which affects the synchronization between the encoder and the decoder. However, this issue is resolved with the help of using the periodic key pictures, which helps the prediction module to quickly resynchronize. The concept of MGS is explained in Figure 7(b).
- **Fine Grain Scalability:** This type of scalability is the most commonly used. In FGS, the output bit rate of the video is continuously adapted by comparing it to the bandwidth available in real time. FGS uses advanced bit plane techniques in which different layers' transport distinct bits for each set of information. This scheme provides data truncation to support improvement in the values of transform coefficients. The concept of FGS is explained in Figure 7(c).

Figure 6. Spatial scalability

Figure 7. SNR scalability over two layers

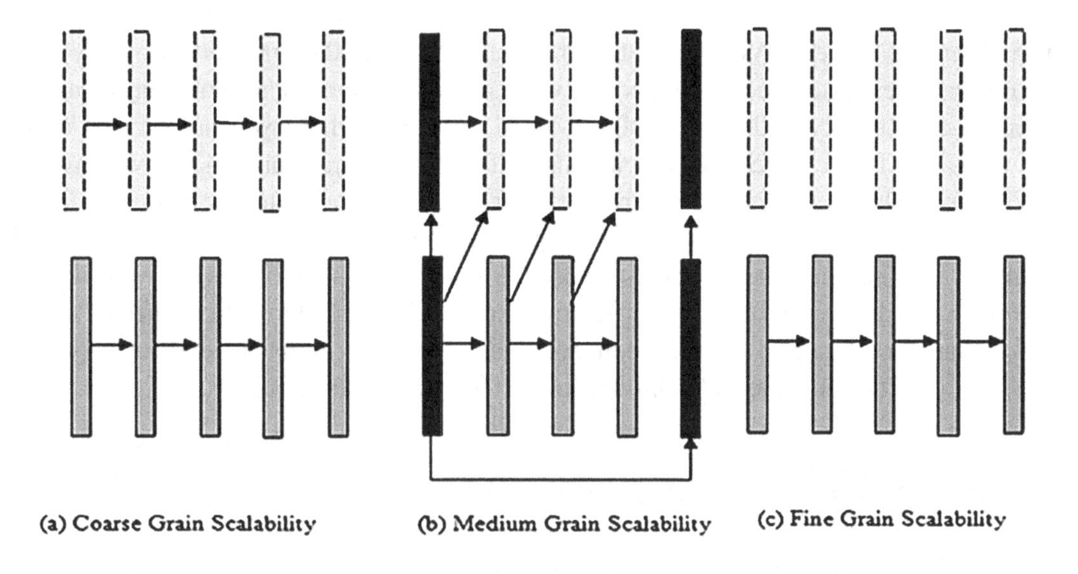

Multiple Descriptive Coding (MDC)

Multiple descriptive coding (MDC) is a type of layered coding that generates multiple independent bit streams or descriptors of a single video stream. The advantage of MDC is that it provides high resilience to packet loss. The quality of the video depends on the number of descriptors received. The basic concept of MDC is well illustrated using Figure 8 in which a media server generates a video with 4 different descriptors and transmits it over the network. At the receiver end, the video is decoded by the receivers with different capabilities such as smart phones or tablets usually have low bandwidths provided by ISP's and processing power so they download only a single descriptor of the video. On the other hand, the devices with larger screens and higher bandwidths download more descriptors. Hence, they have better download rates and the video can be decoded with a higher number of descriptors.

Figure 8. Multiple descriptive coding

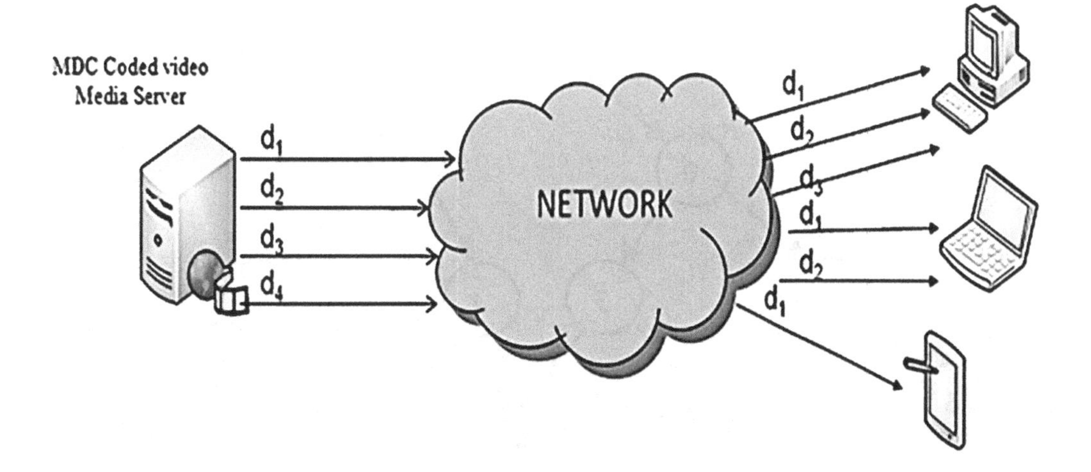

Comparison Between SVC and MDC

The main difference between SVC and MDC techniques is that, in MDC, the quality of video depends on the number of descriptors receive in parallel whereas in SVC, the enhancement layers are used to further improve the quality of the video. Moreover, in SVC, the higher layers are strongly dependent on the lower layers. Hence, it becomes crucial to provide a reliable transmission of lower layers of the video. Whereas MDC doesn't require any priorities or retransmissions. Further it is considered to be more robust as it hardly happens that all the descriptor of the video gets corrupted. Therefore, MDC is widely used in the case where the network is exposed to more churn as it provides error resilience which helps the video to still survive at better quality. However, SVC is still preferred in the networks in which a network is static or centralised control and there is a less chance of churn to enter the network because of its high coding efficiency that still helps to provide the video at better quality.

VIDEO STREAMING OVER P2P NETWORKS

During the last decade, there have been a lot of studies to improve video streaming across P2P networks. In P2P networks, peers join the network and arrange themselves in a form of an overlay. The overlay architectures are categorized as tree based, multi-tree based, mesh based or hybrid overlays (Y. Liu, Y. Guo & C. Liang, 2008). In this section provides the current state of the art in video streaming over P2P networks using non layered video coding techniques. Sections 5 and 6 consider the state of the art for SVC and MDC streaming over a P2P network.

In a tree based overlay, peers organize themselves in a tree shape architecture and form a parent child relationship with each other (H.V. Jagadish, B. C. Ooi & Q. H. Vu, 2005; V. Venkataraman, K. Yoshida & P. Francis 2006) as shown in Figure 9. The media server is located at the root of the tree whereas the peers are located at different locations across the tree. Media content is rooted from the tree root towards the leaf nodes. However, the disadvantage of this approach is that this architecture lacks robustness under peer churn (nodes joining and leaving the overlay) and the leaf nodes do not contribute their upload capacities to the network.

Figure 9. Tree based overlay

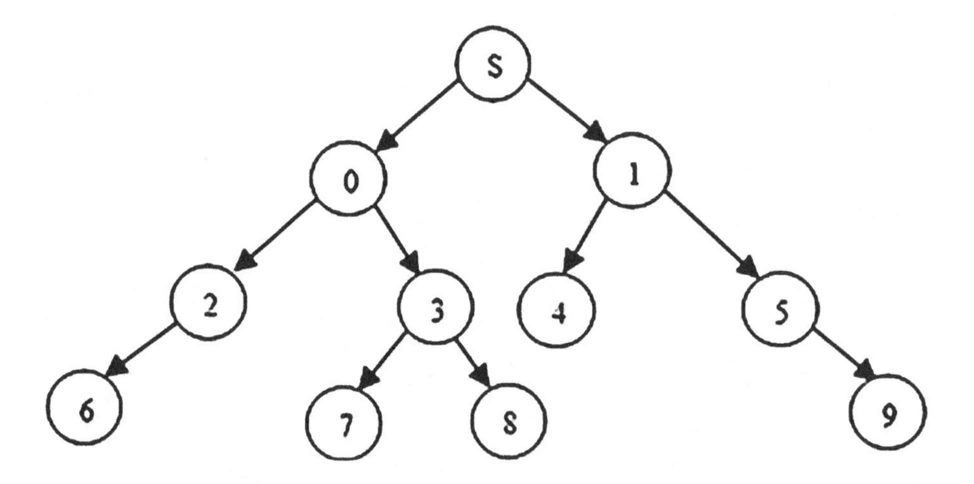

Figure 10. Multi-tree based overlay

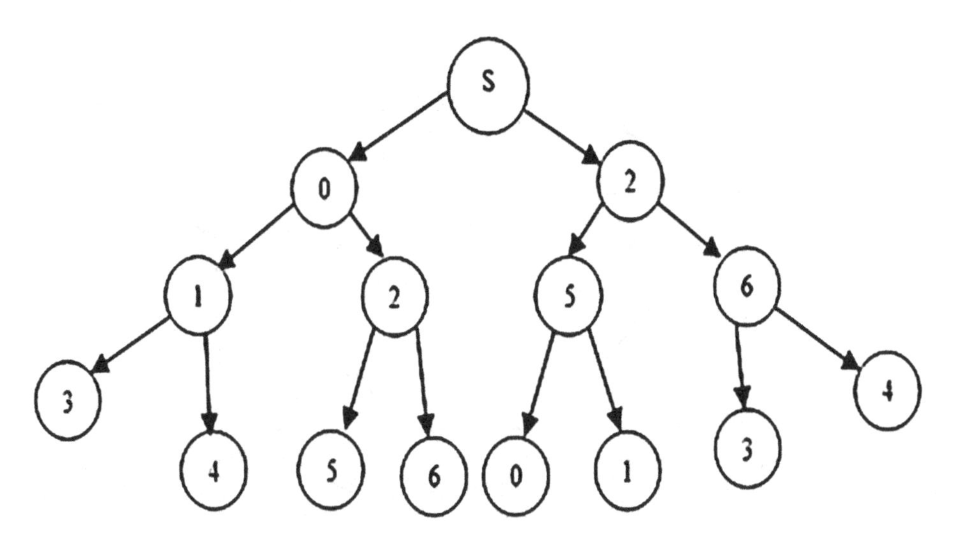

In order to overcome these issues, multi-tree based overlay architectures are proposed (J. Chakareski, S. Han & B. Girod, 2003; Y. Sung, M. Bishop & S. Rao, 2006; N. Magharei, R. Rejaie & Y. Guo, 2007) that divide the video into various sub-streams where each sub-stream is forwarded over one of the sub-trees. Figure 10 shows that the source S divides the video into three sub streams and sends it over its child peers 0 and 2. These peers behave as a parent in one tree to forward the sub stream whereas they act as a child in another tree to download the sub streams. The advantage of this approach is that the upload capacity of the leaf nodes can also be utilized by other peers in the network. However, the churn effect is still a problem.

To overcome this issue, a mesh based overlay architecture is proposed M. Zhang, Y. Xiong, Q. Zhang, L. Sun & S. Yang 2009; Zhang, X, J. Liu, B. Li & Y. S. P. Yum, 2005) that efficiently utilize the peer's bandwidth and improves the overall system performance. In a mesh based approach, whenever a new peer wants to join the network, it makes a request to receive the information about the existing peers in the network. Then based on the information received, the peers form a neighbouring relation with a certain group of peers which shares the common interest. The advantage of mesh based scheme is that it is highly resilient to churn and provides efficient bandwidth utilization as compared to a tree based approach. The concept of mesh based architecture is well illustrated using Figure 11 in which peers receive the chunks using different paths. Therefore, it is being widely used in most of the commercial architectures UUSEE (2015, August) Retrieved from http://www.uusee.com; PPLIVE (2015, August) http://www.pplive.com; PPSTREAM (2015, August) http://www.ppstream.com. However, the disadvantage of mesh based systems is that whenever a peer joins the network it has to forward a number of messages to find its neighbours which produces high overhead.

In order to overcome some existing issues in tree and mesh based overlays, a hybrid overlay architecture is proposed. The hybrid overlay combines the advantages of both mesh and tree based overlays. Figure 12 gives an overview of the hybrid overlay architecture in which a tree is formed over a mesh. There are a number of hybrid overlay architectures available (Jahromi, B. Akbari & A. Movaghar 2010), a hybrid mesh tree overlay structure is implemented that employs the concept of layered streaming to overcome latency and provide resilience in the network. In this overlay design, each peer forms a mesh

Figure 11. Mesh based overlay

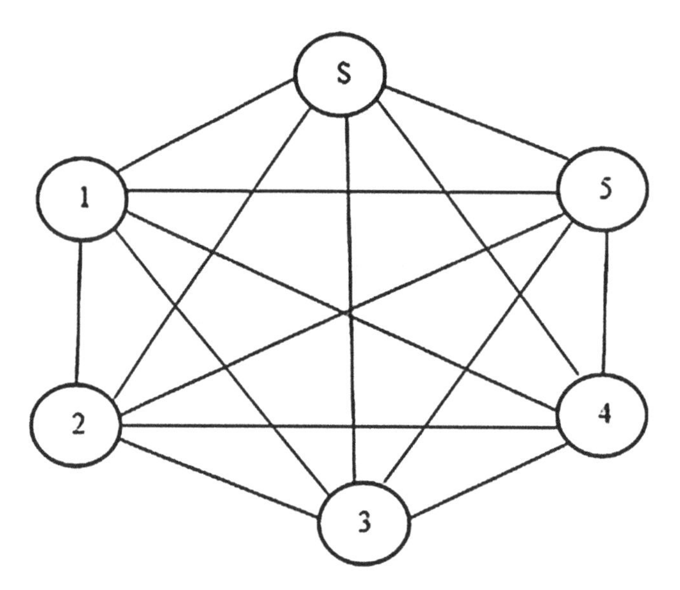

based overlay at the start but after some time a tree overlay is formed by the stable peers. Whenever a node is willing to join a mesh based overlay network, a message is sent to the tracker that searches for the peers with available bandwidth and sends a response message in return to a requested peer who then establishes a connection with the nodes and starts receiving chunks. After a particular threshold time, newly joined peers can be considered as stable peers and they can join a tree based overlay by sending a message to the tracker which finds an appropriate parent for them. The authors proposed two phases of data delivery in this protocol. In the first phase, the base layer and enhancement layer are both transferred using a mesh based scheme and each peer requests chunks of the video from nearby neighbours where as in second phase after peer becomes stable, the base layer is transferred using a tree to reduce delay as it contains the most crucial video content. Whereas, the enhancement layers are requested using mesh based overlay.

In order to overcome the existing issues among different overlay topologies, a number of studies have appeared in the literature. For example, in order to provide tree resilience towards churn in the network, the concept of backup parents is considered (Allani M, Garbinato B & Pietzuch. P 2012; Castro M, Druschel P, Kermarrec A M, Nandi A, Rowstron A & Singh A 2003; Fesci-Sayit M, E. Turhan Tunali & A. Murat Tekalp 2009). The backup parents help to manage better video quality during the node's failure. However, the disadvantage of such an approach is that a single parent is used for the backup which is insufficient to handle high churn. Similarly, (Fesci-Sayit M, E. Turhan Tunali & A. Murat Tekalp 2012), the authors propose the concept of using backup parents' pool in order to provide more resilience.

The authors (Kwon OC & Song H 2013) propose a tree based P2P video streaming method for live video. The proposed method constructs the multicast trees on top of clustered peers. Whenever a peer wants to join the network, it sends a request to the bootstrap server that carries the information of the topology. However, the disadvantage of such an approach is that it can encounter a single point of failure. Similarly, (Kuo JL, Shih CH, Ho CY & Chen YC 2014), the authors consider joining of nodes to the multicast trees based on the round trip time (RTT). The RTT is calculated across root and the joined node. The nodes that have a similar value of RTT are usually placed closer to each other.

Figure 12. Hybrid overlays

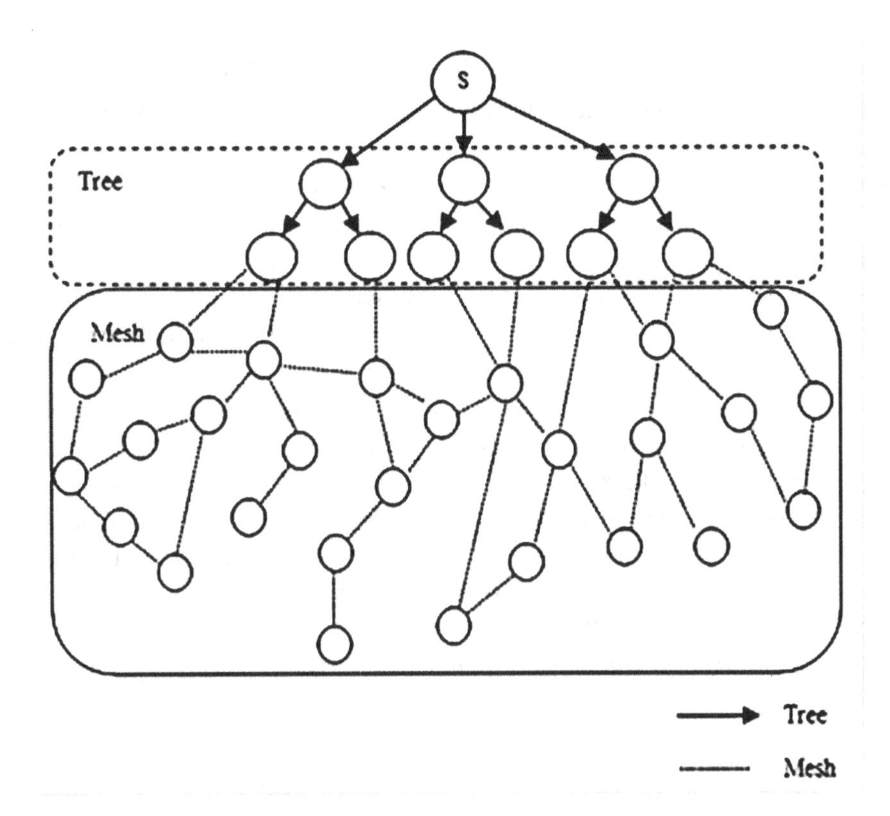

Similarly, the authors (Vinay P, Kumar K, Tamilmani K, Sambamurthy V & and Mohr AE 2005) introduce a multicast streaming system, Chainsaw. This system eliminates the trees concept. In this system, neighbours identify each peer by sending a NOTIFY message related to the new available packets which a peer can request from its neighbours. This overcomes the duplication of data packets which helps to reduce the upload bandwidth consumed in uploading the packets that are available at the neighbouring peers. The experimental results show that the proposed method provides resilience to packet loss and maintains a good QOS with less start-up delay.

Moreover, the authors (Zhang, X, J. Liu, B. Li & Y. S. P. Yum 2005) propose an optimization scheduling method to improve the real time streaming experience based on the concept of Cool Streaming. The authors have used a bandwidth estimation algorithm that monitors the behaviour of dynamic network and estimates the capability of the peer to transfer data based on each data distribution. Furthermore, a zonal request buffer scheme has been introduced that categorizes the buffer into three different zones; urgent zone, common zone and ease zone in order to provide efficient video streaming. The authors found that using a tree based method for sending the control message, reduces the overall overhead by 0.7% as compared to Cool Streaming. Similarly using an optimized scheduling method assures good quality of service by increasing the buffer by 21%.

Furthermore, there are number of studies (Banerjee S, Bhattacharjee B & Kommareddy C 2002; Birrer S & Bustamante F 2005; Tran DA, K. A. Hua & T. T. Do 2004) that were carried out towards providing decentralised methods for streaming the video in tree based overlays. In such methods, peers are organized in the form of clusters and each peer receives the video through the leaders in the cluster known as cluster leaders. The link between a cluster leader and the peers comprise of different multicast trees. However, the disadvantage of such approaches is that the capacity across the leaders is not taken into consideration. However, the authors (Fesci-Sayit M, E. Turhan Tunali & A. Murat Tekalp 2009) show that if the capacity of the cluster leaders is taken into consideration, the QOE across the users is largely improved.

As discussed, an incentive based approach is used to provide better quality of service across a P2P network (X. Shen, H. Yu, J. Buford & M. Akon 2010). These approaches are widely used to overcome free riders (nodes that do not contribute), churn (nodes that may leave and join) and attacks (malicious nodes) in the network. There are a number of incentive based mechanisms that are available such as the authors introduce a reciprocal mechanism that records the history of each node encounters and based on that history, nodes are rewarded (D. Banerjee, S. Saha, S. Sen & P. Dasgupt 2005).

Similarly, the authors have designed a reputation based method that considers a global rank among the peers to provide priorities among peers to receive the requested media segments (A. Satsiou & L. Tassiulas 2010). Moreover, the authors (Y. Chu, J. Chuang & H. Zhang 2004) study a taxation based approach in which peers are given motivation to contribute more towards the system in order to improve the perceived quality received by peers. Furthermore, the authors explain a pricing based mechanism to help manage maximizing the social benefits or incentives for optimal resource allocation across the network (J. Park & M van der. Schaar 2010).

The author (Cohen B 2003) has studied a well-known tit for tat algorithm implemented over Bit Torrent. The proposed method helps to overcome the free riders (Hoong P. & Matsuo H. 2008; Montazeri A, Akbari B & Ghanbari M 2012) across the network. The free riders are considered to be those peers in the network that do not contribute any resources towards the network. According to the proposed method, the node that contributes more towards the network by sharing its resources, receives better quality of the video upon request.

In order to efficiently utilize the bandwidth across the peers in a tree based architecture, authors in (Zhang, X, J. Liu, B. Li & Y. S. P. Yum 2005; Yang S & Wang X 2010) use the concept of incentive based mechanism in which the nodes with higher upload capacities or resources are usually placed closer to root nodes or as a root node. The disadvantage of this mechanism is that the effect of churn across the network is not considered.

(Zhang, X, J. Liu, B. Li & Y. S. P. Yum 2005) the authors propose a data driven overlay network (DONET) for streaming the live media content. In this architecture, each node exchanges information of available data with its nearby neighbours and fetches the required data from them or supplies the available data to other neighbouring peers. Authors focus on three features of DONET design such as: Easy to design as it should not have a complex architecture, efficient in data delivery and robust as available data information is switched swiftly among multiple suppliers.

Furthermore, the authors discuss how a neighbouring relation is formed, how information for data delivery is exchanged and how video data is retrieved and distributed among peers. To provide seamless streaming of media content with low overhead, they proposed a scalable membership management protocol based on gossiping protocol in which a node keeps on sending newly generated messages to random nodes; these nodes then spread the messages in a similar way to other nodes until all nodes retrieve them.

The data delivery concept is partially motivated from gossip but it was not fully utilized as using gossip for streaming can cause redundancy issues due to random pushes. Hence, the authors also design a partner selection algorithm with a low overhead scheduling algorithm that can pull data from the nodes. They also encounter the peers having heterogeneous nature. Each node has a unique identifier e.g. IP address and the node maintains a small list comprises of active nodes in DONET. In a node joining algorithm, each newly joined node contacts the server node, which then selects a deputy node from its membership cache and forwards the new node towards it to make partners. Moreover, each video stream is divided into small segments of uniform length and a buffer map represents the availability of the segments in a buffer. Each node exchanges buffer map with its partners to fetch the video segments.

In order to fetch the available segments from the partners, a scheduling algorithm is proposed that calculates the supplier for each segment. The algorithm monitors the supplier of each segment starting from one potential supplier to many. The scheduler then selects one of the suppliers that have high bandwidth and enough available time. The authors also investigated that a departing node leaves a message having the same format of membership message to inform the partner node about leaving the network. The partner then detects the node failure. Each node that receives the message flushes the departing node from its mCache. Finally, a new partnership relation is formed. The performance results show that DONET has comparatively less overhead which does not grow with the size of the overlay. Furthermore, it was found that under a high dynamic environment, it has high playback continuity and has less end to end delay.

The bandwidth fluctuations result in degradation of playback continuity in a way that video freezes or a portion of video starts skipping. This fluctuation is very severe in delivering the content over a live streaming network. In order to handle content bottlenecks, one way is to degrade the quality of the video or skip some parts of the video using different coding techniques as discussed in Section VI. This means that to transfer only a certain amount of information that helps to recover the lowest quality of the video. However, another solution to such an approach is to use an efficient scheduling technique.

Similarly, the authors present Grid Media to study the performance of a live video streaming system based on users' population, quality, connection heterogeneity and online duration of certain peers in a network (Y. Tang, J. G. Luo, Q. Zhang, M. Zhang & S. Q. Yang 2007). The authors defined a rendezvous point (RP) consisting of content information, IP address and port number of the streaming server. RP helps peers to join overlay, maintain a random part of the active participants and acts as a network administrator. When a new node joins, RP returns peer with the list of peers' already in overlay with their IP addresses and port numbers. Afterwards, a peer calculates the round trip time of each peer in the list and selects peers as its neighbours with minimum trip time as half of its neighbours. However, the other half neighbours are randomly chosen to overcome the overlay division. Each peer in the overlay is provided with a membership list and a neighbours list that keep on updating with time. To provide active nodes within an overlay, overall life time is used that keeps track of the information of message received from a certain peer. If the value of life time increases over the threshold value, then that peer is discarded from the overlay. The updated membership information is exchanged between neighbours for updating their membership lists. The membership list actually provides information that each peer is sharing the same amount of burden, it shares overlay information and video segments with in neighbours. Furthermore, the authors propose a streaming scheduler that takes the responsibility of distributing media segments within the neighbours using a push pull mechanism. Each peer periodically shares buffer maps of video segments and gets the required segments from other peers. In the start, a data driven pull based approach is deployed to request the video segments from the neighbours. However, the disadvantage of this approach is that it causes huge latency. Whereas in a push based approach, the segments are received without making any request, this helps to decrease the transmission delay. However, the disadvantage of this approach is that it can encounter link failure while receiving the segments which may introduce delay.

Better quality content is received if the available streaming rate is high. Furthermore, the higher stream rate also helps to absorb the bandwidth variations caused by churn and the congestion across the network. The authors (Y. Guo, C. Liang & Y. Liu 2008) presented an adaptive queue based scheduling algorithm that can achieve an optimal upload bandwidth rate of peers. In the proposed system, peer side scheduling is performed in a way that each peer maintains streaming content from source and other peers in a playback buffer in the order it needs to be played. Each peer also maintains a forwarding queue which stores the content that needs to be forwarded to other peers. To fully utilize the peer upload bandwidth, it is required that the forwarding queue should always fetch the data. Whenever the forwarding queue becomes empty, a pull signal is sent from the server to request more content. On the server side, it maintains the content and signalling queue. Content queue contains two different dispatchers; F (forward) marked and NF (non-forward) marked dispatcher. If there is a pull signal in a signalling queue, it forwards a chunk of content from content buffer to peer from where a pull signal is originated using F marked dispatcher. If the signalling queue is empty, server forwards a chunk of content from buffer to peers by marking them as NF (Non-forwarding). The authors investigated the algorithm while considering parameters including peer churns, peer bandwidth variations and network congestion to provide an optimal system for real time content streaming.

Similarly, a request peer selection algorithm to maximize the uplink bandwidth utilization across the peers is proposed (N. Liu, Z. Wen, K. L. Yeung & Z. Lei 2012). Each peer in the network monitors the network service response time (SRT) between a neighbor and itself. SRT is measured with respect to the time a data packet request is sent until the requested data packet arrives. Whenever a peer makes a request for a packet, the neighbor with the smallest value of SRT and with fewest data packets will be favored against the potential providers. This happens because smaller SRT involves excessive available capacity and far fewer data packets means less packet requests are received. The authors showed that the proposed peer selection algorithm balances the load across the network as the data packets uploaded by each peer is normalized and the number of repeat requests generated by peers (due to failure) are reduced. They also showed that this algorithm reduces the overall load across the server and improves the quality of service and reduce the startup delay.

In order to measure the streaming capacity of a P2P network, the authors deployed a primal dual algorithm in an undirected graph (Yi C, Baochun L. & Klara N. 2004). However, the disadvantage of such an approach is that the degree bounds (number of peers that can be served at a single time) at each node is not introduced. Later (Shay H & Danny D 2010; Yifeng H & Ling G 2009)), the authors introduced the concept of helper nodes that act as a source of transmission and leads to the deployment of greedy and proximal algorithms for managing the capacity of the P2P networks. However, the proposed algorithms have high overhead complexity which causes huge playback delay, which is not manageable for real-time video distribution.

Similarly, the authors propose an algorithm that determines the maximum streaming capacity that can be achieved in a P2P streaming system (Sudipta S, Shao L, Minghua C, Mung C, Jin L & Philip AC 2011). However, the proposed approach considers the nodes to have an equal bit rate. The authors have used the taxonomy of sixteen different formulations depending on various network scenarios. In each formulation, the authors computed an optimal set of multicast trees to find an optimal P2P streaming capacity. The authors produced a combinatorial optimization approach to solve the streaming capacity problem. The combinatorial problem is converted into a linear optimization problem with exponential variables, which is then solved using a primal dual approach. The authors developed algorithms for single and multiple streaming sessions and found that with the help of smallest price tree construction, each receiving peer is able to achieve a maximum streaming rate.

In a further study, the authors study the advantages of using super peers in the network (B. Yang, B. & Garcia-Molina, H. 2003). According to the authors, a super peer is a peer that works as a server to a certain set of clients. Super peers help to provide the incentives of the centralised approach, load balancing and handles the attacks which affects the distributed network. Furthermore, authors have studied the potential drawback of using seed servers such as cost and complexity of a network.

The authors propose an algorithm that helps to significantly reduce the load across the original server. (Y. Labib, A. E. Sherbini & A. Sabri 2009) propose an enhanced media streaming system that organize the network entities in a structured P2P network to provide big media storage and to dynamically participate in delivering media. Furthermore, multiple sources are used which improves the network resources and reduce the consumption of network bandwidth.

Similarly, the authors (Yi C & Klara N 2003) have introduced the concept of using seed servers in the network. The seed servers are the dedicated servers that handle the requests if the available peers upload capacity is fully utilized. The authors have deployed seed servers in a P2P streaming network for non SVC streams and investigate the optimal utilization of seed servers by evaluating when the servers can be switched on or off.

Correspondingly, the authors (Hong Yi C, Ya Yueh S & Yuan W.L 2012) design a cloud assistive P2P live streaming system that maintains a predefined quality level by renting the helper peers from cloud architecture. The problem is modeled as an optimization problem with an objective to minimize the total cost incurred in renting the cloud resources in order to maintain a desired QOS level. The authors have provided an online heuristic approximate solution that adapts the network dynamics. Authors have used a gossip based aggregation protocol to estimate the upload capacity available in the P2P system and provides provision to get resources from the cloud in order to maintain the QOS at low cost. The simulation results show that their proposed method provides high playback latency with a short playback delay. However, the proposal suffers from the cost of renting the resources from the cloud.

Similarly, the authors have used the concept of dynamically renting the virtual machines from the cloud servers (Seyfabad, M.S & Akbari, B. 2014). The authors have introduced a centralised based method that performs the estimation of the total number of virtual machines required to provide the requested QOS. The experiment results show that the proposed method efficiently distributes the resources across peers and improves the overall system bandwidth.

In another work, the authors studied a mesh based protocol, Fast Mesh, which reduces the source to peer delay while considering the requirement of streaming bandwidth (D. Ren, YT. H. Li & S.-H. G. Chan 2009). The proposed protocol supports the network comprises of super nodes, proxies or content distribution networks. The authors study a minimal delay multipath tree problem and proposed a centralised heuristic that can be used over a small network. Furthermore, the authors propose a distributed algorithm in which peers select parents based on power factor that is obtained through the ratio of throughput and delay. The simulation results show that the proposed method reduces delay and load across the servers.

VIDEO STREAMING OVER P2P NETWORKS USING SVC

As discussed earlier, SVC is a type of layer coding that divides a single video stream into different sub streams or layers based on the resolution, frame rate or fidelity level of the video. The sub streams are

Table 1. Qualitative comparison for single layer P2P streaming techniques

Approach	Cod. Tech.	Objectives	Advantages	Disadvantages	Multi. Sender Sources
(H. Jagadish et.al 2005; V. Venkataraman et.al 2006) tree based architecture	S.L	Media streaming	Imp. scalability	Churn cannot be handled	No
(J. Chakareski et.al 2003; Y. Sung et.al 2006; N. Magharei et.al 2007) multi-tree architecture	S.L	Handle leaf nodes capacities	Upload cap. Of leaf nodes is utilized	Churn effect	No
(Zhang, X. et.al 2005; M. Zhang et.al 2009) mesh based architecture	S.L	Utilize the peer's b/width to imp. Sys. Perform.	Resilient to churn and prov. b/width utilization	High overhead complexity	No
(Jahromi et al. 2010) hybrid based architecture	S.L	Exploit the prop. of both tree & mesh	Improve sys. Performance. Imp. Latency & provide resilience	High overhead complexity	No
(Fesci-Sayit M et.al 2012) concept of backup parents' pool	S.L	In order to provide more resilience	Reduce quality degradation	Single node failure	No
(Kwon OC 2013) multi cast tree over clustered peers	S.L	Construct multi cast tree over clustered peer	Maintain quality	Single node failure	No
(Kuo JL et.al 2014) RTT over multi cast trees	S.L	Build the overlay	Eff. Management of resources	Not appropriate solution	No
(Vinay, P. et.al 2005) rand. Picking strategy & override algo.	S.L	Introduce a multicast streaming	Reduce cons. of upload b/width; provide resilience to packet loss & main. QOS	Upload b/width is not eff. utilized	No
(Zhang, X. et.al 2005) Optimization scheduling method, b/width estim. algo	S.L	Imp. real time streaming exp.	Red. overhead, assures good QOS	Flash crowd effects are not studied	No
(Kwon OC 2013; Banerjee S et.al 2002; Tran DA et.al 2004) cluster based tree overlays	S.L	Improve QOE	Better Quality is rec. across peers	Cluster leaders capacity is not cons.	No
(Fesci-Sayit M 2009) Cluster based tree overlays	S.L	Improve QOE	QOE improved. Cap. Across leaders are cons.	Churn is not studied	No
(X. Shen et. al 2010) incentive based approach	S.L	Provide better QOS across network	Overcome free riders and handle churn	Low capabilities nodes have to cont. more to rec. more	No
(D. Banerjee et.al 2005) reciprocal mechanism	S.L	Monitors history	Improve QOS	No app. sol. History changes freq.	No
(Satsiou, L. et.al 2010) Reputation based approach	S.L	Priorities across peers to rec. data	Nodes with more rep. rec. data first	Less prior. Nodes are not cons.	No
(Y. Chu et.al 2004) study a taxation based approach	S.L	To imp. received quality	Quality across peers is imp.	Flash crowd is not cons.	No
(J. Park et.al 2010) pricing based mechanism	S.L	Optimal resource allocation	Resource alloc. Improved	Churn is not studied	No
(Cohen B 2003) tit for tat algorithm	S.L	Over. free riders	Rec better quality on request	Churn is not studied	No
(Fesci-Sayit M 2009) score based app.	S.L	Share content based on connections.	High score rec. more content.	Res. Are not eff. Utilized.	No
(Zhang, X. et.al 2005; Yang S, et.al 2010) concept of incentive based mechanism	S.L	Higher upload cap. nodes are placed closer to root	Eff. B/width utilization in tree based network	Churn is not studied	No
(Zhang, X. et.al 2005) Membership man. Protocol & partner sel.& scheduling algo.	S.L	Data exchange across nodes	Less over. , high playback cont. & low delay	B/width fluctuation reduce quality	No

continued on following page

Table 1. Continued

Approach	Cod. Tech.	Objectives	Advantages	Disadvantages	Multi. Sender Sources
(Tang et al. 2007) data driven pull approach	S.L	Req. packets from neighbours	Better quality rec. Red. churn & congestion	Latency issues	No
(Guo et al. 2008) Adaptive queue based scheduling algo.	S.L	Achieve optimal upload bandwidth of peers	Provide better streaming rate	Free riders are not studied.	No
(Nianwang, L et.al 2012) a request peer selection algorithm	S.L	Max. uplink b/width util. across peers	Load bal. red. & imp. quality of video	No content adaption	No
(Yi C et.al 2004), a primal dual algorithm in an undirected graph	S.L	Meas. the stream. capacity of a tree based network	Max. the throughput	The degree bounds at each node is not introduced	No
(Shay H et.al 2010; Yifeng H et.al 2009) deployed greedy and proximal algorithms	S.L	Managing the capacity of the P2P networks	Improves capacity received	Overhead complexity which causes huge playback delay	Helper Nodes
(Yifeng H et.al 2009) algorithm to provision resources for streaming	S.L	Monitor network performance	Offer high video quality & red. the effect of churn	Nodes with equal bit rates	Helper Nodes
(Sudipta S et.al 2011) the taxonomy of sixteen diff. problems	S.L	Opt. set of multicast trees to find the P2P stream. capacity	Rec. peer is able to achieve max. stream rate	Eff. util. of upload cap. at each peer is not considered	Helper Nodes
(Beverly Yang et.al 2003) Design a net. Comp.on super peer	S.L	To provide load bal. & handling attacks	Load is balanced	Cost & complexity	Super Peers
(Labib, Y. et.al 2009) enhanced media streaming system	S.L	Provide big media storage	Imp. Net. res. & red. consumption of b/width	Cost	No
(Yi C et.al 2003) deployed seed servers in a P2P network	S.L	Opt. util. of seed servers by det. switched on/off	Imp. Capacity & red. impact of peers failure	Intro. delay and suffer from low quality	Seed Servers
(Hong-Yi C et.al 2012) a cloud assistive opt. prob. to maintain a predefined quality	S.L	Min the total cost incurred in renting the cloud resources	Provide high playback lat. with short playback delay	Suffers from cost of renting cloud res.	Cloud Servers
(Seyfabad, M.S. et.al 2014) introduced a centralised based method	S.L	No. of virtual machines req. to provide QOS	Imp. Overall sys b/width	Cost & complexity	Virtual Machines
(Dongni Ren et.al 2009) min. delay multipath tree & distributed algo.	S.L	Supports net. comp. of super nodes, proxies or CDN.	Red. delay & load across servers	Small network	Proxy, CDN

divided into one base layer and several enhancement layers. The base layer carries the basic quality of the video whereas enhancement layers further improve the quality of the base layer stream.

The authors (H. Hu, Y. Guo & Y. Liu 2011) present a streaming design to provide efficiency, fairness and incentives with in a layer P2P streaming system. In this design, video content is distributed over mesh and a tracker acts as a bootstrapping node for the system. The design parameters for layered P2P streaming are layer subscription, chunk scheduling and topology adoption. To determine the number of layers required by a peer, a layer subscription algorithm with an adaptive increase or decrease and exponential back off is proposed. In this algorithm, whenever a peer joins a streaming session, an initial number of layers are subscribed to it. If a peer receives all the layers within the timeframe and the neighbouring peers have more number of layers to share, peer increases its layer subscription rate. However, if the top subscription layer becomes non decode able for a peer, it reduces the assigned number of layers. Furthermore, each peer shares their buffer maps with the neighbouring peers to monitor the chunk availability.

The authors also present a chunk scheduling algorithm that decides how to issue and request the chunks from the neighbouring peers. During a chunk request, chunks with high importance are requested first. Whereas in chunk serving, each peer maintains two FIFO queues labelled as entitled and excess queues for each neighbour. Entitled queue are retrieved first and excess queues are served if a peer has an excessive bandwidth. Furthermore, each peer periodically contacts the tracker to receive the list of active neighbouring peers in the network and maintains a present out degree. If the number of neighbours increases the present out degree, peer cancels its connections with some of its neighbouring peers whereas if the present out degree gets lower, it starts making connections with more neighbours. The authors show that the propose method provide the balance between social welfare and individual utility within a layered P2P system. However, it is still unpredictable to judge the fairness of the system under a given incentive mechanism.

The authors consider the behavior of live P2P multicast session over a large network (S. Agarwal, J. P. Singh, A. Mavlankar, P. Baccichet & B. Girod 2008). According to the authors, in order to efficiently distribute the video across the requesting nodes, it is necessary to encounter high bandwidth, high peer churn and the low peer persistence. Moreover, the authors monitor the quality of service (QOS) of the most popular content and correlate the monitored quality against the peer behaviors so that the better performance strategies can be provided.

Similarly, the authors (A. Riabov, Z. Liu & L. Zhang 2004) propose a system that monitors the available bandwidth among the nodes and consider multicast trees scenario to disseminate the content across nodes. Furthermore, SVC is used to maximize the video quality received with minimum possible delay while keeping the upload bandwidth of peers into consideration. Authors consider that if the peers with better available bandwidth are placed closer to the source or root node of the tree the overall performance of the system improves.

In (A. Abdelhalim, T. Ahmed, H. Walid-Khaled & S. Matsuoka 2012), the authors study a P2P streaming network using SVC in order to provide an efficient video streaming and video sharing system. They extended the famous bit-torrent protocol by combining it with SVC to support live content delivery with different quality levels. In order to achieve a better QoS, they organized the overlay based on grouping the peers with the same capacity together. Furthermore, the high capacity peers are placed closer to the source node. The authors showed that the proposed system has better QoS received and performs better than the existing single layer techniques.

In a P2P live streaming system, the overall bandwidth across the network automatically scales up based on bandwidth contribution of the peers in a system. To video stream efficiently, each peer is required to download the video within a certain playback interval. Hence, it is important to manage a right balance between the bandwidth supplied and bandwidth demanded. The authors (H. Luan, K. W. Kwong, Z. Huang & D. H. K. Tsang 2008) address this issue by proposing a system that automatically adapts the network towards full bandwidth utilization. A link level homogeneous network is designed that have identical bandwidth value. The advantage of identical bandwidth is that video flowing through an overlay will not encounter any issues, and guaranteed downloading rates can be achieved. Moreover, depending upon the peer downloading rate, the server adjusts video playback rate to provide quality video by fully utilizing the network bandwidth. In the proposed system, raw video is generated at the source which is then forwarded to the media server. The media server encodes the video using SVC. The compression rate of the video is estimated based on monitoring the downloading rate of peers which then helps to select an appropriate playback rate. After the video compression, a channel coding scheme is applied before it is broadcasted on the P2P network. On the other hand, at the receiver, a reverse operation is performed in order to decode the video to be played at the given playback rate.

The authors (Yi C., Liang D. & Yuan X. 2007) propose an algorithm for online selection of peers in order to provide throughput maximization under peer churn. The algorithm can be upgraded to multi parent streaming from a single peer streaming. Authors have also implemented an admission control mechanism by which the peers are rejected if the desired throughput is not achieved.

Moreover, the authors (Z. Liu, Y. Shen, K. W. Ross, S. S. Panwar & Y. Wang 2009), propose a LayerP2P system that combines layer coding with mesh overlay. The authors then provide a tit for tat strategy in which more incentives are given to those peers which contributes more towards the network. The video source encodes a video into different layers which is further broken down into small chunks known as layered chunks. The layer chunks are then distributed using a mesh based overlay. Whenever a peer joins the network, it obtains a complete list of its neighbouring peers similar to single layer streaming. Afterwards, the tit for tat strategy is used by which peers allocate more bandwidth to those peers that have high contribution of the upload bandwidth. Furthermore, the proposed system provides a viewable quality to peers if the sender peer's bandwidth falls below the total supply bandwidth. The authors have also considered an incentive based approach using SVC. In the proposed approach, a peer requests the base layer chunks firstly based on their upload capacities. Similarly, the authors (Liu Z, Shen Y, Panwar SS, Ross KW & Wang Y 2007) have introduced a probability based resource sharing. The probability is proportional to the upload capacity of the node.

In a different work, the authors (X Xiao, Y Shi, Y Gao & Q Zhang 2009) propose a new data scheduling approach called LayerP2P to stream the layered video in a P2P network. According to the authors, the proposed method achieves high throughput, high delivery ratio, low useless packets ratio and low subscribing jitter. Furthermore, the absent blocks are requested using a three stage mechanism; free, decision and remedy stage. In a free stage, the network is modelled as a minimum cost network flow model in order to schedule the data for achieving high throughput. The decision stage considers the total number of layers subscribed to a specific window in order to achieve high delivery ratio, less jitter and less useless packet ratio. Finally, the remedy stage carries the blocks with the most urgent playback time; the missed blocks are then requested through peers using multiple users.

In (M. Moshref, R. Motamedi, H. R. Rabiee & M. Khansari 2010), the authors have introduced a new video streaming protocol, LayeredCast. The proposed protocol uses the incentives of both mesh and tree based approaches and provide a multi service network core to clients. The experimental results show that the proposed method provides better QoS. The tree based approach is used to push the base layer across all the requesting nodes whereas the enhancement layers are pulled using the mesh based approach.

The authors (Muge S, Sercan D, Yagiz K & E. Turhan T. 2015) study a new approach for dynamic construction and maintenance of a tree based method for streaming live video in a P2P network. The authors divide the overlay multicast tree into hierarchical clusters so that it becomes easy to change the position of peers located under small trees. The joining of node comprises of two phases; firstly, node joins the cluster and then afterwards it joins the tree within the particular cluster based on the node's upload capacity. This joining of nodes reduces the message complexity because of its simplicity. Furthermore, authors have considered different factors while the construction of multicast tree such as dynamics, capacity awareness, incentive mechanism and scalable utilization. The authors have also introduced the concept of streaming leaders list, in this case if a leader fails or leave the network, an alternate leader is available. Each streaming leader is responsible to provide management to the part of the tree within their clusters. Moreover, a data dissemination algorithm is provided that maintains the backup parents list to provide tree resilience and maintains better quality. The experiment results show that the proposed method provides high QoE among peers, improve playback latency and reduce the duration of video pauses.

In this work (Hao H, Yang G & Yong L 2010), a taxation based scheme is deployed that provides fairness among peers requesting scalable video streams and have variable upload and download bandwidths. Similarly, (Mohamed H & Cheng HH 2008), the authors have considered a rate distortion model for SVC video using a fine grained scalable method in order to maximize the perceived video quality. The proposed model is more suitable as compared to the assumption that all layers have equal bit rates. However, in both models, different SVC coded packets of a video are streamed from each peer that carries data for multiple layers. If one of the peers fails or leaves the network, the video cannot be streamed and peers start starving until a new peer is available, which produces an excessive delay.

The authors (Lahbabi Y, Ibn Elhaj EH & Hammouch A 2014)) propose a quality adaption mechanism over the P2P network that uses SVC. The quality adaption mechanism comprises of two different phases. In the first phase, a layer level initialization (LLI) strategy is introduced that selects an initial quality of the video based on the peer's static resources such as power, screen resolution and the available bandwidth. When the streaming starts, the second phase is initiated in which layer level adjustment (LLA) algorithm is used. The algorithm helps to quickly adapt the quality of video based on various dynamic parameters such as memory, energy consumption, block availability and throughput. However, the disadvantage of the proposed method is that the degree bound at each node is not introduced that helps to improve the quality of the video received.

The authors (Medjiah S, Ahmed T & Boutaba R 2014) address the problem of the quality bottleneck in adaptive SVC streaming. The authors investigated the problem as a joint optimization problem of overlay formation, data distribution and content adaption in order to maximize the quality of experience whereas avoiding the quality bottleneck. In the scheduling strategy, the authors aimed to consider the effect of neighbor's departure on the received video quality. According to them, the data requests are forwarded to peers based on what they can handle with the available upload capacity. Furthermore, in order to avoid the quality bottleneck problem, the authors form an overlay with more stable neighbors based on their lifetime duration. However, it is suggested that even with proper scheduling and an overlay design, the network can still be affected with different network conditions such as bandwidth fluctuation. Therefore, the authors have considered using a smoothing function to overcome bandwidth fluctuations and provide better quality of experience. The experimental results showed that the proposed method reduces the quality bottlenecks, increase the churn tolerance and efficiently utilizes the bandwidth.

The investigators (Bradai A, Ahmed T, Boutaba R & Ahmed R 2014) have considered an efficient bandwidth allocation technique to allocate the sender peers upload bandwidth to the receiver peers. In the proposed method, the peer's upload bandwidth is allocated based on the quality level requested by the requesting peer. The authors have used an auction based game that distributes the bandwidth across the requesting peers, where the sender peers sell their upload bandwidth based on the bids from the requesting peers. The overall goal of their proposed work is to provide benefit to high priority peers whereas ensure the minimum quality for all the peers. The authors have combined the bandwidth allocation technique with an efficient scheduling mechanism that takes the advantage of allocated bandwidth with respect to the layer dependency and the playback deadline of the data packets. The simulation results showed that the proposed technique improves quality of the video, bandwidth utilization under heterogeneous peers.

The authors (Xuguang L, Nanning Z, Jianru X, Xiaoguang W & Bin G 2007) presented a data driven overlay network for streaming live media content using SVC. Peers forward data according to their upload bandwidths. A centralized server is used with an efficient scheduler that handles the requests from peers and serves them according to their capacities. However, the disadvantage of this approach is that it can only be used over the small P2P networks.

In different works (Yi C & Klara N 2003; Reza R & Antonio O 2003), the authors have studied the resource allocation problem in P2P streaming using SVC and the seed servers. In the first work, the authors proposed an algorithm that runs on each peer separately in order to request SVC layers from a given set of heterogeneous senders (each sender has a different outbound capacity), however it is assumed that all the layers have equal bit rates and offer equal video quality. Moreover, each peer sends SVC layers depending upon its outbound capacity and the receiver receives layers according to its inbound capacity. If all the peers are not able to overcome the requirement of layers of receiving peers, then the remaining layers are served using seed servers.

Similarly, in the second work, the receiver continuously sends periodic messages for requesting the packets from each sender. The receiver determines a set of senders that can increase the system throughput and then finds the maximum number of layers that can be sent. The authors have implemented a congestion control mechanism in which each sender sends the packets to receivers upon request.

In another work (Shabnam M & Mohamed H 2010), the authors proposed a P2P system that uses SVC and network coding (NC). SVC helps to support heterogeneous peers in the network whereas NC handles the peer dynamics and maximizes the network throughput. NC enables to perform single operations on the video packets before they are forwarded. The forwarding of packets allows peers to share partial information with the destination node. The receiver can receive the whole video after receiving all the necessary partial information from the peers. The proposed model uses three different entities; trackers, sources and peers. Trackers identify the peers who are watching the same video in the network. Source nodes initialize the video stream in the network and provide extra capacity in the network if the available upload capacity at the peers is fully utilized. Source nodes perform NC on video data before forwarding them to distribute in the network. The results show that the proposed system improves average video quality, average streaming rates reduces the effects of churn and manages flash crowd.

The authors propose an efficient scheme to manage seed server resources in a P2P network (Kianoosh M & Mohamed H 2013). The scheme uses an adaptive layer streaming and monitors the peers' contribution according to their upload bandwidths. Furthermore, the authors consider the problem of capacity management across seed servers and provide a capacity allocation algorithm. The algorithm helps to deliver seed server resources in order to maximize a system wide utility function i.e. overall video quality received by the peers. The results show that when seed servers are introduced in the network, the overall quality of the video is improved. However, the drawback of this approach is that it introduces a certain cost at the seed servers.

The authors (Hong Yi C, Ya Yueh S & Yuan W.L 2012) suggest a cloud based P2P live video streaming platform (CloudPP) that uses cloud servers as peers to develop a P2P streaming network using SVC. The authors have designed a tree based network using SVC so that all the requests are served using the minimum number of cloud servers. The working of the proposed method is like whenever a new client joins the system, it searches for a cloud server using a breadth first search method starting from the SVC base layer tree. If the cloud server has enough available bandwidth to maintain the streaming quality requested by the new client, the client is added as its child. However, if an existing cloud server is unable to provide the required streaming quality to the client, the system boots a new cloud server and the existing client redirects its entire client to the new cloud server and makes it free to serve more clients. Similarly, when a peer leaves a network, the parent cloud server shuts down if no other children are connected to it that reduces the overall cost.

Table 2. Qualitative comparison for multilayer SVC P2P video streaming techniques

Approach	Cod. Tech.	Objectives	Advantages	Disadvantages	Multi. Sender Sources
(Hoa Hu et al. 2011) Layer subs., chunk sched. & topology adoption algo.	SVC	To provide eff. fairness and incentives	Improve QOS, Red. Effect of free rider	B/width is not efficiently utilized	No
(Riabov et.al 2004) multi-cast tree to disseminate content	SVC	To max. video quality received with min. delay	Overall system performance is improved.	Do not support heterogeneous peers	No
(A. Abdelhalim et.al 2012)	SVC	Live video at diff. qualities	Achieves better QOS	Priority to nodes with more capacity	No
(Luan et al. 2008) Stream. system	SVC	Full b/width util. of network	Guarantee downloads rates.	Receiver Sync. problem	No
(Cui et.al 2007) an admission control mechanism	SVC	To provide throughput maximization.	Impr. Throughput of net. Reduce churn	Poor Utilization of b/ width at peer	No
(Zhengye Li. et.al 2009) LayerP2P	SVC	To provide more incentive to peers that cont. more	Improve video quality	Flash crowd is not studied	No
(Zhengye Li. et.al 2007), incentive based approach	SVC	Reduce skipping of content	Reduce latency	More overhead complexity	No
(Xiao et al. 2009) mech. to request missing blocks	SVC	To achieve high throughput & low delay	Provide high quality with min. delay	High overhead complexity	No
(M. Moshref et. al 2010) introduced Layered Cast protocol	SVC	Incentives of both mesh and tree based approaches	Provides better QOS.	Increase complexity	No
(Sayit et.al 2015) Data diss. algo. with stream. leaders list	SVC	To construct & maintain trees	Red. message compl. and churn. Provide high QOE	B/width is not eff. utilized.	No
(Hao H et.al 2010) taxation based scheme	SVC	Provide fairness among peers	Balance b/w efficiency & fairness	Utilizes most of the capacities of peers	No
(Mohamed H et.al 2008) rate distortion scheme	SVC	Max. perceived video quality	Improves quality of video	Handle heterogeneous network	Multi. Senders
(Lahbabi Y et.al 2014) link layer initialization &adaption mech.	SVC	Adapts quality of video over various dynamic parameters	Devices retrieve qualities based on resources	Effect of different network conditions flash crowd &churn	No
(Medjiah S et.al 2014) opt. prob. of overlay form., data distr. & adaption	SVC	Max. QOE while avoid quality bottleneck	Red. Quality bottleneck, incr. churn tol. & utilize b/width	Overhead complexity &Effect of flash crowd	No
(Bradai A et.al 2014) auction based game to distribute bandwidth	SVC	Incentives to priority peers & ensure min. quality for all peers	Improves quality, bandwidth util. across network	Upload capacity of peers is not efficiently utilized	No

continued on following page

Table 2. Continued

Approach	Cod. Tech.	Objectives	Advantages	Disadvantages	Multi. Sender Sources
(Xuguang L et.al 2007) data driven overlay net. for stream. live media	SVC	Node handles req. according to their capacities	Adapts quickly to heterogeneous nature of peers	Only for small P2P networks	No
(Yi C et.al 2003) algo. runs at each peer sep. to req. layers from sender peers &intro. seed servers	SVC	Peers send layers dep. on outbound cap. and receives based on inbound capacity	Max. deliver quality while min. server & network load	Dynamics of b/width variation is not discussed.	Seed Servers
(Reza R et.al 2003) congestion control mech. to send packets to receivers upon req.	SVC	Det. set of senders that incr. throughput &finds senders that max. no. of layers rec.	Improves system throughput and the video quality rec.	Doesn't support variable bit rate and peer's dynamics	Multi. Senders
(Shabnam M et.al 2010) proposed a P2P system	SVC-NC	Support hetero. Peers & hand. peer dynamic	Imp. Video quality, avg. rate, red. churn eff. & manage crowd	Upload b/width across peer is not utilized	Seed Servers
(Kianoosh M et.al 2013) adaptive layer streaming scheme	SVC	Use seed server res. to max. system utility	Reduce cost of stream. seed server resources	Upload capacity at peers is not utilized	Seed Servers
(Hong-Yi C et.al 2012) cloud network that uses breadth first search method to handle req.	SVC	Tree based approach to serve the req. using min. cloud servers	Reduces the overall cost of the network	Resources across the peers are not utilized	Cloud Servers

VIDEO STREAMING OVER P2P NETWORKS USING MDC

Multi-Layer Descriptive Coding (MDC) was introduced in the early 1980s by researchers (]. By the mid-1990s, MDC has become an important mechanise to reduce the propagation errors in video delivery over networks. As discussed, MDC is a type of layer coding technique that generates multiple layers or descriptors of a single video stream. The advantage of MDC is that the video can be decoded at the receiver at different qualities and it provides high resilience to packet loss (A. El-Gamal & T.M. Cover 1982; N S Jayant 1981; V.A. Vaishampayan 1993) Below, a brief description of some streaming algorithms that use MDC are provided. These include SEACAST and CoopNet.

The authors (VK. Goyal 2001) propose an algorithm to provide better QoS across peers using MDC. The proposed algorithm distributes the data by quickly adapting the variable bandwidth across peers. However, the disadvantage of this method is that the transmission loss occurs due to unavailability of signals at the decoder. In another work S. Zezza, E. Magli, G. Olmo & M. Grangetto 2009), the authors propose SEACAST, a P2P streaming protocol for streaming live media content. The proposed protocol provides flow control and application-layer error control using MDC techniques as discussed (T. Tillo, M. Grangetto & G. Olmo (2008). The flow control states that the network assures a constant flow of data through the overlay with low start-up latency and with a considerable packet loss. The advantage of SEACAST is that it started using new techniques like RTP, UDP, and RTSP/SDP to establish sessions and provide flow control across them. On the other hand, the error control technique helps to reduce the error within the prediction loop.

The authors (Jaganathan & Eranti, J. 2011) use MDC to provide error resilience in order to handle the lost frames at the receiver end. The authors propose an algorithm that uses spatial temporal correlation to reconstruct the video signal from lost descriptors. In the algorithm, if both the descriptions are received, the decoder will efficiently reconstruct the signal. However, if any one of the descriptions is received, then the decoder reconstructs the signal using spatial-temporal smoothness measure in the received description.

Furthermore, the authors introduce CoopNet that uses the concept of cooperative networking to allocate the streaming content. It comprises of a central tree management protocol that helps to provide redundancy in both network path using multiple trees and for the diverse distribution of data using MDC. The protocol helps to reduce the effect of churn in the network. Furthermore, scalable feedback method is used that manage the effectiveness of trees by monitoring the physical and logical topology (Venkata N. Padmanabhan, Helen J. Wang & Philip A. Chou 2003). Similarly, the authors used multicast tree to stream the MDC video comprises of different descriptors. In the propose methods, different descriptors of the video are forwarded using different trees (J. D. Mol, D. H. P. Epema & H. J. Sips 2007; Noh J & Girod B 2012).

COMPARING S.L, SVC AND MDC PROPOSALS

Table 1, Table 2 and Table 3 show the qualitative analysis of the existing video streaming techniques that use S.L, SVC and MDC to improve the video quality received across the peers. The analysis shows that in a P2P network with heterogeneous peers, S.L coding usually achieves low video quality whereas SVC and MDC maintain better video quality due to their quality adaption property. Furthermore, the

Table 3. Qualitative comparison for multilayer MDC P2P video streaming techniques

Approach	Cod. Tech.	Objectives	Advantages	Disadvantages	Multi. Sender Sources
(Goyal et. al 2001) Algorithm that distrib. data by adapt. Var. b/ width at peers	MDC	To provide better streaming quality	Provide better QOS	transmission loss occurs due to unavailability of signals	No
(Zezza et. al 2009; Tillo et al. 2008) SEACAST protocol	MDC	To provide flow control & app. layer error control	Flow & error control. Low latency	Churn & Flash crowd is not studied.	No
(Jaganathan et. al 2011) propose an algo. that uses spatial temporal correlation	MDC	To reconstructs video signal from lost descriptors	Pro. error resilience to handle lost frame at rec. end	Bandwidth is not efficiently utilized	No
(V. N. Padmanabhan et.al 2003) CoopNet that uses cooperative networking	MDC	Allocate streaming content	Reduces effect of churn	Handling flash crowd	No
(J. D. Mol et.al 2007; Noh J. et.al 2012) multicast tree	MDC	Stream MDC coded video over diff. paths	Provide error resilience	Flash crowd are not considered	No

techniques are differentiated based on the available source such as peers, helper nodes, multiple senders, seed servers and cloud severs. These video streaming sources help to increase the overall network capacity and maintain the better video in a highly dynamic network such as network with churn or flash crowds.

CONCLUSION AND FUTURE RESEARCH DIRECTIONS

The chapter presented a basic understanding of P2P networks and its applications and further investigated the use of different video streaming techniques to stream video over P2P networks. The proposed video streaming techniques help to provide better QoS among peers and reduces the effect of churn in the network. Moreover, the use of different incentive based mechanisms help to overcome the free riders in the network. However, the disadvantage of these techniques is that they do not provide any adaptive solution to stream the video under the heterogeneous nodes in the network which introduces playback delay and skipping of the video content.

To overcome such issues, different video coding techniques have been discussed in Section 5 related to P2P video streaming using SVC and Section 6 covers different approaches for P2P video streaming using MDC. Theses coding techniques help to improve the average video quality received at the receivers. Furthermore, the most important advantage of using video coding techniques in P2P networks is that it can quickly adapt the video quality based on receiver capabilities and adapt to heterogeneous network conditions. However, there are certain limitations to each approach as in SVC; the video segment is encoded into different layers such as one base layer and several enhancement layers. Each layer is decoded one after another starting from the lower layer such that higher layers depend on lower layers. Hence, it is required to provide reliable transmission of a base layer so that video can be played at the receiver. Hence if the base layer is not received there is no use in receiving the higher quality layers. However, in the case of MDC, it allows more scalability while encoding the video into different layers or descriptors. This is because MDC does not require any priorities or retransmissions. Further it is considered to be robust as it hardly happens that all the descriptors of the video get corrupted. Therefore, MDC is widely used in the case where the network is exposed to more churn as it provides error resilience which helps the video to still survive at better quality. On the other hand, SVC is used for the networks which are static or centralized control and there is a less chance that a network is affected due to churn.

Given that video is such an integral part of all future communications, further work in this area is still required. We believe the early work presented by the authors (Raheel, M.S.; Raad, R & Ritz, C 2014, M. S. Raheel, R. Raad, & C. Ritz, 2015) can be further built on. The main focus is to exploit the properties of layering and split the layers across multiple sources, hence reducing the required upload capacity from each node and hence in this way encouraging more nodes to participate in the streaming function. Further work is still required in synchronisation of the streams from multiple sources. Further work is also required to efficiently distribute the layers of the video in the first place and further work is required into the layer discovery algorithms to bring the video together at the receiver.

REFERENCES

Akbari & Movaghar. (2010). A hybrid mesh-tree peer-to-peer overlay structure for layered video streaming. *5th International Symposium on Telecommunications,* 706 – 710.

Abdelhalim, A., Ahmed, T., Walid-Khaled, H., & Matsuoka, S. (2012) Using Bit-torrent and SVC for efficient video sharing and streaming. *IEEE Symp on Comput. Commun, 12*, 537–543.

Agarwal, S., Singh, J. P., Mavlankar, A., Baccichet, P., & Girod, B. (2008). Performance and Quality-of-Service Analysis of a Live P2P Video Multicast Session on the Internet. *Quality of Service, 2008. IWQOS 2008. 16th International Workshop on*, 11-19. doi:10.1109/IWQOS.2008.7

Ahmed, T., Buridant, G., & Mehaoua, A. (2001). Encapsulation and marking of MPEG-4 video over IP differentiated services. *ISCC, 1*, 346–352.

Ahmed, T., Mehaoua, A., & Buridant, G. (2001). Implementing MPEG-4 video on demand over IP differentiated services. *GLOBECOM, 1*, 2489–2493.

Ahmed, T., Nafaa, A., & Mehaoua, A. (2003) An Object-Based MPEG-4 Multimedia Content Classification Model for IP QOS Differentiation. *IEEE Symposium on Computers and Communications*. doi:10.1109/ISCC.2003.1214260

Allani, M., Garbinato, B., & Pietzuch, P. (2012). Chams: Churn-aware overlay construction for media streaming. *Peer-to-Peer Networking and Applications*, *5*(4), 412–427. doi:10.1007/s12083-012-0161-7

Anh, T. N., Baochun, L., & Frank, E. (2010). Chameleon: adaptive peer-to peer streaming with network coding.*Proceedings of IEEE INFOCOM*, 1–9.

Banerjee, D., Saha, S., Sen, S., & Dasgupt, P. (2005). Reciprocal Resource Sharing in P2P Environments. *Proceedings of ACM AAMAS05*.

Banerjee, S., Bhattacharjee, B., & Kommareddy, C. (2002). Scalable application layer multicast.*Proceedings of the 2002 Conference on Applications, Technologies, Architectures, and Protocols for Computer Communications*, 205–217.

Birrer, S., & Bustamante, F. (2005). Resilient peer-to-peer multicast without the cost. *Proceedings 12th Annual Multimedia Computing and Networking Conf.*, 113–120.

Bittorrent. (2015) Available at http://www.bittorrent.com/

Bradai, A., Ahmed, T., Boutaba, R., & Ahmed, R. (2014). Efficient content delivery scheme for layered video streaming in large-scale networks. *Journal of Network and Computer Applications*, *45*, 1–14. doi:10.1016/j.jnca.2014.07.004

Castro, M., Druschel, P., Kermarrec, A. M., Nandi, A., Rowstron, A., & Singh, A. (2003). Split-stream: High-bandwidth multicast in cooperative environments. *SIGOPS Operating System Review*, *37*(5), 298–313. doi:10.1145/1165389.945474

Chakareski, J., Han, S., & Girod, B. (2003). *Layered coding vs. multiple descriptions for video streaming over multiple paths* (A. C. M. Multimedia, Ed.). doi:10.1145/957013.957100

Chu, Y., Chuang, J., & Zhang, H. (2004). A Case for Taxation in Peer-to-Peer Streaming Broadcast. Proc of ACM SIGCOMM '04 Workshops.

Cohen, B. (2003). Incentives build robustness in bit torrent. *Proceedings ACM SIGCOMM Workshop Economics of Peer-to-Peer Systems*, 1–5.

Dingledine, R., Freedman, M., & Molnar, D. (2001). Free Haven. In Peer-to-Peer: Harnessing the Power of Disruptive Technologies. O'Reilly & Associates.

El-Gamal, A., & Cover, T. M. (1982). Achievable rates for multiple descriptions. *IEEE Transactions on Information Theory*, *28*(06), 851–857. doi:10.1109/TIT.1982.1056588

Fesci-Sayit, M., Turhan Tunali, E., & Murat Tekalp, A. (2009) Bandwidth-aware multiple multicast tree formation for P2P scalable video streaming using hierarchical clusters. ICIP'09. IEEE, 945–948 doi:10.1109/ICIP.2009.5414021

Fesci-Sayit, M., Turhan Tunali, E., & Murat Tekalp, A. (2012). Resilient peer-to peer streaming of scalable video over hierarchical multicast trees with backup parent pools. *Signal Processing Image Communication*, *27*(2), 113–125. doi:10.1016/j.image.2011.11.004

Freenet. (2015). Available at http://freenet.sourceforge.net/

German, S., Markus, W., & Herwig, U. (2006). Search Methods in P2P Networks: A Survey. Lecture Notes in Computer Science, 3473, 59-68.

Guha, S., Daswani, N., & Jainet, R. (2006). An experimental study of the Skype peer-to-peer VoIP system. *Proceedings of the 5th International Workshop on Peer-to-Peer Systems.*

Guo, Y., Liang, C., & Liu, Y. (2008). Adaptive Queue-based Chunk Scheduling for P2P Live Streaming. *Proceedings of 7th International IFIP-TCP Networking conference on Adhoc and sensor networks, wireless networks, next generation internet*, 433-444.

Hao, H., Yang, G., & Yong, L. (2010). Mesh-based peer-to-peer layered video streaming with taxation. *Proceedings of ACM Workshop on Network and Operating System Support for Digital Audio and Video NOSSDAV'10*, 27–32.

Hareesh, K., & Manjaiah, D. H. (2011). Peer to Peer live streaming and video on demand design issues and its challenges. *International Journal of Peer to Peer Networks*, *02*(04), 1–11.

Hong Yi, C., Ya Yueh, S., & Yuan, W. L (2012). Cloud PP: A novel cloud-based P2P live video streaming platform with SVC technology. *8th International Conference on Computer Technology and Information Management*, 64–68

Hoong, P., & Matsuo, H. (2008). Palms: A reliable and incentive-based P2P live media streaming system. Lecture Notes in Electrical Engineering, 4, 51–66. doi:10.1007/978-0-387-74938-9_5

Hu, H., Guo, Y., & Liu, Y. (2011). Peer to Peer Streaming of Layered Video: Efficiency, Fairness and Incentive. *IEEE Transactions on Circuits and Systems for Video Technology*, *21*(08), 1013–1026. doi:10.1109/TCSVT.2011.2129290

Jagadish, H. V., Ooi, B. C., & Vu, Q. H. (2005). BATON: a balanced tree structure for peer to-peer networks.*Proceedings of the International Conference on Very Large Databases*, 661–672

Jaganathan, & Eranti, J. (2011). High quality of service on video streaming in P2P networks using FST-MDC. *International Journal of Multimedia and Its Applications, 3*(2), 33-43.

Jayant, N. S. (1981). Sub-sampling of a DPCM speech channel to provide two self-contained half-rate channels. *The Bell System Technical Journal*, *60*(04), 501–509. doi:10.1002/j.1538-7305.1981.tb03069.x

Kan, G. (2001). Gnutella. In *Peer-to-Peer: Harnessing the Power of Disruptive Technologies* (pp. 94–122). O'Reilly & Associates.

Kianoosh, M., & Mohamed, H. (2013). Capacity management of seed servers in peer to peer streaming systems with scalable video streams. *IEEE Transactions on Multimedia*, *15*(1), 181–194. doi:10.1109/TMM.2012.2225042

Klingberg, T., & Manfredi, R. (2002). *Gnutella 0.6.* Technical report. Available at: http://rfcgnutella.sourceforge.net/src/rfc-0_6-draft.html

Kofler, I., Seidl, J., Timmerer, C., Hellwagner, H., Djama, I., & Ahmed, T. (2008). Using MPEG-21 for cross-layer multimedia content adaptation. *Signal Image Video Process.*, *2*(4), 355–370. doi:10.1007/s11760-008-0088-x

Kuo, J. L., Shih, C. H., Ho, C. Y., & Chen, Y. C. (2014). *Advanced bootstrap and adjusted bandwidth for content distribution in peer-to-peer live streaming*. Peer-to-Peer Networking and Applications.

Kwon, O. C., & Song, H. (2013). Adaptive tree-based P2P video streaming multicast system under high peer-churn rate. *Journal of Visual Communication and Image Representation*, *24*(3), 203–216. doi:10.1016/j.jvcir.2012.11.004

Labib, Y., Sherbini, A. E., & Sabri, A. (2009). A Clustered P2P Proxy-Assisted Architecture for On Demand Media Streaming. *Computer Technology and Development, ICCTD '09. International Conference on, 2*, 95-100.

Lahbabi, Y., Ibn Elhaj, E. H., & Hammouch, A. (2014) Quality adaptation using scalable video coding (SVC) in peer-to-peer (P2P) video-on demand (VoD) streaming. *International conference on Multimedia Computing and Systems ICMCS'14*, 1140–1146.

Leszek, C. (2006). Scalable Video Coding for Flexible Multimedia Services. *IEEE Transactions on Circuits and Systems for Video Technology, 18*(7).

Liu, Wen, Yeung, & Lei. (2012). Request-peer selection for load-balancing in P2P live streaming systems. *IEEE Conference on Wireless Communications and Networking*, 3227–3232.

Liu, Y., Guo, Y., & Liang, C. (2008). A survey on peer-to-peer video streaming systems. *Journal of Peer-to-Peer Networking and Applications*, *1*(1), 18–28. doi:10.1007/s12083-007-0006-y

Liu, Z., Shen, Y., Panwar, S. S., Ross, K. W., & Wang, Y. (2007). P2P video live streaming with MDC: Providing incentives for redistribution. ICME. IEEE, 48–51.

Liu, Z., Shen, Y., Ross, K. W., Panwar, S. S., & Wang, Y. (2009). Layer P2P: Using layered video Chunks in P2P live streaming. *IEEE Transactions on Multimedia*, *11*(7), 1340–1352. doi:10.1109/TMM.2009.2030656

Luan, H., Kwong, K. W., Huang, Z., & Tsang, D. H. K. (2008). P2P live streaming towards best quality video. 5th IEEE Consumer Communications and Networking Conference (IEEE CCNC), 458-460.

Magharei, N., Rejaie, R., & Guo, Y. (2007). Mesh or multiple-tree: A comparative study of live P2P streaming approaches. *Proceedings of IEEE INFOCOM*.

Magi. (2015). Available at http://www.endeavors.com

Medjiah, S., Ahmed, T., & Boutaba, R. (2014). Avoiding quality bottlenecks in P2P adaptive streaming. *IEEE Journal on Selected Areas in Communications*, *32*(4), 734–745. doi:10.1109/JSAC.2014.140406

Mohamed, H., & Cheng, H.H. (2008). Rate-distortion optimized streaming of fine grained scalable video sequences. *ACM Transactions on Multimedia Computing, Communications and Applications*, *4*(1), 2:1–2:28.

Mol, J. D., Epema, D. H. P., & Sips, H. J. (2007). The orchard algorithm: Building multicast trees for P2P video multicasting without free-riding. *IEEE Transactions on Multimedia*, *9*(8), 1593–1604. doi:10.1109/TMM.2007.907450

Montazeri, A., Akbari, B., & Ghanbari, M. (2012). An incentive scheduling mechanism for peer-to-peer video streaming. *Peer-to-Peer Networking and Applications*, *5*(3), 257–278. doi:10.1007/s12083-011-0121-7

Muge, S., Sercan, D., Yagiz, K., & Turhan, T. (2015). Adaptive, incentive and scalable dynamic tree overlay for P2P live video streaming. *Springer Journal of Peer to Peer networking and Applications*. DOI 10.1007/s12083-015-0390-7

Napster. (2015). Available at http://www.napster.com/

Noh, J., & Girod, B. (2012). Time-shifted streaming in a tree-based peer-to-peer system. *Journal of Communication*, *7*(3), 202–212.

Padmanabhan, V. N., Wang, H. J., & Chou, P. A. (2003). Resilient Peer-to-Peer Streaming.*International conference on network protocols (IEEE ICNP 03)*, 16-27.

Park, J., & van der. Schaar, M. (2010). A game theoretic analysis of incentives in content production and sharing over P2P networks. *IEEE Journal of Selected Topics in Signal Processing*, *4*(4), 704–717. doi:10.1109/JSTSP.2010.2048609

PPLive. (2015). Available at http://www.pplive.com/

PPStream. (2015). Available at http://www.ppstream.com

Raheel, Raad, & Ritz. (2015, August). Achieving maximum utilization of peer's upload capacity in P2P networks using SVC. *Springer Peer-to-Peer Networking and Applications*, 1-21.

Raheel, M. S., Raad, R., & Ritz, C. (2014). Efficient utilization of peer's upload capacity in P2P networks using SVC. *IEEECommunications and Information Technologies (ISCIT),201414th International Symposium on*, 66-70. doi:10.1109/ISCIT.2014.7011871

Ratnasamy, Francis, Handley, Karp, & Shenker. (2001). *A scalable content-addressable network*. Technical report, ACM SIGCOMM'01.

Ren, D., Li, Y. T. H., & Chan, S.-H. G. (2009). Fast-Mesh: A Low-Delay High-Bandwidth Mesh for Peer-to-Peer Live Streaming. *IEEE Transactions on Multimedia, 11*(8), 1446–1456. doi:10.1109/TMM.2009.2032677

Reza, R., & Antonio, O. (2003). PALS: Peer-to-peer adaptive layered streaming.*Proceedings of ACM International Workshop on Network and Operating System Support for Digital Audio and Video NOSS-DAV'03*, 153–161.

Riabov, A., Liu, Z., & Zhang, L. (2004). Overlay multicast trees of minimal delay. *Distributed Computing Systems, 2004. Proceedings. 24th International Conference on*, 654-661. doi:10.1109/ICDCS.2004.1281633

Satsiou, A., & Tassiulas, L. (2010). Reputation-Based Resource Allocation in P2P Systems of Rational Users. *IEEE Transactions on Parallel and Distributed Systems, 21*(4), 476–479. doi:10.1109/TPDS.2009.80

Seyfabad, M. S., & Akbari, B. (2014). CAC-live: Centralized assisted cloud P2P live streaming. *Electrical Engineering (ICEE), 2014 22nd Iranian Conference on*, 908-913.

Shabnam, M., & Mohamed, H. (2010). Live peer to peer streaming with scalable video coding and network coding.*Proceedings of the first annual ACM SIGMM conference on Multimedia systems MM-Sys'10*, 123-132.

Shay, H., & Danny, D. (2010). *On the Role of Helper Peers inP2P Networks. Parallel and Distributed Computing* (A. Ros, Ed.). InTech. doi:10.5772/9450

Shen, X., Yu, H., Buford, J., & Akon, M. (2010). *Handbook of Peer-to-Peer Networking*. Springer Publishing Company. doi:10.1007/978-0-387-09751-0

SopCast. (2015). Available at http://www.sopcast.com/

Stoica, I., Morris, R., Liben, N.D., Karger, D.R., Kaashoek, M.F., Dabek, F., & Balakrishnan, H. (2003). Chord: A Scalable Peer-to-Peer Lookup Protocol for Internet Applications. *IEEE/ACM Transaction of Networking, 11*(1), 17–32.

Sudipta, S., Shao, L., Minghua, C., Mung, C., Jin, L., & Philip, A. C. (2011). Peer to peer streaming capacity. *IEEE Transactions on Information Theory, 57*(08), 5072–5088. doi:10.1109/TIT.2011.2145630

Sung, Bishop, & Rao. (2006). Enabling contribution awareness in an overlay broadcasting system. *Proceedings of ACM SIGCOMM.*

Tang, Y., Luo, J. G., Zhang, Q., Zhang, M., & Yang, S. Q. (2007). Deploying P2P networks for large-scale live video-streaming service. *IEEE Communications Magazine, 45*(6), 100–106. doi:10.1109/MCOM.2007.374426

Tillo, T., Grangetto, M., & Olmo, G. (2008). Redundant slice optimal allocation for H.264 multiple description coding. *IEEE Transactions on Circuits and Systems for Video Technology, 18*(01), 59–70. doi:10.1109/TCSVT.2007.913751

Tran, D. A., Hua, K. A., & Do, T. T. (2004). A peer-to-peer architecture for media streaming. *IEEE Journal on Selected Areas in Communications, 22*(1), 121–133. doi:10.1109/JSAC.2003.818803

UUSee. (2015). Available at http://www.uusee.com/

Vaishampayan, V. A. (1993). Design of multiple description scalar quantizers. *IEEE Transactions on Information Theory*, *39*(03), 821–834. doi:10.1109/18.256491

Venkataraman, V., Yoshida, K., & Francis, P. (2006). Chunkyspread: Heterogeneous Unstructured Tree-Based Peer-to-Peer Multicast. *Proceedings of the 2006 IEEE International Conference on Network Protocols*, 2-11. doi:10.1109/ICNP.2006.320193

Vinay, P., Kumar, K., Tamilmani, K., Sambamurthy, V., & Mohr, A. E. (2005). Chainsaw: Eliminating Trees from Overlay Multicast. *4th International Workshop on Peer-to-Peer Systems,*127-140.

Vivek, K. G. (2001). Multiple description coding: Compression meets the network. *IEEE Signal Processing Magazine*, *18*(5), 74–93. doi:10.1109/79.952806

Waldman, M., Cranor, L., & Rubin, A. (2001). Publius. In *Peer-to-Peer: Harnessing the Power of Disruptive Technologies* (pp. 145–158). O'Reilly & Associates.

Wang, X. (2002). BuddyWeb: a P2P-based collaborative web caching system. *Proceedings of the Web Engineering and Peer-to-Peer Computing, Networking Workshops*, 247–251.

Wien, M., Schwarz, H., & Oelbaum, T. (2007). Performance analysis of SVC. *IEEE Transactions on Circuits and Systems for Video Technology*, *17*(9), 1194–1203. doi:10.1109/TCSVT.2007.905530

Xiao, X., Shi, Y., Gao, Y., & Zhang, Q. (2009). Layer P2P: A new data scheduling approach for layered streaming in heterogeneous networks. IEEE INFOCOM Proceedings, 603-611.

Xuguang, L., Nanning, Z., Jianru, X., Xiaoguang, W., & Bin, G. (2007). A Peer-to-peer architecture for efficient live scalable media streaming on Internet. *Proceedings of ACM Multimedia Conference*, 783–786.

Yang, B., B. & Garcia-Molina, H. (2003). Designing a super-peer network, *Proceedings.19th International Conference on Data Engineering*, 49-60.

Yang, S., & Wang, X. (2010). An incentive mechanism for tree-based live media streaming service. *Journal of Networks*, *5*(1), 57–64. doi:10.4304/jnw.5.1.57-64

Yi, C., Baochun, L., & Klara, N. (2004). On achieving optimized capacity utilization in application overlay networks with multiple competing sessions. *ACM Symposium on Parallelism in Algorithms and Architectures.*

Yi, C., & Klara, N. (2003). Layered peer-to-peer streaming. *Proceedings of ACM International Workshop on Network and Operating System Support for Digital Audio and Video*, 162–171.

Yi, C., Liang, D., & Yuan, X. (2007). Optimizing P2P streaming throughput under peer churning. *IEEE Global Telecommunications Conference*, 231-235.

Yi, C. T., Jianzhong, S., Mohamed, H., & Sunil, P. (2005). An analytical study of peer-to peer media streaming systems. *ACM Trans. Multimedia Computer, Communication, Application*, *1*(4), 354–376.

Yifeng, H., & Ling, G. (2009). Improving the streaming capacity in P2P VoD systems with helpers. *Proceedings of ICME*, 790–793.

Zezza, S., Magli, E., Olmo, G., & Grangetto, M. (2009). SEACAST: a protocol for peer-to-peer video streaming supporting multiple description coding. *Proceedings of the 2009 IEEE international conference on Multimedia and Expo*, 1586-1587. doi:10.1109/ICME.2009.5202819

Zhang, M., Xiong, Y., Zhang, Q., Sun, L., & Yang, S. (2009). Optimizing the throughput of data driven peer-to-peer streaming. *IEEE Transactions on Parallel and Distributed Systems*, *20*(1), 97–110. doi:10.1109/TPDS.2008.69

Zhang, X., Liu, J., Li, B., & Yum, Y. S. P. (2005). DONet/CoolStreaming: a data-driven overlay network for live media streaming. Proc. of IEEE INFOCOM'05, 2102-2111.

Zhao, B., Huang, L., Stribling, J., Rhea, S. C., Joseph, A. D., & Kubiatowicz, J. D. (2004). Tapestry: A Resilient Global-scale Overlay for Service Deployment. *IEEE Journal on Selected Areas in Communications*, *22*(1), 41–53. doi:10.1109/JSAC.2003.818784

Zhengye, L., Yanming, S., Keith, W. R., Shivendra, S. P., & Yao, W. (2009). Layer P2P: Using layered video chunks in P2P live streaming. *IEEE Transactions on Multimedia*, *11*(07), 1340–1352. doi:10.1109/TMM.2009.2030656

Zhijie, S., Jun, L., Roger, Z., & Athanasios, V. V. (2011). Peer to Peer media streaming: Insights and new developments. *Proceedings of the IEEE*, *99*(12), 2089–2109.

Chapter 10
Quantifying QoS in Heterogeneous Networks:
A Generalized Metric-Based Approach

Farnaz Farid
Western Sydney University, Australia

Seyed Shahrestani
Western Sydney University, Australia

Chun Ruan
Western Sydney University, Australia

ABSTRACT

The heterogeneous-based 4G wireless networks will offer noticeable advantages for both users and network operators. The users will benefit from the vibrant coverage and capacity. A vast number of available resources will allow them to connect seamlessly to the best available access technology. The network operators, on the other hand, will be benefited from the lower cost and the efficient usage of the network resources. However, managing QoS for video or voice applications over these networks is still a challenging task. In this chapter, a generalized metric-based approach is described for QoS quantification in Heterogeneous networks. To investigate the efficiency of the designed approach, a range of simulation studies based on different models of service over the heterogeneous networks are carried out. The simulation results indicate that the proposed approach facilitates better management and monitoring of heterogeneous network configurations and applications utilizing them.

INTRODUCTION

A heterogeneous communication network provides transparent and self-configurable services across wireless local area networks (WLANs), Wireless Metropolitan Area Networks (WMANs) and Wireless wide area networks (WWANs). Primarily, heterogeneous networks were visioned as an integration of IEEE 802.11 WLANs and 3G/2.5G/2G/B3G all these cellular technologies, mobile WiMAX being

DOI: 10.4018/978-1-5225-2113-6.ch010

the major player in the middle. However, the recent advancement of LTE-advance has added one more new technology in the picture, which would play a fundamental role in this integrated architecture and will form the 4th generation (4G) or next-generation of wireless networks. The heterogeneous wireless access, the exclusive all IP-based architecture and the advanced mobility support are the key drivers of this generation (Hossain, 2008).

All the technologies, which are behind the heterogeneous networks poses the characteristics that complement each other (Zahran, Liang, & Saleh, 2006). 3G and 2G-based cellular communication technologies are well-known for wide area coverage, complete mobility, and roaming. However, traditionally these technologies offer low bandwidth, and expensive data traffic solutions (Chalmers, Krishnamurthi, & Almeroth, 2006). LTE technology is developed in response to overcoming the limitations of the conventional cellular technologies. On the other hand, WLANs provide high data rate at low cost, but with limited coverage, whereas WiMAX delivers last mile mobile broadband access and backhaul for WLANs (Fangmin, Luyong, & Zheng, 2007).

Hence, from network designing, planning, and troubleshooting perspectives, a QoS evaluation method is required that can bring these domains into a common platform. Generally speaking, the key performance parameters for these technologies are not directly comparable. For example, the delay ranges of UMTS and WLAN are entirely different. Therefore, a high value of delay measured from a WLAN may not be considered elevated in a UMTS environment. On the other hand, the applications running over them have the same QoS characteristics regardless of the communication technology they are utilizing. Hence, in this context, an application-based QoS evaluation approach is more applicable than a communication technology-based performance evaluation approach. The most current studies aim to evaluate the QoS of each application or radio access network individually. This method is useful in case of networks that use one type of communication technology. However, for heterogeneous networks, the presence of multiple types of applications and access technologies make the assessment of the overall performance challenging.

This chapter proposes a generalized metric based QoS evaluation method to unify multiple performance evaluation parameters into a single measurement metric. The metrics measure the performance of different entities in a network such as applications and radio access networks (RANs) and represent the measurement with a single value. This approach is evaluated using various simulation scenarios. The particular focus is on multimedia-based applications such as video conferencing (VC), video streaming (VS) and voice. These applications have distinct characteristics, necessitating special attention to their QoS evaluation methods.

Background

A conventional heterogeneous network constitutes of different communication technologies and diverse types of applications. Figure 1 shows a typical heterogeneous network scenario. In this network, there are several radio access networks, such as UMTS, LTE, and WLAN. Each of these access networks has several active applications; for example, UMTS has a few active voice calls and a VC session. Some users of VC session are in the WLAN and LTE network. The QoS evaluation of this type of network can be very complex in some situations. For instance, to measure the QoS of the UMTS network considering the performance of both voice calls and VC session is a challenging task. If service providers or network planners want to evaluate the QoS of the overall network configuration involving the performance of UMTS, LTE and WLAN will be a difficult case. The method proposed in this chapter evaluates the

Figure 1. A typical heterogeneous network scenario

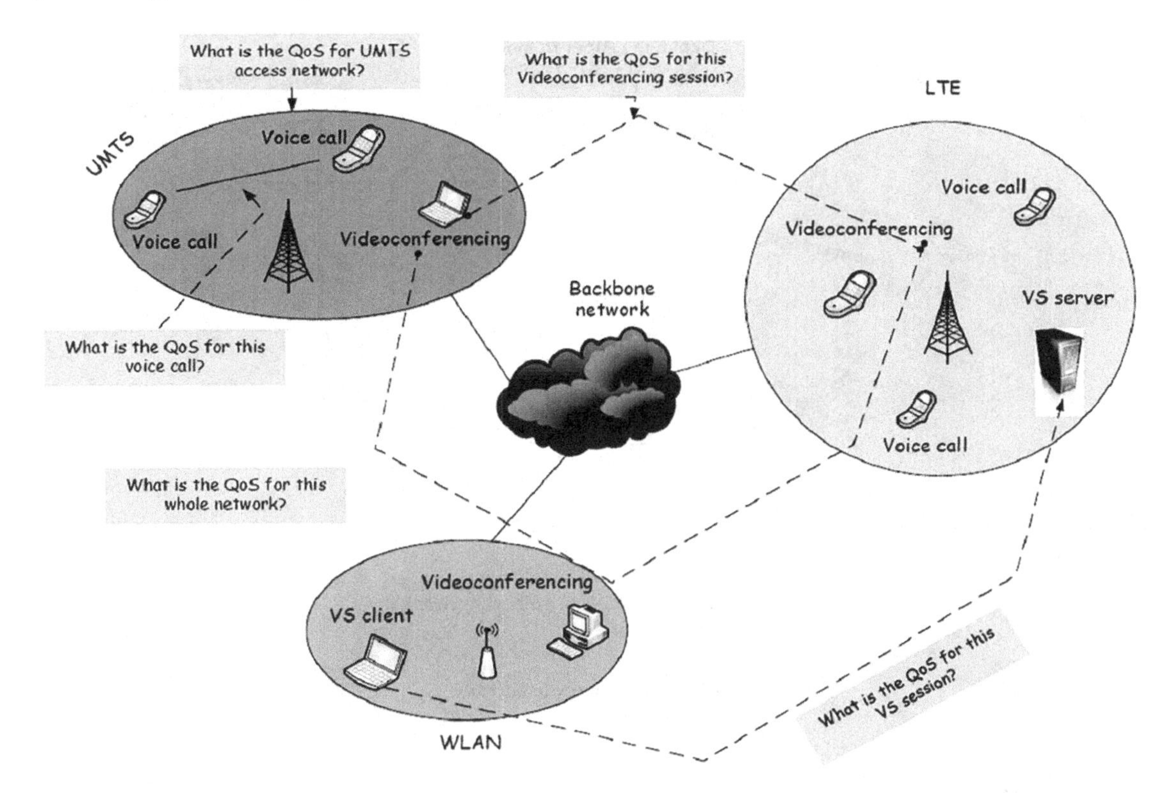

QoS of the applications, and the radio access networks individually and derives a unified network configuration QoS metric for the whole network. Therefore, this QoS evaluation method can handle the above-discussed situation efficiently.

Figure 2 shows a scenario of a heterogeneous network-based service model. For example, there is a network-based distance education model in a developed country *A*. Another developing country *B* wants to deploy the same service model. The question is in the network designing and planning stage, how to analyze the satisfactory QoS level for the users in country *B*, for such a model. A unified metric can facilitate the evaluation in such cases. The goal of this unified metric is to include various constraints of the network to measure the network QoS with a single numerical value.

The third case demonstrates a scenario where service operators want to deploy a health service-centric network in a rural area *B*. The scenario is depicted in Figure 3. They are considering several available technologies, such as 3G and WiMAX. While planning the configuration, they would consider the number of active users for the applications, relevant to health service and non-relevant to health service. It would be better from the planning perspective if they the achievable QoS level with such constraints and the ideal QoS level for this type of network beforehand. To analyze these situations, an analytical method that can evaluate the QoS level with a unified metric would be useful.

The unified QoS metric can also be helpful in selecting a suitable access network. Although, some classical access selection algorithms use an integrated QoS metric for this purpose, in this case, the considered situations are different, though. Figure 4 shows the scenario of such an access network selection. For example, there are three networks: X, Y, and Z. These networks have the following specifications:

Figure 2. Deploying the network-based service model of a developed country in a developing country

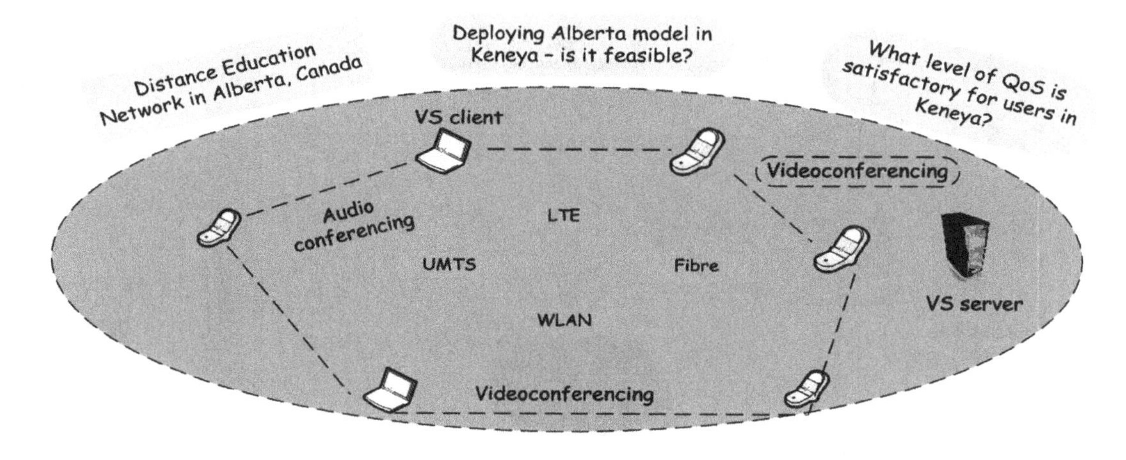

Figure 3. Deploying a new network

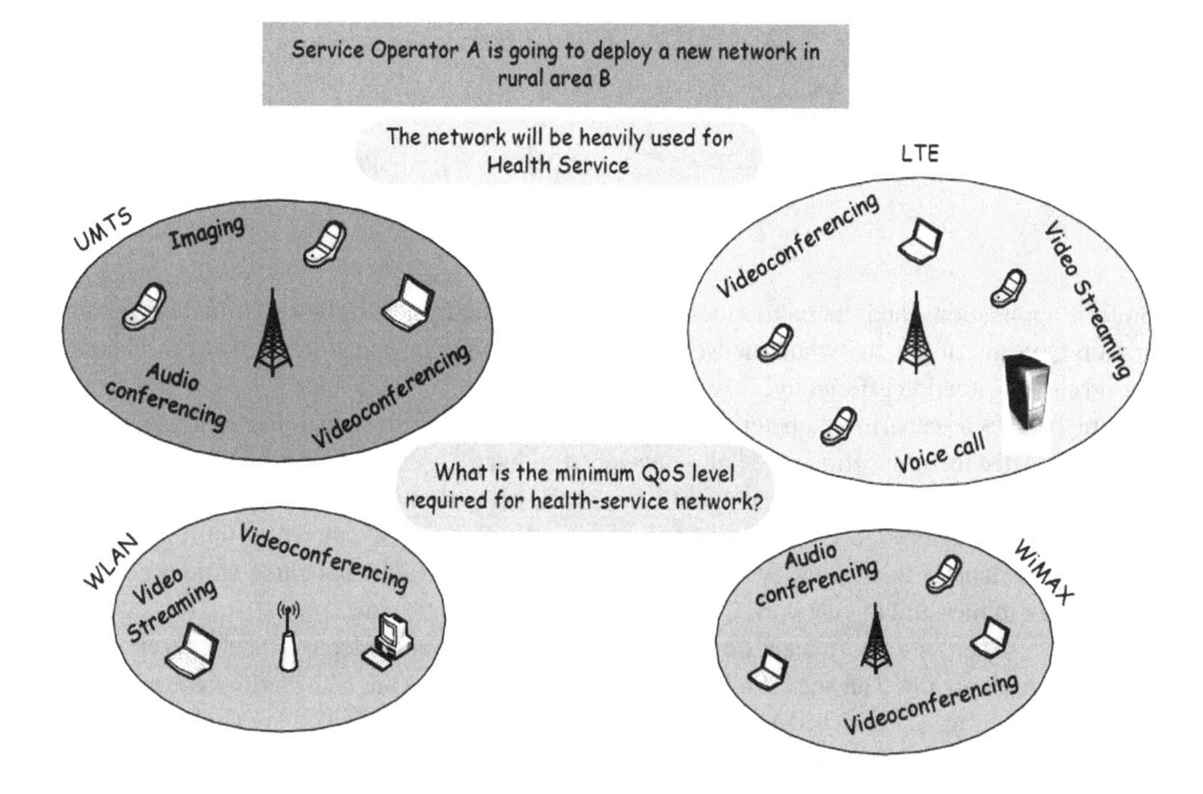

- Network X:
    - Radio Access Network: UMTS, WLAN
    - Active users:
        - Numbers of VC users: 5
        - Numbers of VS clients: 25
        - Numbers of Voice calls: 50

- Network Y:
  - Radio Access Network: LTE, WLAN
  - Active users:
    - Numbers of VC users: 10
    - Numbers of VS clients: 20
    - Numbers of Voice calls: 40
- Network Z:
  - Radio Access Network: LTE, WLAN, UMTS
  - Active users:
    - Numbers of VC users: 30
    - Numbers of VS clients: 40
    - Numbers of Voice calls: 40

Now if a user *A* wants to join a video lecture, which network is the most suitable one from their perspective? Alternatively, another user *B* in a rural area *K* wants to join a VC session with a doctor in an urban area *J*, and needs to choose the most suitable network from their standpoint? Clearly, these matters relate to QoE and its measurement, as well. To manage these situations, a QoS evaluation method that can evaluate the overall network RAN and application performance individually with single QoS metrics would be useful.

QoS evaluation in heterogeneous networks has been an active area of research. Different studies have taken it from different perspectives. The major studies can be divided into a few groups; these are roaming across heterogeneous networks, resource utilization, adoption, resource allocation, and multimedia QoS issues across heterogeneous networks. Roaming across heterogeneous networks can be further divided into call admission control and vertical handover (VHO).

An end-to-end QoS framework for heterogeneous wireless networks is proposed in (Xia, Gang, & Miki, 2004). The authors integrate a three-pane network architecture of a unified based hierarchical policy management framework. A game-theoretic framework for radio resource management is proposed

Figure 4. Access network selection

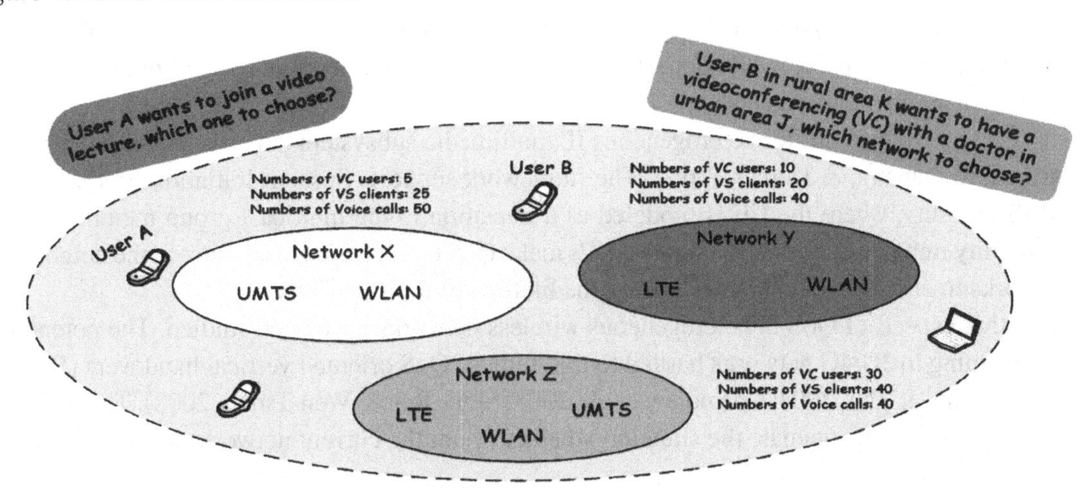

based on allocated bandwidth and the capacity reservation threshold to maintain a certain level of QoS performance of the ongoing connections in the heterogeneous networks (Niyato & Hossain, 2008).

One of the key focuses of many research works related to QoS provision for heterogeneous networks is about resource allocation to mobile users on a call or connection level. Probabilities of the new call blocking and existing call dropping are two main QoS metrics that are widely studied (Levine, Akyildiz, & Naghshineh, 1997; Niyato & Hossain, 2008). The call admission algorithm determines whether an incoming call can be admitted or not in the network, based on the efficient allocation of the limited bandwidth maintaining the guaranteed users' QoS requirements at the same time. A call is forced to terminate if the target cell does not have enough bandwidth to support the connection while opting for a handover. Therefore, a network that has some buffer bandwidth for the expected handovers is likely to have lower connection dropping probability and hence better QoS (Hasib & Fapojuwo, 2008). A distributed resource application algorithm for heterogeneous networks is proposed in (Ahluwalia & Varshney, 2007). The algorithm enables each network base station or access point to perform its resource allocation rather than the central resource manager. Most of the other available resource management algorithms, in this case, use a central resource manager (Xuebing, Tao, Daiming, Guangxi, & Liu, 2010).

Bandwidth reservation techniques are another widely studied topic for QoS provision in heterogeneous wireless networks. Some studies have suggested applying the expert knowledge of the traffic pattern of a particular area and the estimation of channel occupancy time distribution for bandwidth reservation (Lu & Koutsakis, 2011). To reduce the call dropping probability and the call blocking probability, and to maximize the bandwidth utilization, the previous movements of the mobile users are used for bandwidth reservation (Rashad, Kantardzic, & Kumar, 2006). The local and global mobility profiles for mobile users are generated to predict the future path of a mobile user.

Due to the unpredictable nature of wireless networks, the mobile applications, especially the multimedia applications should be able to adapt to changing QoS condition. As a result, the QoS provision of multimedia applications over heterogeneous networks is another extensively researched topic. A novel quality-aware the adaptive concurrent multipath transfer solution is proposed in (Changqiao, Tianjiao, Jianfeng, Hongke, & Muntean, 2013). The main idea behind this algorithm is the continuous observation and analyze of each path's data handling capability to take a decision on the data delivery adaptation. Some of the studies also consider QoE-driven adoption scheme for multimedia applications (Changqiao et al., 2013).

A QoS-aware routing algorithm is proposed in (Jammeh et al., 2012). The authors propose this algorithm according to the analysis of a policy-based QoS supporting infrastructure. In a dynamic network environment, the policy-based QoS supporting system enables the operators to accommodate policies to meet ever-changing requirements.

A mobile QoS framework for heterogeneous IP multimedia subsystem (IMS) is presented in (Vasu, Maheshwari, Mahapatra, & Kumar, 2012). The framework supports Session Initiation Protocol (SIP)-based IMS mobility, where the UE is modeled as a transition in the multicast group membership. To reduce mobility impact on service guarantees, UEs make QoS reservations in advance at the neighboring IMS networks to ensure QoS guarantee during the lifetime of their sessions.

Many other aspects of QoS in heterogeneous wireless networks have been studied. The potential of seamless roaming in 3G/4G networks has led to the study of QoS oriented vertical handovers (Chuanxiong, Zihua, Qian, & Zhu, 2004; Garmonov et al., 2008; Shun-Ren & Wen-Tsuen, 2008). The main idea behind these studies is to evaluate the situation whether or not the current network can satisfy the QoS requirements of the considered application. If the current network is unable to meet the QoS requirements of the considered application, then the Mobile Terminal (MT) moves to a suitable network.

Resource management solutions in integrated networking architecture comprising various wireless communication networks such as GPRS and UMTS, IEEE 802.11 and WiMAX are also widely studied. An efficient load-balancing based policy framework for resource management in a loosely-coupled cellular/WLAN integrated is presented in (Dong & Maode, 2012). The policy uses a two-phase control strategy. It uses call assignment to provide a statistical quality of service guarantee during the admission phase. Then, the dynamic vertical handover during the traffic service phase is used to minimize the performance variations. A generic reservation-based QoS model for the integrated cellular and WLAN networks is proposed in (Wei, Weihua, & Yu, 2007). The model ensures the delivery of adaptive real-time flows for end users by taking the advantage of high data-rate WLAN systems and the wide coverage area of cellular networks.

Other topics that are studied related to QoS in heterogeneous networks are signaling protocol, hardware robustness, and energy efficiency. Effective signaling protocols ensure the seamless exchange of information between wireless terminals and various components of the integrated network (Chang, 1996; Wang, Min, Mellor, Al-Begain, & Guan, 2005). The optimization of power control and base station assignment can affect the resource management and QoS in wireless networks (Douros & Polyzos, 2011; Papavassiliou & Tassiulas, 1996).

The QoS analysis that considers different aspects of a heterogeneous network is necessary to evaluate and enhance the application-based performance. To address this necessity, several studies (Alshamrani, Xuemin, & Liang-Liang, 2011; Karimi & Fathy, 2010) have examined the application-centric performance of these networks. Nevertheless, the establishment of an integrated QoS metric for this type of configuration as a whole is still a challenging task (Luo & Shyu, 2011). Most of the existing research discussed above focuses on the partial QoS evaluation of a heterogeneous network by deriving the QoS level of a single access network and a single application present within a heterogeneous environment. Also, different studies have come up with various performance metrics for QoS evaluation of these networks. The proposed approach in this chapter focuses on combining the various performance metrics together to come up with a unified QoS metric to evaluate the QoS of heterogeneous network configuration as a whole. It integrates the application QoS level and the network QoS level in heterogeneous wireless networks under various traffic and mobility scenarios.

THE GENERALIZED METRIC-BASED QoS ANALYSIS

In this section, the steps of the proposed generalized metric-based QoS evaluation method are discussed in detail. The approach treats the heterogeneous network as two separate entities, which are the active applications and RANs. The overall network configuration consists of both of these entities and works as an overlaying layer of these two entities. Figure 5 shows the conceptual diagram. A heterogeneous wireless network N is considered, which has $j\left(j = 1, 2, 3, \ldots, n\right)$ number of radio access networks. Each of this access networks contains $i\left(i = 1, 2, 3, \ldots, m\right)$ number of applications.

The key QoS-related parameters of each application and their acceptable ranges have been identified in (Farid, Shahrestani, & Ruan, 2013). The proposed approach applies those ranges to calculate the application, the RAN and the network configuration QoS metric.

Figure 5. QoS analysis: entities of a network

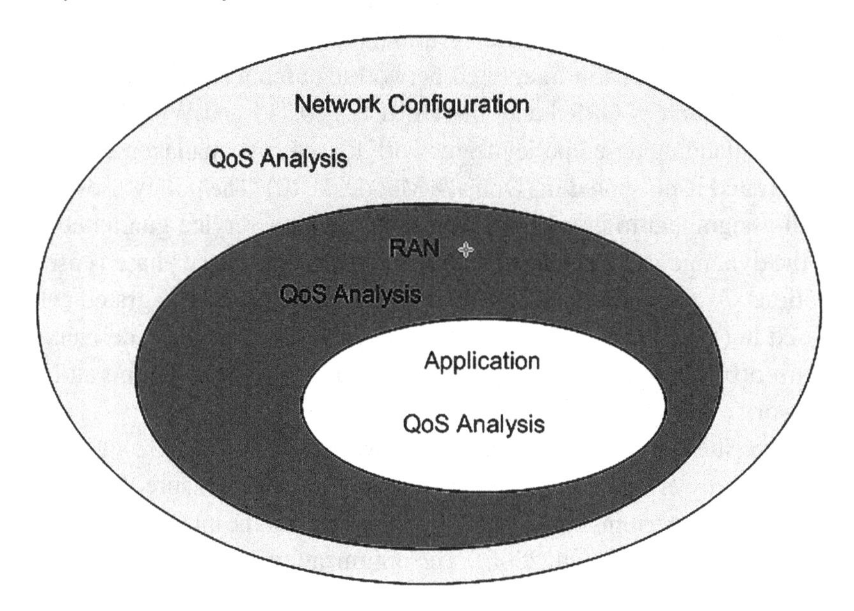

The variables, which are used for this method, are specified as follows:

- QoS-related parameter: QP
- Application: A
- Radio Access Network: R
- Network Configuration: N
- A QoS-related parameter related to any Application A, in any RAN, R: $QP_{k,A,R}$,where k is the index for the QoS-related parameters and $k=\{1,2,\ldots,p\}$.
- The upper bound for the benchmark range of QP_k in any application A: $UB_{QP_{k,A}}$
- The lower bound for the benchmark range of QP_k in an application: $LB_{QP_{k,A}}$
- The weight of a QoS-related parameter QP_k in any application A: $W_{QP_{k,A}}$
- The weight of any application A, in a *RANR*: W_A^R
- The weight of any RAN, R in a network N: W_R^N
- The performance metric for a QoS-related parameter QP_k : $P_{QP_{k,A}^R}$
- The real-time performance metric for a QoS-related parameter QP_k : $RTP_{QP_{k,A}^R}$
- The application QoS metric for an application A, in a RAN, R: $QoSAM_A^R$
- The radio access network QoS metric for any RAN, R in a Network N: $QoSRM_R^N$
- Network Configuration QoS Metric: $QoSCM_N$

Figure 6. Workflow for the proposed QoS evaluation method

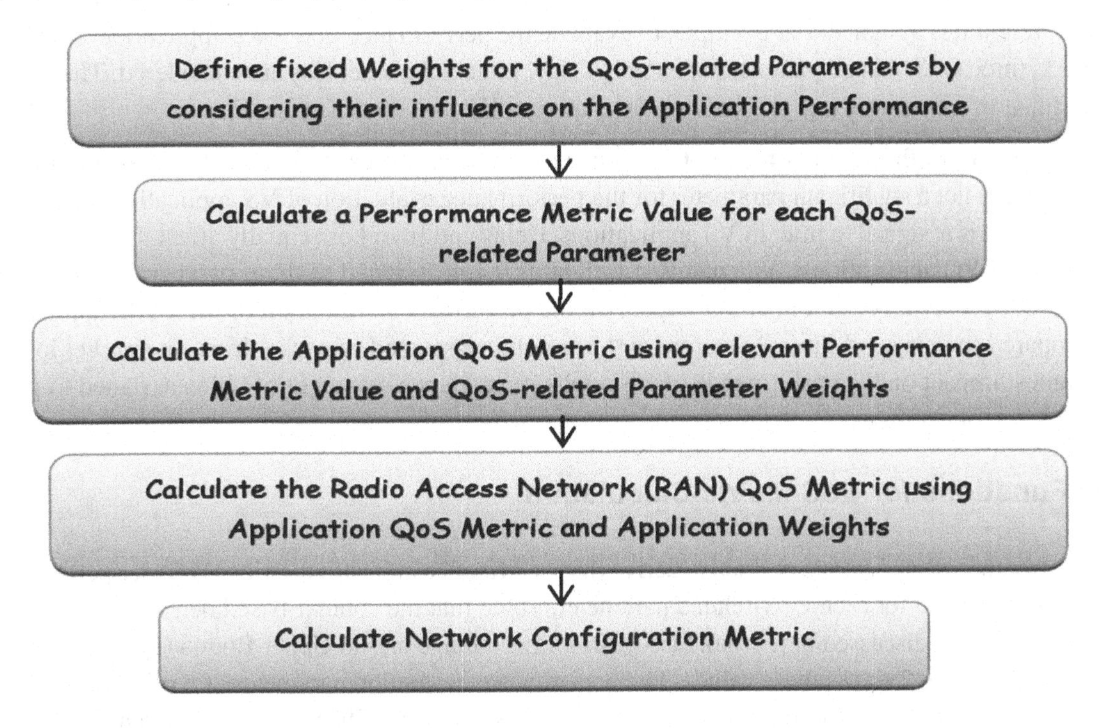

The steps of the proposed method are presented in Figure 6.

1. In the first step, the weights for the key QoS-related parameters are determined for each application. These weights are defined based on their impacts on the application performance. For example, jitter plays a crucial role in the performance evaluation of voice and video conferencing applications; however, it does not affect the performance of video streaming applications. As a result, while assigning weight, the jitter is assigned a higher value for the VC and voice applications.
2. In the second step, a performance metric is calculated for each application-related QoS parameter.
3. These performance metrics and the weights of the QoS-related parameters are used to compute the QoS metric for each application.
4. Then the RAN QoS metric is derived using the application QoS metrics and the weight of each application.
5. The final network configuration QoS metric is calculated using the RAN QoS metrics and their weights.

The following sections present each of these steps in more detail.

Calculate the QoS-related Parameter Weights

The weights are determined based on the performance-related impacts that each QoS-related parameter has on an application. The key QoS-related parameters for the performance evaluation of the voice, VC and VS have been discussed in (Farid et al., 2013). In this chapter, to establish the weights of those parameters, a scale that ranges between 0 and 1 is defined where 1 represent the highest importance level and 0 depicts the lowest. The more impact the parameter has on any application performance, the higher weight it is assigned. For example, to evaluate the performance of a voice application in a UMTS network, three QoS-related parameters, namely delay, jitter and packet loss are considered. The weights are defined in such a way that $W_{QP_{\text{Delay,Voice}}} + W_{QP_{\text{Jitter,Voice}}} + W_{QP_{\text{Packet Loss, Voice}}} = 1$.

Table 1 shows the weights for the QoS-related parameters of each considered application in this work. Jitter is not a significant parameter for the performance evaluation of VS applications. Therefore, a zero weight is assigned to jitter in VS applications. Delay and Jitter both equally affect the performance of voice and VC applications. As a result, a weight of 0.4 is assigned to these two parameters. On the other hand, packet loss has a minimal impact on the performance evaluation of these two applications as compared to delay and jitter. Therefore, a 0.2 weight is assigned to packet loss. The packet loss has the highest impact on the performance of VS application. Hence, a weight of 0.8 is assigned to packet loss in VS application.

The Functions for QoS Metric Calculation

In this step, three functions are defined to derive the performance indicator metrics. At first, a performance metric is calculated for each QoS-related parameter. To do that the context-based benchmark ranges are applied which are discussed in (Farnaz Farid, 2012). Each range has an Upper Bound (UB) and a Lower Bound (LB). Table 2 shows these values. The performance evaluation parameter of a network can be divided into two categories; these are the benefit and the cost category. The parameters, which are preferred to have a maximum value, such as throughput, falls within the benefits category. On the other hand, the parameters, which are preferred to have minimal costs, are regarded as the cost type parameter. Delay, jitter, and packet loss all these QoS-related parameters fall within the cost category. In the proposed method, only the cost-based parameters are used for the performance evaluation of any application.

The function to derive the parameter-related performance metric uses a real-time performance metric, and the upper and lower bounds of the specific parameter. The real-time performance metric value is referred to as $RTP_{QP^{R}_{k,A}}$. This value is collected for each $QP_{k,A,R}$ from an ongoing session of an application A in a RAN R at some time interval. The function is as follows:

Table 1. Weights or different QoS parameters

QoS metric	Application	Weight
Packet loss	Voice	0.2
	VS	0.8
	VC	0.2
Delay	Voice	0.4
	VS	0.2
	VC	0.4
Jitter	Voice	0.4
	VS	0
	VC	0.4

Table 2. Context-based ranges

Applications		Packet loss (%)		Delay(msec/sec)		Jitter(msec)	
		LB	UB	LB	UB	LB	UB
Voice	Rural	3	20	150 msec	400 msec	30	75
	Urban	1	3	100 msec	150 msec	1	30
VS	Rural	3	5	5 sec	10 sec	n/a	n/a
	Urban	1	3	1 sec	5 sec	n/a	n/a
VC	Rural	2.5	5	100 msec	400 msec	30	75
	Urban	1	2.5	100 msec	150 msec	1	30

$$P_{QP^R_{k,A}} = \begin{cases} 0 & , \quad RTP_{QP^R_{k,A}} \geq UB_{QP_{k,A}} \\ 1 & , \quad RTP_{QP^R_{k,A}} \leq LB_{QP_{k,A}} \\ \dfrac{UB_{QP_{k,A}} - RTP_{QP^R_{k,A}}}{UB_{QP_{k,A}} - LB_{QP_{k,A}}} & , \quad UB_{QP_{k,A}} > RTP_{QP^R_{k,A}} > LB_{QP_{k,A}} \end{cases} \tag{1}$$

For example, to evaluate the performance of an ongoing VC session in the UMTS access network, three QoS-related parameters, namely delay, jitter and packet loss are considered. The real-time performance metric for each of these parameters is expressed as $RTP_{QP^{U}_{D,VC}}$, $RTP_{QP^{U}_{J,VC}}$, and $RTP_{QP^{U}_{PL,VC}}$,where D, J, and PL refer to as delay, jitter, and packet loss respectively, and U refers to as UMTS network.

If $RTP_{QP^{U}_{D,VC}} > UB_{QP_{D,VC}}$ then $P_{QP^{U}_{D,VC}} = 0$

If $RTP_{QP^{U}_{J,VC}} < LB_{QP_{J,VC}}$ then $P_{QP^{U}_{J,VC}} =1$

If $UB_{QP_{PL,VC}} > RTP_{QP^{U}_{PL,VC}} > LB_{QP_{PL,VC}}$ then

$$P_{QP^{U}_{PL,VC}} = \frac{UB_{QP_{PL,VC}} - RTP_{QP^{U}_{PL,VC}}}{UB_{QP_{PL,VC}} - LB_{QP_{PL,VC}}}$$

If the real-time performance metric value of any parameter is greater than or equal to the acceptable upper bound of that parameter, the parameter related performance metric value is calculated as zero. In the above calculation example, it is assumed that the delay in the VC application at some time interval shows a greater value than the acceptable upper bound of the delay. If this real-time performance metric value of any parameter is lower than or equal to the acceptable lower limit of that parameter, then this metric value is one. This is the highest possible value for any performance metric. In the above example, jitter has the highest performance value. If this real-time metric value lies between the upper and lower bound of any parameter, then the performance metric is calculated using the normalization function. In the above example, packet loss has a value between the acceptable lower and upper bound of packet loss.

Next, the performance metric for each active application presented in an access network is calculated using the application metric calculation function. This function combines the weights and the performance metric of the QoS-related parameters together. With $W_{QP_{k,A}}$ corresponding to the weight of the QoS-related parameter, QP_k in an application A and $P_{QP^R_{k,A}}$ representing the performance metric for the same this will be:

$$QoSAM_A^R = \sum_{k=1}^{p} P_{QP^R_{k,A}} W_{QP_{k,A}} \tag{2}$$

where $k=\{1,2,\ldots,p\}$ and $1 = \text{Delay}, 2 = \text{Jitter etc}$. For example, there is a VC session running in a UMTS network. The application QoS metric for this VC session is evaluated as:

$$QoSAM_{\text{VC}}^{\text{UMTS}} = P_{QP^{\text{UMTS}}_{\text{D,VC}}} W_{QP_{\text{D,VC}}} + P_{QP^{\text{UMTS}}_{\text{J,VC}}} W_{QP_{\text{J,VC}}} + P_{QP^{\text{UMTS}}_{\text{PL,VC}}} W_{QP_{\text{PL,VC}}}$$

where $\text{D} = \text{Delay}, \text{J} = \text{Jitter}$ and $\text{PL} = \text{Packet Loss}$.

Then the RAN metric calculation function is applied to derive the performance metric for each available RAN in the network. The function combines the values of the application QoS metrics and application weights to derive the QoS metric for a specific RAN. Suppose, there are m numbers of applications $(A_1 A_2, \ldots\ldots, A_m)$ in a RAN R and the weight for each of this application is defined as $(W^R_{A_1}, W^R_{A_2} \ldots\ldots, W^R_{A_m})$. An equal weight is distributed for the applications present in the access network to for normalization. The function is as follows:

$$QoSRM_R^N = \sum_{i=1}^{m} QoSAM^R_{A_i} W^R_{A_i} \tag{3}$$

where $i = \{1,2, \ldots,m\}$. For example, in the LTE-based access network, there are two applications VC and voice. In this case, the RAN QoS metric in the LTE network is derived as:

$$QoSRM^{\text{N}}_{\text{LTE}} = QoSAM^{\text{LTE}}_{\text{VC}} W^{\text{LTE}}_{\text{VC}} + QoSAM^{\text{LTE}}_{\text{Voice}} W^{\text{LTE}}_{\text{Voice}}$$

where $A_1 = \text{VC}$, $A_2 = \text{Voice}$ and R=LTE

The final function computes the network configuration performance. This function amalgamates the values of RAN performance metrics to derive the unified QoS metric for the network configuration. Each RAN is assigned an equal weight for normalization purpose. Suppose, there are n number of radio access networks $(R_1 R_2, \ldots\ldots, R_n)$ in a network N, and the weight for each of this RAN is defined as

$(W_{R_1}^{N}, W_{R_2}^{N} ...., W_{R_n}^{N})$. In this case, the weights are assigned in such a way that $\sum W_{R_j}^{N} = 1$. The function is expressed as follows:

$$QoSCM_N = \sum_{j=1}^{n} QoSRM_{R_j}^{N} W_{R_j}^{N} \tag{4}$$

where $j=\{1,2,3....,n\}$. For example, in a network N, there are two RANs: UMTS and LTE denoted as U and L respectively, the network configuration QoS Metric is calculated as:

$$QoSCM_N = QoSRM_{\mathrm{U}}^{\mathrm{N}} W_{\mathrm{U}}^{\mathrm{N}} + QoSRM_{\mathrm{L}}^{\mathrm{N}} W_{\mathrm{L}}^{\mathrm{N}}$$

where $R_1 = \mathrm{UMTS}$ and $R_2 = \mathrm{LTE}$

SIMULATION SCENARIOS

This section discusses the setups for the simulation scenarios, which have been designed to evaluate the performance of the proposed QoS evaluation method. Primarily, the simulation scenarios are divided into three major categories, and they are:

- UMTS/UMTS-WiMAX-based network scenarios
- LTE-UMTS/LTE-based network scenarios

Each of these categories has several sub-scenarios. It is assumed for each scenario that there is a network N with j numbers of RANs, and each RAN has i numbers of active applications.

Table 3. Simulation parameters for the UMTS network

Layers	Parameters				
Application		QoS class		Payload size	
	Voice calls	Conversational		224 bits	
	VS	Background		Variant	
	VC	Conversational		576 bytes	
RLC	Mode	Unacknowledged			
	Timer MRW (msec)	900			
	Timer Discard (msec)	7500			
	MAX MRW	6			
	MAX DAT	4			
PHY	Channel type	DCH			
		Conversational		Background	
		Bit rate (Kbps)	TTI (msec)	Bit rate (Kbps)	TTI (msec)
		64	10	144	20

UMTS-WiMAX-Based Heterogeneous Network Scenarios

A UMTS-WiMAX and a UMTS only network are considered with Voice, VC, and VS applications in these scenarios. The UMTS and the WiMAX model of OPNET are used for the simulations as these models are well established. The rural outdoor environment is chosen for these simulations. Some of the simulation parameters are stated in Table 3. A detail of these simulation scenarios has been presented in (Farid et al., 2013). The highest bitrate chosen is 144 Kbps as in a rural area usually the bandwidth is lower than the UMTS standard bandwidth of 2 Mbps. In the first phase of the simulations, a UMTS network in the context of a rural area is simulated. The number of voice calls is used from a range of 8 to 20. The values of packet loss, delay, and jitter are evaluated in each scenario. The scenarios are run for 13 times with different seed values and 95% confidence interval. In the second phase of the simulations, a VS application is added to the network. Two separate video codecs, H.263, and MPEG-4, are used for the VS client. The VS server is placed in the WiMAX environment. The performance of the voice application is evaluated with the presence of different codec-based video streaming clients. This assessment is conducted to identify the optimal number of voice and streaming users for this sort of network settings. In the third stage of simulations, a VC session is added to the voice and VS sessions.

IP-based calls are used for voice communications. 7.4 kbit/s mode Adaptive Multi-Rate (AMR) speech codec is used for these calls with an activity factor of 0.5. A typical AMR packet runs for 20 msec with a payload size of 224 bits. VS applications are the traditional mediums for distance education. The video codec used, in this case, are the H.263 and MPEG-4 as these two are the most popular video codec for mobile devices (Yao, 2012). Two video trace files encoded with H.263 and MPEG-4 are used for simulations. The trace files are collected from (TKN). The parameters for video trace files are specified in Table 4. The Group of Pictures (GoP) structure is denoted using GgBb. Here, g defines

Table 4. Streaming trace specification of H.263 and MPEG-4

Codec	Performance metrics	
	Metric	Value
H.263	QP	5
	Resolution	QCIF (177*144)
	Encoder	tmn encoder (Version 2.0/3.2)
	Bit rate	16 kbit/s
	USE of PB-frames	ON
	Length	3603.28 sec
	No. Frames	19436
MPEG-4	Layer	Single Layer
	Encoder	MOMUSYS MPEG-4
	Frame Size	QCIF 176x144
	No. Frames	89998
	Number of GoP	7499
	INTRA PERIOD	12,IBBPBB,PBBPBB
	QP	5

Figure 7. UMTS-based network scenario

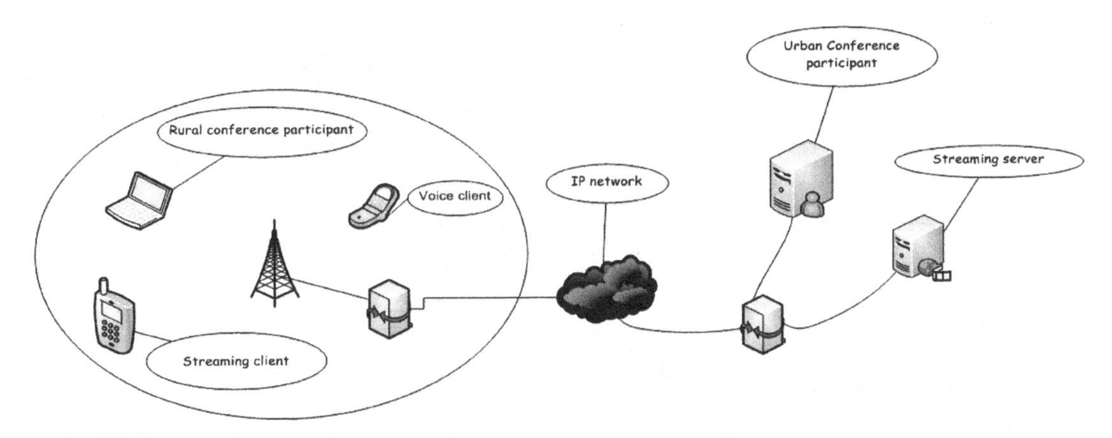

the total number of frames in a GoP, and b indicates the number of B frames between successive I or P frames. B frames are the bi-directionally predictive coded frames, I frames are intra-coded frames and P are predictive coded frames. The VC session uses a frame rate of 15 fps. The packet size used for the VC application is 576 bytes. The FDD version of UMTS is considered in these simulation scenarios. A free space path loss model is used, which is the most suitable model for a rural environment. The throughput-based admission control algorithm is used and in the downlink, the other-cell interference factor is set to 0.65. Figure 7 shows the simulation setup scenario.

LTE-UMTS-Based Heterogeneous Network Scenarios

In these scenarios, LTE-UMTS and LTE only network are considered with Voice, VC, and VS applications. In this case, also the LTE model of OPNET is used for the simulations. The environments chosen for these simulations are also the rural outdoor environment. Some of the simulation parameters are stated in Table 5.

In the first phase of the simulations, the LTE-based network is also simulated in the context of a rural area with various numbers of voice application users starting from 8 to 20. Like UMTS-based scenarios, the values of packet loss, delay, and jitter are evaluated, and they are run for 13 times with different seed values and 95% confidence interval. The application settings in the LTE network are same as the UMTS network.

In the second phase of the simulations, LTE-UMTS-based networks in the context of a rural-urban area are simulated with various numbers of voice calls, VS, and VC sessions. In some of the scenarios, the VS server is placed in the LTE RAN. For some of them, the voice calls are initiated between the UMTS and LTE network.

SIMULATION RESULT ANALYSIS

This section presents the result analysis from the above-discussed simulation scenarios. Figure 8 shows the packet loss for voice calls in the presence of MPEG-4 and H.263 VC client. The voice calls experience a better performance in the presence of H.263 clients. The simulation results clearly indicate that

Table 5. Simulation parameters for LTE

Layers	Parameters		
Application		QoS class	Payload size
	Voice calls	Interactive Voice	224 bits
	VS	Best Effort	Variant
	VC	Interactive Voice	576 bytes
PHY -UL	Modulation and Coding Scheme	9	
	Timer MRW (msec)		
	Base Frequency	1920MHz	
	Bandwidth	20 MHz	
PHY-DL	Base Frequency	2110 MHz	
	Bandwidth	20 MHz	
EPS Bearer	QoS Class Indetifier	GOLD	
	Uplink Guranteed Bit Rate (bps)	64 Kbps	
	Downlink Guranteed Bit Rate (BPS)	64 Kbps	
	UL Maximum Bit Rate	64 Kbps	
	DL Maximum Bit Rate	64 Kbps	
	Aoolocation Retention Priority	1	
	QoS Class Indetifier	Silver	
	UL Guranteed Bit Rate (bps)	384 Kbps	
	DL Guranteed Bit Rate (bps)	384 Kbps	
	UL Maximum Bit Rate (bps)	384 Kbps	
	DL Maximum Bit Rate (bps)	384 Kbps	
	Allocation Retention Priority	4	
	QoS Class Indetifier	Bronze	
	UL Guranteed Bit Rate (bps)	384 Kbps	
	DL Guranteed Bit Rate (bps)	384 Kbps	
	UL Maximum Bit Rate (bps)	384 Kbps	
	DL Maximum Bit Rate (bps)	384 Kbps	
	Allocation Retention Priority	5	

the voice calls experience less packet loss in the presence of an H.263-based VS client. The figure also shows that H.263 VS client with ten simultaneous voice calls experiences a higher packet loss than other situations. This is due to poor coverage experienced by some of the users in that particular scenario. As some of the callers moved far away from the base station, the statistics showed more packet loss compared to other scenarios.

Table 6 presents the performance of a VS client while using H.263 and MPEG-4 codec. Regarding packet loss, H.263 codec shows a better performance under the UMTS RAN than MPEG-4 codec does. In the case of delay, the results indicate the same behavior. The simulation results also demonstrate that the network experiences better performance with one H.263 codec-based VS client and twelve simulta-

Figure 8. Comparison of packet loss for voice clients in the presence of different VC clients

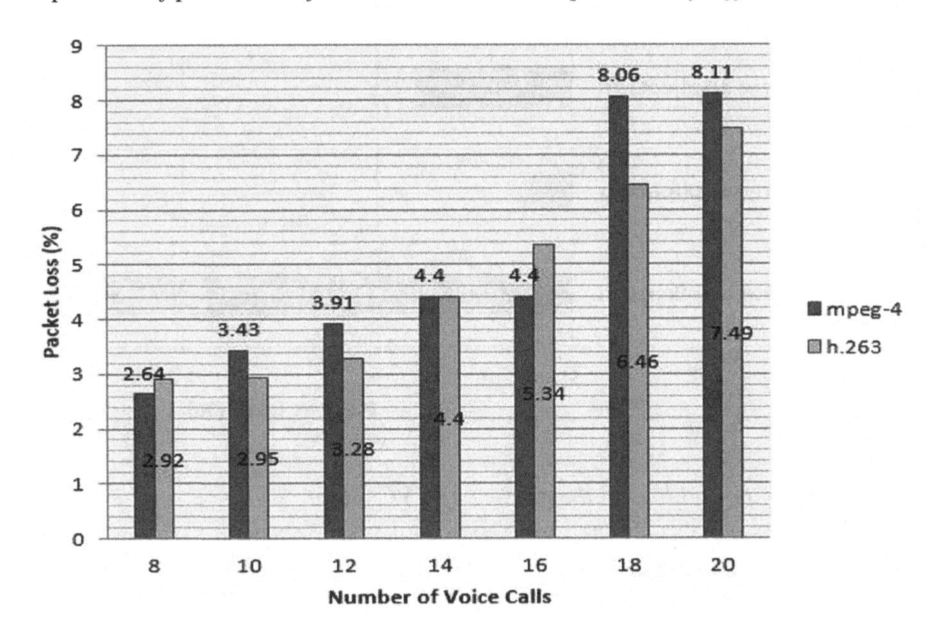

Table 6. Packet loss for streaming client in the presence of different number of voice clients

Number of Voice calls	Number of streaming client	Packet loss for different codec	
		MPEG-4	H.263
8	1	5.58	1.82
10	1	5.56	1.85
12	1	5.75	1.89
14	1	5.93	1.89
16	1	5.79	1.91
18	1	5.62	1.94
20	1	5.75	2.02

neous voice calls regarding packet loss. However, regarding the delay, the platform shows an acceptable performance with one H.263-based VS client and twenty simultaneous voice calls. On the other hand, the network shows a different behavior in the presence of the MPEG-4-based VS client. With the MPEG-4-based client, the optimal number is ten simultaneous voice calls and one VS client.

Figure 9 shows the comparisons of packet loss for different applications when voice, VS and VC applications are present on the network. VC influences the performance of voice calls to a great extent. Figure 10 compares the percentage of packet loss for voice and streaming client with and without the presence of video conferencing client. The results indicate that the simultaneous existence of the VC client affects the performance of voice calls and streaming applications to a noticeable extent.

Table 7 shows the data for the LTE network with a different number of voice calls. The voice calls demonstrate a better performance compared to the UMTS network.

Figure 9. Comparison of packet loss for different applications

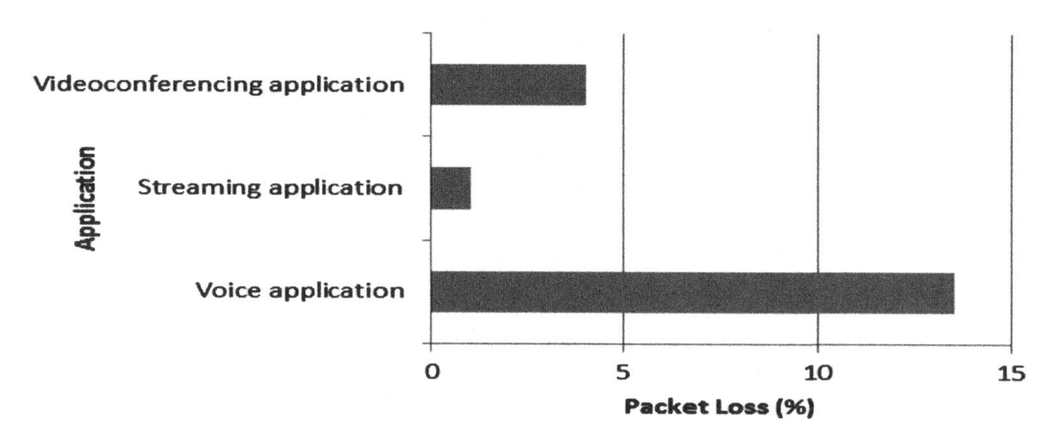

Figure 10. Comparison of packet loss in the presence of VC client

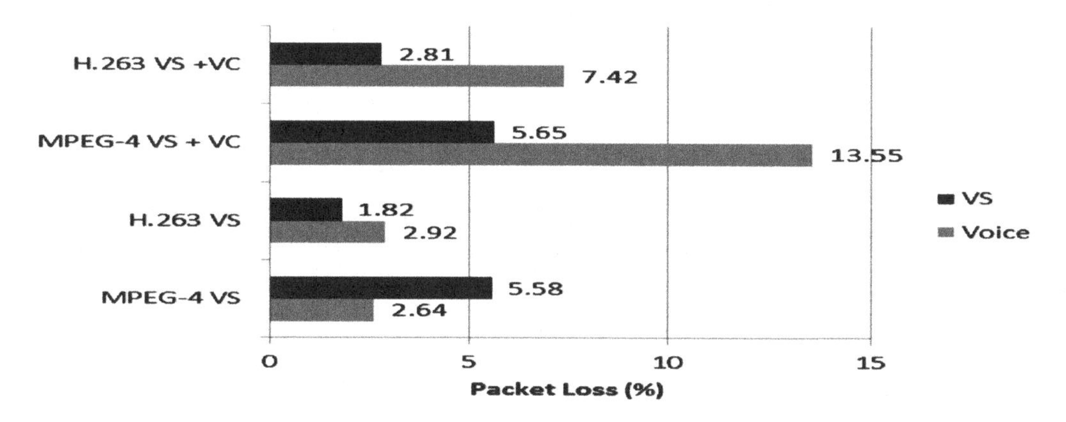

Table 7. Data analysis for LTE network

Number of voice calls	End-to-end delay (msec)	Packet loss (%)	Jitter (msec)
8	0.092	0.02	20
10	0.092	0.06	24
12	0.093	0.08	28
14	0.094	0.12	30
16	0.095	0.15	32
18	0.099	0.18	34

Figure 11 shows the comparison of packet loss for VS client when the LTE and the WiMAX-based servers are used for the VS session. The VS client is placed in the UMTS network. Figure 12 shows the comparison of end-to-end delay for the same. If the server side uses the LTE technology, the VS client experiences less packet loss and delay compared to the WiMAX-based network.

Figure 11. Comparison of packet losses for streaming client in LTE and WiMAX

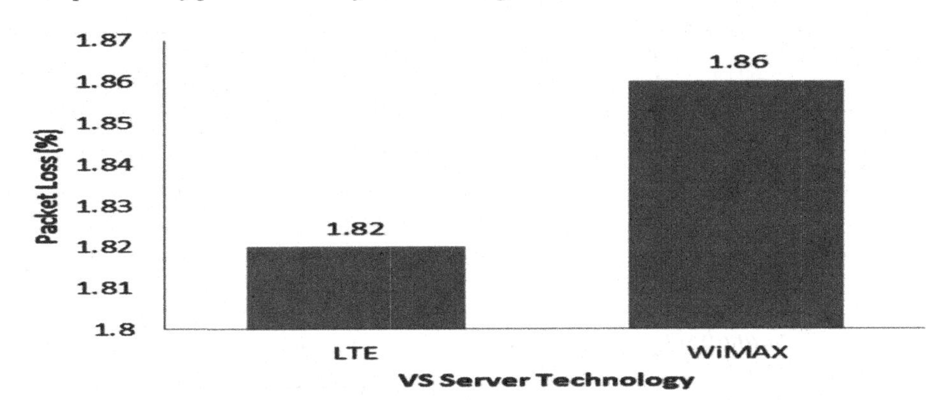

Figure 12. Comparison of end-to-end delay for streaming client in LTE and WiMAX

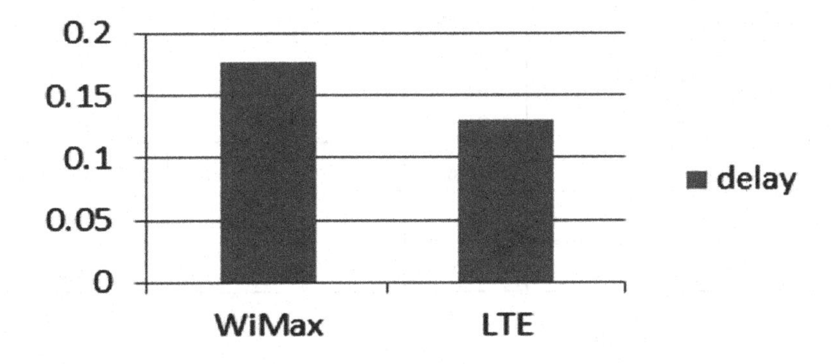

PERFORMANCE ANALYSIS USING THE PROPOSED APPROACH

The simulation result analysis in the previous section demonstrates that the assessment of network performance by investigating the value of each QoS-related parameter individually is a challenging task. Different communication technologies and varying numbers of users in the picture makes it even more complicated. The analysis of various scenarios shows that in some cases of VC session, although the value of end-to-end delay is acceptable, the amount of packet loss is below the acceptable level. On the other hand, VC and voice conferencing show inconsistent performance under different communication technologies.

The proposed unified metric-based method can handle these complexities of QoS analysis by using the QoS metric evaluation functions stated earlier in this chapter. The ranges of these QoS metrics are first interpreted. As stated earlier, there are three QoS metrics to evaluate the network performance. The first one is the application QoS metric, the second one is the RAN QoS metric, and the final one is the network configuration QoS metric. A scale that ranges between 0 and 1 is used for each QoS metric; 1 being the highest and zero being the lowest level. Table 8 shows these scales.

Table 9 outlines the QoS analysis results of the UMTS-based scenarios in the presence of a different number of voice conferencing participants. The results presented in Table 9 are illustrated in the following calculations in detail. In all these calculations, the following denotations are used:

Table 8. Interpretation of QoS metric values

QoS Metric	Value Range	Interpretation		
		Good	Average	Poor
Application QoS Metric	0 - 1	0.8-1	0.6-0.8	0-0.6
Access Network QoS Metric	0 - 1	0.8-1	0.6-0.8	0-0.6
Network Configuration QoS Metric	0 - 1	0.8-1	0.6-0.8	0-0.6

Table 9. QoS metrics from analysis of voice calls

Num-ber of voice calls	Delay (msec)	PL (%)	J (msec)	$P_{QP^R_{k,A}}$	$P_{QP^R_{k,A}}$	$P_{QP^R_{k,A}}$	$QoSAM^R_A$	Overall QoS
8	202	2.06	20	0.792	1	1	0.92	Good
10	206	2.90	22	0.776	1	1	0.9104	Good
12	209	2.91	25	0.764	1	1	0.91	Good
14	210	2.94	30	0.76	1	1	0.90	Good
16	213	2.98	45	0.748	1	0.67	0.77	Average
18	227	3.23	68	0.692	1	0.04	0.54	Poor

UMTS – U

Voice application –V

End-to-end delay – D

Jitter – J

Packet loss – PL

For eight simultaneous calls, the participants experience an average end-to-end delay of 202 msec, an average packet loss of 2.06% and an average jitter of 20 msec. The performance metric of these parameters for this scenario using Equation 1 is calculated as:

$$\mathrm{m}_{\mathrm{d(voice)}} = \frac{400-216}{400-216} = 0.74$$

$$P_{QP^{\mathrm{U}}_{\mathrm{J,V}}} = 1$$

$$P_{QP^{\mathrm{U}}_{\mathrm{PL,V}}} = 1$$

The application QoS metric using Equation 2 is calculated as:

$$QoSAM_{\mathrm{V}}^{\mathrm{U}} = 0.792 \times 0.4 + 1 \times 0.4 + 1 \times 0.2 = 0.9168 \ \mathrm{QoS}_{\mathrm{voice}}.$$

In this case, the RAN QoS metric and the network configuration metric have the same values as the application QoS metric. This is because the network has single application and single RAN. The results of other simulation scenarios have been calculated in a similar way.

Next, the QoS is analyzed for the next set of UMTS-WiMAX-based scenarios. As these scenarios have different radio access networks in the both ends of the connection, to distinguish between them, the Equation 1 is expressed in the following form:

$$P_{QP_{k,A}^{R,RS}} = \begin{cases} 0 & , \quad RTP_{QP_{k,A}^{R,RS}} \geq UB_{QP_{k,A}} \\ 1 & , \quad RTP_{QP_{k,A}^{R,RS}} \leq LB_{QP_{k,A}} \\ \dfrac{UB_{QP_{k,A}} - RTP_{QP_{k,A}^{R,RS}}}{UB_{QP_{k,A}} - LB_{QP_{k,A}}} & , \quad UB_{QP_{k,A}} > RTP_{QP_{k,A}^{R,RS}} > LB_{QP_{k,A}} \end{cases} \tag{5}$$

where *RS* is the technology on the other side of the connection.

The Equation 2 is expressed in the following form at one end of the connection:

$$QoSAM_{A}^{R,RS} = \sum_{k=1}^{p} P_{QP_{k,A}^{R,RS}} W_{QP_{k,A}} \tag{6}$$

The function for the RAN QoS that is the Equation 3 is expressed as:

$$QoSRM_{R,RS}^{N} = \sum_{i=1}^{m} QoSAM_{A_i}^{R,RS} W_{A_i}^{R} \tag{7}$$

The function to calculate the network configuration metric at the one end of the connection that is Equation 4 is expressed as:

$$QoSCM_{N} = \sum_{j=1}^{n} QoSRM_{R_j,RS}^{N} W_{R_j}^{N} \tag{8}$$

The Equation 8 is used in the cases when there is one access technology on the other side of the connection. If there is a combination of multiple access technologies in the both sides of the connection, then the Equation 8 is expressed as:

$$QoSCM_N = \sum_{l=1}^{q}\sum_{j=1}^{n} QoSRM_{R_j,RS_l}^{N} W_{R_j}^{N} \tag{9}$$

To calculate the performance metric, application QoS metric, RAN QoS metric, and network configuration QoS metric for the other side of the connection the above equations can be expressed as:

$$P_{QP_{k,A}^{RS,R}} = \begin{cases} 0 & , \quad RTP_{QP_{k,A}^{RS,R}} \geq UB_{QP_{k,A}} \\ 1 & , \quad RTP_{QP_{k,A}^{RS,R}} \leq LB_{QP_{k,A}} \\ \frac{UB_{QP_{k,A}} - RTP_{QP_{k,A}^{RS,R}}}{UB_{QP_{k,A}} - LB_{QP_{k,A}}} & , \quad UB_{QP_{k,A}} > RTP_{QP_{k,A}^{RS,R}} > LB_{QP_{k,A}} \end{cases} \tag{10}$$

$$QoSAM_A^{RS,R} = \sum_{k=1}^{p} P_{QP_{k,A}^{RS,R}} W_{QP_{k,A}} \tag{11}$$

$$QoSRM_{RS,R}^{N} = \sum_{i=1}^{m} QoSAM_{A_i}^{RS,R} W_{A_i}^{RS} \tag{12}$$

$$QoSCM_N = \sum_{j=1}^{n} QoSRM_{RS_l,R_j}^{N} W_{RS_l}^{N} \tag{13}$$

If there are multiple access networks on the both sides of the connection, then the Equation 13 is expressed as:

$$QoSCM_N = \sum_{j=1}^{n}\sum_{l=1}^{q} QoSRM_{RS_l,R_j}^{N} W_{RS_l}^{N} \tag{14}$$

For eight simultaneous voice calls and one VS session, in the UMTS network, the voice calls experience an average end-to-end delay of 216 msec, an average packet loss of 2.92%, and an average jitter of 40 msec. In all the following calculations, the UMTS technology is expressed as U and the WiMAX technology is expressed as W. The performance metric of these parameters in voice calls for this scenario using Equation 1 is calculated as:

$$\mathrm{m}_{\mathrm{d(voice)}} = \frac{400-216}{400-216} = 0.74$$

$$P_{QP_{\mathrm{J,V}}^{\mathrm{U}}} = {}^{75-40}\!/_{75-30} = 0.77$$

$$P_{QP_{\mathrm{PL,V}}^{\mathrm{U}}} = 1$$

The application QoS metric using Equation 2 is calculated as:

$$QoSAM_{\mathrm{V}}^{\mathrm{U}} = 0.74 \times 0.4 + \ 0.77 \times 0.4 + \ 1 \times 0.2 = 0.804$$

The weights are applied from Table 1.

The performance metrics for VS application are calculated using the Equation 5.

$$P_{QP_{\mathrm{D,VS}}^{\mathrm{U,W}}} = 1$$

$$P_{QP_{\mathrm{PL,VS}}^{\mathrm{U,W}}} = 1$$

The application QoS metric using Equation 6 is calculated as:

$$QoSAM_{\mathrm{VS}}^{\mathrm{U,W}} = 1 \times 0.5 + \ 1 \times 0.5 = 1$$

Table 10. QoS metric values for mixed traffic

Numbers of active calls/sessions		$QoSAM_A^R$		$QoSRM_R^N$	Overall QoS
V	VS	V	VS		
8	1	0.804	1	0.8824	Good
10	1	0.64	1	0.784	Average
12	1	0.57	1	0.74	Average
14	1	0.52	1	0.71	Average
16	1	0.488	1	0.69	Average
18	1	0.43	1	0.66	Average

Table 11. QoS-related parameter values for mixed traffic

Number of active calls/sessions		End-to-end delay (msec)		Packet loss (%)		Jitter (msec)
V	VS	V	VS	V	VS	V
8	1	216	176	2.92	1.82	40
10	1	221	179	2.95	1.85	58
12	1	224	180	3.28	1.86	65
14	1	228	180	4.41	1.89	68
16	1	230	180	5.34	1.92	70
18	1	233	180	6.46	1.94	99

The RAN QoS metric using Equation 7 is calculated as:

$$QoSRM^{N}_{U,W} = QoSAM^{U}_{V} \times W^{U}_{V} + QoSAM^{U,W}_{VS} \times W^{U}_{VS} = 0.8824$$

The weight for the applications is set as 0.5. Table 10 and 11 show the QoS-related parameter values and the overall QoS measurement of this network. Figure 13 displays the effects of VS application on the performance of voice application. Due to the presence of VS application in the network, the performance of voice application drops by 11.6% when there are eight simultaneous voice calls and one VS session on the network. When there are ten simultaneous voice calls, this performance decreases by 27.04%, and when there are twelve voice calls, it drops by 34%. With 18 simultaneous voice calls, the voice calls experience poor performance even without the presence of VS client. The Voice application QoS metric is rated as poor in this context.

Figure 14 shows the effects of different VS codec on the performance of the network. If the VS Session uses the MPEG-4 codec when there are nine active users in the network, the performance degrades by 35.24% compared to applying H.263 codec. When there are eleven active users, this performance degrades by 29%. When there are thirteen active users, this performance degrades by 26.8%. With the numbers of increased active users in the network, the performance difference between H.263 and MPEG-4 network becomes less. Because, in both cases, the number of increased active users affects the performance of the network.

Table 12 shows the QoS Metrics for the LTE network with a different number of voice calls. Table 13 presents the comparison of application QoS Metrics for the UMTS and LTE access networks. The LTE access network demonstrates a better performance for voice calls than the UMTS access network. Figure 15 shows the comparisons of RAN QoS Metric for the UMTS and LTE access networks. For 20 voice calls, the LTE network experiences 30% better performance than the UMTS network. When there are few numbers of voice calls, the performance does not vary that much between these two networks.

Figure 13. Effects of VS Application on the performance of voice calls

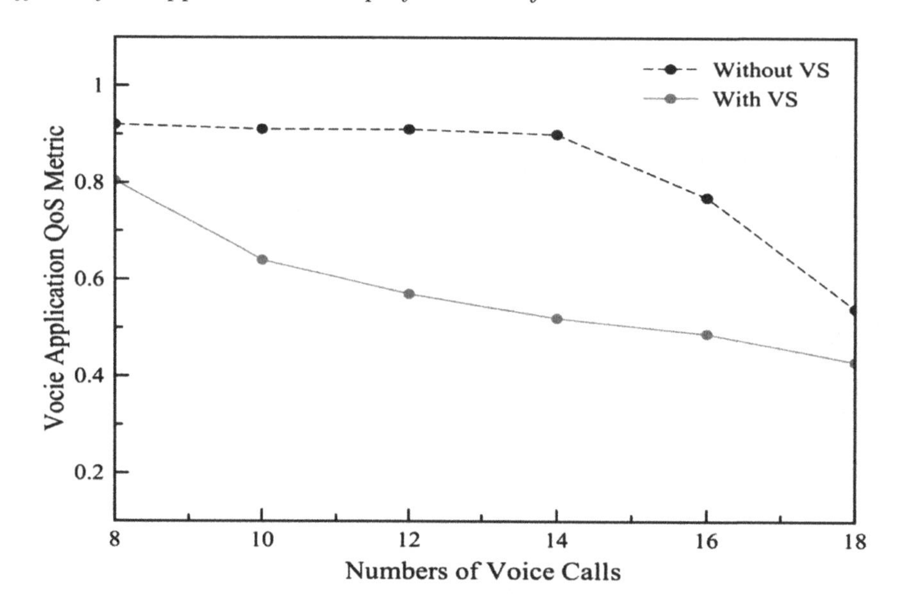

Figure 14. Effects of different VS Codec on the performance of network

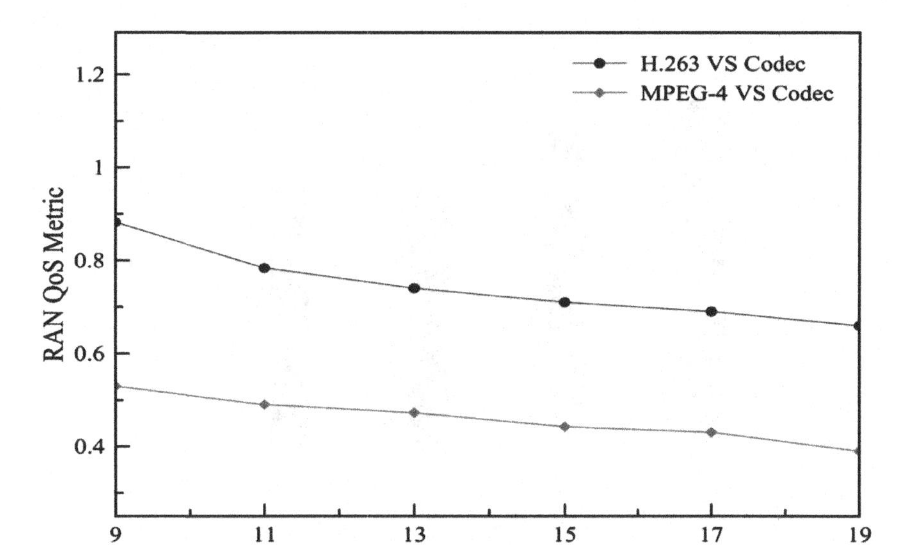

Table 12. QoS metrics for LTE traffic

Num-ber of voice calls	End-to-end delay (msec)	Packet loss (%)	Jitter (msec)	$P_{QP^R_{k,A}}$	$P_{QP^R_{k,A}}$	$P_{QP^R_{k,A}}$	$QoSAM^R_A$	Overall QoS
8	0.092	0.02	20	1	1	1	1	Good
10	0.092	0.06	24	1	1	1	1	Good
12	0.093	0.08	28	1	1	1	1	Good
14	0.094	0.12	30	1	1	1	1	Good
16	0.095	0.15	32	1	1	1	1	Good
18	0.099	0.18	34	1	1	1	1	Good

Table 13. Comparisons of QoS metrics for LTE and UMTS network

Radio Access Network	Applications	Number of users		Application QoS Metric	
UMTS	Voice, VS	V	VS	V	VS
		8	1	0.804	1
		10	1	0.64	1
		14	1	0.52	1
		20	1	0.40	1
LTE	Voice, VS	8	1	1	1
		10	1	1	1
		14	1	1	1
		20	1	0.85	1

Figure 15. Comparisons of RAN QoS metric for LTE and UMTS network

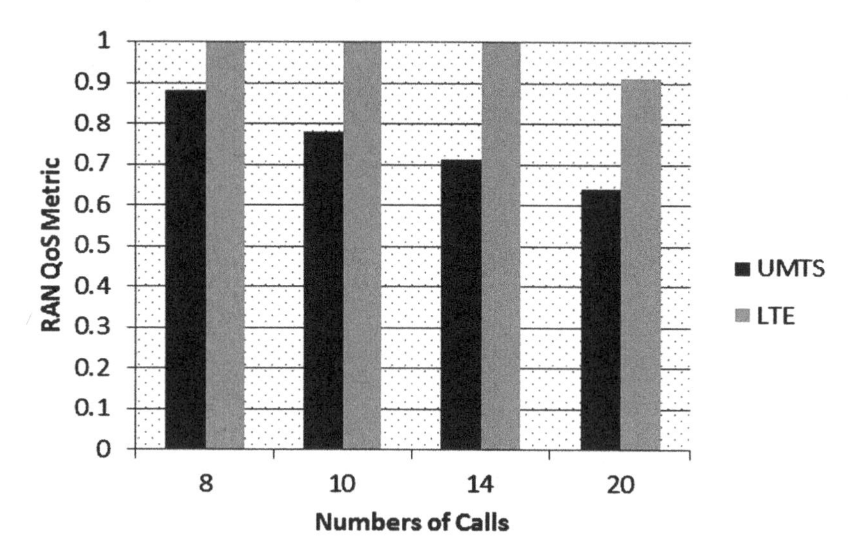

CONCLUSION

In this chapter, a generalized metric-based QoS evaluation method has been proposed and evaluated. Through different simulation scenarios and result analysis, the efficiency of this approach is presented. This approach can be applied in the network planning and designing stage to evaluate the QoS of heterogeneous network-based service models. The values of RAN QoS metrics, which are calculated using the simulation results, show that when the network is congested, the LTE-based networks perform better compared to the UMTS-based networks. On the other hand, with a few of number users, the UMTS-based networks almost have the same performance as the LTE-based network. Such observations, which are derived using the values of RAN QoS metrics, could be used in recommending the LTE-based networks for a densely populated rural area and the UMTS-based networks for a less densely populated rural area. It is also possible to use this method for QoS management and monitoring. When the codec of the VS session has been changed from H.263 to MPEG-4, the performance of the network degrades by 35.24%. Thus, the unified application, RAN, and network QoS metrics can remove the overhead of the service providers to analyze each dynamic of the network individually for QoS evaluation. Using the proposed QoS analysis method, it was easy to identify that the WLAN in combination with the WiMAX technology is more suitable for VC services than the UMTS RAN or the UMTS-WiMAX network when there are an increased number of users. Whereas, using the traditional analysis method, the values of delay, jitter and packet loss are compared separately for the performance evaluation of VC sessions in each RAN. The end-users are also left with the options to choose the most suitable network, according to their requirements by simply analyze these metric values. They can also investigate if they are getting the services they are paying for.

REFERENCES

Ahluwalia, P., & Varshney, U. (2007). Managing end-to-end quality of service in multiple heterogeneous wireless networks. International Journal of Network Management,John Wiley & Sons. *Inc.*, *17*(3), 243–260. doi:10.1002/nem.639

Alshamrani, A., Xuemin, S., & Liang-Liang, X. (2011). QoS Provisioning for Heterogeneous Services in Cooperative Cognitive Radio Networks. Selected Areas in Communications. *IEEE Journal on*, *29*(4), 819–830. doi:10.1109/JSAC.2011.110413

Chalmers, R. C., Krishnamurthi, G., & Almeroth, K. C. (2006). Enabling intelligent handovers in heterogeneous wireless networks. *Mobile Networks and Applications, Springer-Verlag*, *11*(2), 215–227. doi:10.1007/s11036-006-4474-8

Chang, D.-Y. (1996). Applications of the extent analysis method on fuzzy AHP. *European Journal of Operational Research*, *95*(3), 649–655. doi:10.1016/0377-2217(95)00300-2

Changqiao, X., Tianjiao, L., Jianfeng, G., Hongke, Z., & Muntean, G. M. (2013). CMT-QA: Quality-Aware Adaptive Concurrent Multipath Data Transfer in Heterogeneous Wireless Networks. Mobile Computing. *IEEE Transactions on*, *12*(11), 2193–2205. doi:10.1109/TMC.2012.189

Chuanxiong, G., Zihua, G., Qian, Z., & Zhu, W. (2004). A seamless and proactive end-to-end mobility solution for roaming across heterogeneous wireless networks. Selected Areas in Communications. *IEEE Journal on*, *22*(5), 834–848. doi:10.1109/JSAC.2004.826921

Dong, M., & Maode, M. (2012). A QoS Oriented Vertical Handoff Scheme for WiMAX/WLAN Overlay Networks. Parallel and Distributed Systems. *IEEE Transactions on*, *23*(4), 598–606. doi:10.1109/TPDS.2011.216

Douros, V. G., & Polyzos, G. C. (2011). Review of some fundamental approaches for power control in wireless networks. *Computer Communications*, *34*(13), 1580–1592. doi:10.1016/j.comcom.2011.03.001

Fangmin, X., Luyong, Z., & Zheng, Z. (2007). Interworking of Wimax and 3GPP networks based on IMS. *Communications Magazine, IEEE*, *45*(3), 144–150. doi:10.1109/MCOM.2007.344596

Farid, F., Shahrestani, S., & Ruan, C. (2013). *QoS analysis and evaluations: Improving cellular-based distance education*. Paper presented at the Local Computer Networks Workshops (LCN Workshops), 2013 IEEE 38th Conference on.

Farnaz Farid, S. S. (2012). *Quality of service requirements in wireless and cellular networks: application-based analysis*. Paper presented at the International Business Information Management Association Conference, Bercelona, Spain.

Garmonov, A. V., Seok Ho, C., Do Hyon, Y., Ki Lae, H., Yun Sang, P., Savinkov, A. Y., & Kondakov, M. S. (2008). QoS-Oriented Intersystem Handover Between IEEE 802.11b and Overlay Networks. Vehicular Technology. *IEEE Transactions on*, *57*(2), 1142–1154. doi:10.1109/TVT.2007.906347

Hasib, A., & Fapojuwo, A. (2008). Analysis of Common Radio Resource Management Scheme for End-to-End QoS Support in Multiservice Heterogeneous Wireless Networks. Vehicular Technology. *IEEE Transactions on*, *57*(4), 2426–2439. doi:10.1109/TVT.2007.912326

Hossain, E. (2008). *Heterogeneous Wireless Access Networks: Architectures and Protocols*. Springer.

Jammeh, E., Mkwawa, I., Khan, A., Goudarzi, M., Sun, L., & Ifeachor, E. (2012). Quality of experience (QoE) driven adaptation scheme for voice/video over IP. *Telecommunication Systems*, *49*(1), 99–111. doi:10.1007/s11235-010-9356-5

Karimi, O. B., & Fathy, M. (2010). Adaptive end-to-end QoS for multimedia over heterogeneous wireless networks. *Computers & Electrical Engineering*, *36*(1), 45–55. doi:10.1016/j.compeleceng.2009.04.006

Levine, D. A., Akyildiz, I. F., & Naghshineh, M. (1997). A resource estimation and call admission algorithm for wireless multimedia networks using the shadow cluster concept. *Networking, IEEE/ACM Transactions on, 5*(1), 1-12. doi: 10.1109/90.554717

Lu, Q., & Koutsakis, P. (2011). Adaptive Bandwidth Reservation and Scheduling for Efficient Wireless Telemedicine Traffic Transmission. Vehicular Technology. *IEEE Transactions on*, *60*(2), 632–643. doi:10.1109/TVT.2010.2095472

Luo, H., & Shyu, M.-L. (2011). Quality of service provision in mobile multimedia - a survey. *Human-centric Computing and Information Sciences*, *1*(1), 1–15. doi:10.1186/2192-1962-1-5

Niyato, D., & Hossain, E. (2008). A Noncooperative Game-Theoretic Framework for Radio Resource Management in 4G Heterogeneous Wireless Access Networks. Mobile Computing. *IEEE Transactions on*, *7*(3), 332–345. doi:10.1109/TMC.2007.70727

Papavassiliou, S., & Tassiulas, L. (1996). Joint optimal channel base station and power assignment for wireless access. *IEEE/ACM Transactions on Networking*, *4*(6), 857–872. doi:10.1109/90.556343

Rashad, S., Kantardzic, M., & Kumar, A. (2006). User mobility oriented predictive call admission control and resource reservation for next-generation mobile networks. *Journal of Parallel and Distributed Computing*, *66*(7), 971–988. doi:10.1016/j.jpdc.2006.03.007

Shun-Ren, Y., & Wen-Tsuen, C. (2008). SIP Multicast-Based Mobile Quality-of-Service Support over Heterogeneous IP Multimedia Subsystems. Mobile Computing. *IEEE Transactions on*, *7*(11), 1297–1310. doi:10.1109/TMC.2008.53

TKN. (2012). *MPEG-4 and H.263 Video Traces for Network Performance Evaluation*. Retrieved from http://trace.eas.asu.edu/TRACE/ltvt.html

Vasu, K., Maheshwari, S., Mahapatra, S., & Kumar, C. S. (2012). QoS-aware fuzzy rule-based vertical handoff decision algorithm incorporating a new evaluation model for wireless heterogeneous networks. *EURASIP Journal on Wireless Communications and Networking, 2012*(322). doi: 10.1186/1687-1499-2012-322

Wang, X. G., Min, G., Mellor, J. E., Al-Begain, K., & Guan, L. (2005). An adaptive QoS framework for integrated cellular and WLAN networks. *Computer Networks*, *47*(2), 167–183. doi:10.1016/j.comnet.2004.07.003

Wei, S., Weihua, Z., & Yu, C. (2007). Load balancing for cellular/WLAN integrated networks. *IEEE Network, 21*(1), 27–33. doi:10.1109/MNET.2007.314535

Xia, G., Gang, W., & Miki, T. (2004). End-to-end QoS provisioning in mobile heterogeneous networks. *Wireless Communications, IEEE, 11*(3), 24–34. doi:10.1109/MWC.2004.1308940

Xuebing, P., Tao, J., Daiming, Q., Guangxi, Z., & Liu, J. (2010). Radio-Resource Management and Access-Control Mechanism Based on a Novel Economic Model in Heterogeneous Wireless Networks. Vehicular Technology. *IEEE Transactions on, 59*(6), 3047–3056. doi:10.1109/TVT.2010.2049039

Yao, L. (2012). Measurement and Analysis of an Internet Streaming Service to Mobile Devices. *IEEE Transactions on Parallel and Distributed Systems, 99*, 1–1.

Zahran, A. H., Liang, B., & Saleh, A. (2006). Signal threshold adaptation for vertical handoff in heterogeneous wireless networks. *Mobile Networks and Applications, 11*(4), 625–640. doi:10.1007/s11036-006-7326-7

KEY TERMS AND DEFINITIONS

Generalized Metric-Based QoS Analysis: A QoS analysis method, which combines different parameters from the application, network, and user-level to quantify the QoS of a network.

Heterogeneous Network: A Heterogeneous Network consists of multiple radio access technologies serving various traffic flows. It provides transparent and self-configurable services across wireless local area networks (WLANs), Wireless Metropolitan Area Networks (WMANs) and Wireless wide area networks (WWANs).

LTE: In contrast to the circuit-switched cellular model, the LTE supports only packet-switched services. The aim of LTE is to provide seamless Internet Connectivity between the User Equipment and the Packet Data Network.

Network-Based Service Models: Any socio-economic service such as education or health care service provided through various communication technologies.

Quality of Service (QoS): The term QoS combines the application-level, network-level, and user-level requirements of a network. It deals with both human experiences and technological concepts. From a human perspective, QoS is essentially a subjective matter, and hence hard to define. On the other hand, from the network perspective, it can be quantified through a range of parameters such as delay, jitter, and packet loss.

UMTS: UMTS is the third generation technology based on Wideband Code Division Multiple Access (WCDMA). Most of the GSM/GPRS mobile operators use UMTS as the 3G technology because it has the essential infrastructure to migrate from GSM/GPRS.UMTS is managed by the 3rd Generation Partnership Project (3GPP).

Unified Metric: The unified metric uses the concept of unified metric measurement functions for network QoS evaluation. This function considers all the critical performance parameters to quantify the network and application QoS with a single numerical value.

Chapter 11
Advanced Retransmission Protocols for Critical Wireless Communications

Salima El Makhtari
Abdelmalek Essaadi University, Morocco

Ahmed El Oualkadi
Abdelmalek Essaadi University, Morocco

Mohamed Moussaoui
Abdelmalek Essaadi University, Morocco

Hassan Samadi
Abdelmalek Essaadi University, Morocco

ABSTRACT

This chapter provides background about Hybrid Automatic Repeat reQuest (HARQ) protocols. First, the critical situations that may be faced by wireless communication systems especially cellular mobile technologies in case of very noisy radio channels are introduced. Particularly, the chapter introduces the HARQ protocols, their main constituent components as well as some related application areas. Then, the state-of-the-art of HARQ protocols is presented. The next section explains the three basic ARQ protocols. Then, the different HARQ types are detailed. Then, a mathematical model of type II HARQ based on Rayleigh fading channel is provided. This analytical analysis is followed by a discussion of the throughput which is one of the most interesting metrics used to measure the performance of HARQ systems. The readers can find in the next section a description of the HARQ systems architecture where 3GPP LTE is used to illustrate and explain how such systems operate. Finally, the last section concludes the chapter.

INTRODUCTION

In modern wireless communication systems, especially digital cellular mobile phone technologies, demands for high data rates and reliability are increasing dramatically. However, ensuring that the data transmitted at high speed, over noisy wireless channels, is correctly received by the end user becomes more and more challenging. In despite of the availability of different channel estimation models and methods, the errors induced in case of severe noise and fading remain critical and may cause great damage. Many solutions have been developed to deal with poor transmission channels quality. To illustrate: Forward Error Correction (FEC) codes, Adaptive Modulation and Coding (AMC), Automatic Repeat Request (ARQ) and Hybrid Automatic Repeat reQuest (HARQ).

DOI: 10.4018/978-1-5225-2113-6.ch011

This chapter provides information on HARQ protocols for readers. These solutions are a powerful type of feedback-based communications and play a key role in enhancing reliability of wireless transmission links. HARQ protocols are cross-layer techniques and the result of combining classical ARQ schemes, FEC codes and error detection techniques such as Cyclic Redundancy Check (CRC) in addition of packets soft combining solutions. They enable reliable communication and improve the throughput performance over time-varying fading channels by leveraging both FEC and ARQ to their high potential when used together and mitigating their individual drawbacks. HARQ protocols continue to attract the intensive attention of researchers and have been adopted by many wireless communication systems, particularly the most recent cellular mobile technologies like as IEEE 802.16m (Institute of Electrical and Electronics Engineers [IEEE], 2011), IEEE 802.16e (IEEE, 2006) and the most recent 3GPP cellular standards such as High Speed Downlink Packet Access (HSDPA) (Third Generation Partnership Project, 2009), Long Term Evolution (LTE) (Third Generation Partnership Project, 2008) and Long Term Evolution Advanced (LTE-A) (Third Generation Partnership Project, 2010).

Background

ARQ techniques have served for years to deal with channel errors. However, the throughput of ARQ-based systems decreases as the channel error rate increases since more retransmissions are required. Thus, the situation becomes critical in front of the huge demand in terms of both reliability and throughput. The authors of Wozencraft and Horstein (1960) and Wozencraft and Horstein (1961) have introduced a combination of ARQ, FEC and error detection codes in order to overcome the drawbacks of ARQ schemes when used individually. This technique is referred to as Hybrid Automatic Repeat reQuest (HARQ). It takes advantage of both FEC and ARQ. Particularly, in 1997, Narayanan and Stuber (1997) suggested a combination of ARQ and turbo-codes. Turbo-codes are channel coding and decoding algorithms which have been introduced by the authors of Berrou et al. (1993). Turbo-codes are based on soft decision iterative decoding of parallel codes and known to provide high coding gains achieving near-Shannon limit performance (Berrou et al., 1993).

HARQ protocols are recognized as a powerful solution used to deal with poor channel quality. In addition of using ARQ in conjunction with FEC, another key technique which is soft combining of packets is also involved in. So, instead of discarding the erroneous data packets, the communication system preserves them in order to combine them with the new packets that will be retransmitted.

A simple technique of erroneous packets combining at the receiver, called Chase Combing (CC), has been introduced in 1985 (Chase, 1985). Essentially, CC consists of retransmitting a full copy of the data packet originally sent and found to be corrupted. Later on, more sophisticated combination protocols, called Incremental Redundancy (IR), have been developed. As a result of deploying soft combining techniques, when the data packet is found to be in error, the receiver requests a retransmission instead of sending erroneous decoded data packets to upper layers. In that way, HARQ schemes are more reliable than using FEC alone. Moreover, HARQ systems have some advanced options such as the possibility to adjust the code rates according to the need. Consequently, they can increase the overall throughput and be more efficient than ARQ schemes.

However, the complexity of wireless communication systems increases with nonstop. Thus, enhancing HARQ performance continuously is required and vital. Researchers have addressed many aspects of HARQ systems. To illustrate, Hong Chen Maunder and Hanzo (2013) worked on the complexity reduction of the turbo coded HARQ schemes. Also, as discussed by the authors of Zhiping Shi et al. (2002), several schemes to combine packets before or after the turbo decoding have been proposed. In addition, the authors of Yidong Lang et al. (2013) suggested transmitting the sequence resulting from applying a XOR function to two erroneous received packets instead of retransmitting these packets individually. Besides, two HARQ complexity reduction strategies have been proposed by the authors of Hong Chen et al. (2013) for a system based on multiple components turbo codes: early stopping strategy which is used to stop the iterative decoding process when the mutual information improvements become less than a given threshold, and deferred iteration technique which is used to delay the iterative decoding process until the receiver confidently estimates that it has received enough information for successful decoding. Furthermore, Holland et al. (2005) proposed a scheme using soft combining at symbol level.

ARQ SCHEMES

ARQ is an error control mechanism which was developed to deal with transmission errors: when the received data packet is found to be in error, the receiver sends a negative acknowledgment (NACK) to the transmitter requesting the retransmission of the erroneous packet. Theoretically, the retransmission continues until the received data is error free. Practically speaking, the retransmission is stopped once a defined maximum number of retransmissions is reached or a configured time-out is expired. Basically, there are three different types of ARQ schemes: Stop-and-Wait (SW) (Marcille et al, 2012), Go-Back-N (GBN) as discussed by the authors of Zhang (2013), Kwatra (2013), and Shiann-Tsong Sheu et al. (2013) and Selective Repeat (SR) as illustrated by the authors of Alsebae et al. (2013).

Stop-and-Wait Scheme

In Stop-and-Wait ARQ, the transmitter sends only one packet at a time and waits for the related acknowledgment from the receiver. If the received packet is found to be corrupted, the receiver sends back a negative acknowledgment (NACK) to the transmitter. Otherwise, it sends back a positive acknowledgment (ACK). Figure 1 illustrates the Stop-and-Wait process. The transmitter sends a packet P1 and waits for the receiver's feedback. The received packet P1 is detected in error, so the receiver discards it and sends back a NACK. Accordingly, the transmitter resends the same previous packet P1. Typically, this process continues until the packet P1 is error free or a pre-defined maximum number of retransmissions is achieved. Assuming that P1 is error free after the first retransmission as shown in Figure 1, then the transmitter sends the next packet P2. The receiver sends an ACK response to the transmitter since P2 is correctly received and the transmitter sends the next packet P3 and so on.

Figure 1. Stop-and-Wait ARQ scheme

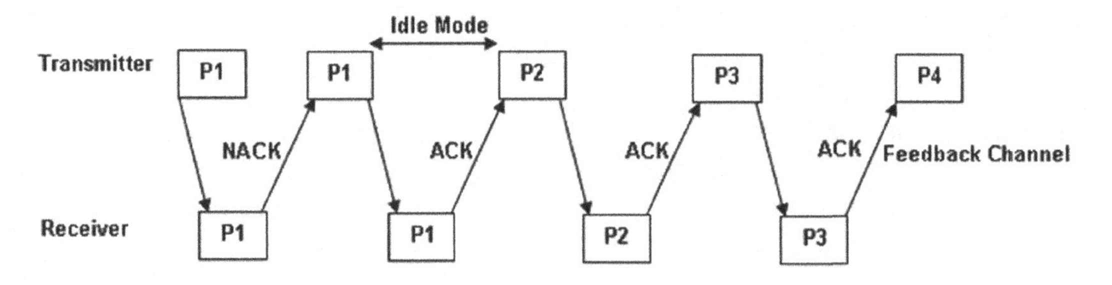

Stop-and-Wait is the simplest ARQ scheme. Nevertheless, it requires the transmitter to remain in an idle state until the transmitted packet is positively acknowledged by the receiver. As a result, the communication system's throughput decreases as the waiting time duration becomes longer. Another drawback of SW technique is that the transmitted packet could be lost or damaged before reaching the receiver, which means that the transmitter could wait for an acknowledgment of a packet that has never been received. In general, to solve this issue, a common solution is used. It consists of defining a time interval beyond which a transmitted packet is considered lost. This time-out technique is implemented at the transmitter. In addition, SW requires to save a copy of the transmitted packet in a buffer at the transmitter until the related positive acknowledgment is received. Moreover, the ACK/NACK feedback channel can be itself corrupted by noise and induces by consequence the receiver's response distortion. In this way, the transmitter may get a wrong ACK/NACK feedback and resend data while the packet originally transmitted was positively acknowledged by the receiver in reality. This kind of critical situations creates confusion for both transmitter and receiver.

Go-Back-N Scheme

The transmitter sends packets continuously to the receiver without waiting for an acknowledgment. At the receiver, when a given packet is detected in error, this packet and all the subsequent packets are discarded and a NACK is sent back to the transmitter. After receiving this feedback, the transmitter resends the concerned packet and all the next ones. Figure 2 depicts the G-Back-N scheme.

Go-Back-N solution is more efficient than SW. It reduces the waiting time of the transmitter. However, this scheme increases the memory size needed at the transmitter since it requires storing more than one packet until their related positive acknowledgments are received. Besides, packets error free could be lost at the receiver since this last one discards the packet found in error as well as all succeeding ones.

Selective-Repeat Scheme

In order to avoid retransmitting the error-free packets, Selective-Repeat scheme uses a packets numbering mechanism in such way to resend only the packets negatively acknowledged or those for which the time-out has expired. Figure 3 describes SR scheme. The transmitter sends a set of packets P1, P2, P3

Figure 2. Go-Back-N ARQ scheme

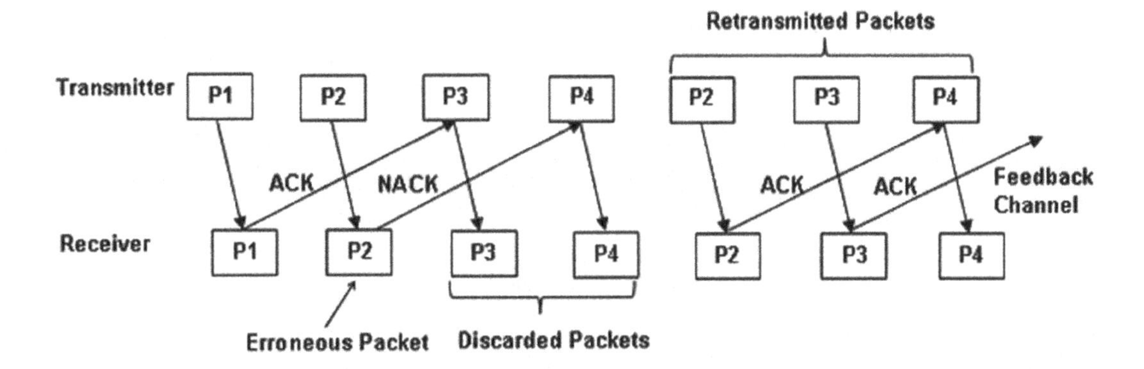

and P4 and stores in memory a copy of these packets until they are positively acknowledged. At the receiver, P2 is found to be corrupted and a NACK is sent back to the transmitter which will resend this packet and continue to transmit the remaining packets P5 and P6. Assuming that P3 and P4 are error-free, these packets will be buffered at the receiver until P2 is correctly received. The receiver is able to reorder packets using the number associated to each one indicating its appropriate location within the received packets sequence.

Thanks to packets numbering and reordering capabilities, SR technique is more efficient than SW and GBN techniques, it removes the waiting time at the transmitter and improves the throughput. However, this solution is complex and requires large memories. On the one hand, it is needed to preserve the transmitted packets at the transmitter until they are positively acknowledged. On the other hand, packets error-free should be buffered at the receiver until those completing the sequence and detected in error are also received correctly. Then the full sequence can be delivered. The situation becomes even critical when several retransmissions are required to get a given packet error-free since the number of packets to be memorized at the receiver becomes higher.

HARQ PROTOCOLS

Practically, ARQ schemes have some limitations. They induce additional delay, decrease the throughput in case of SW technique or are highly complex in case of SR scheme. Also, the system can suffer from congestion when increasing the number of retransmissions in case of high error rates. In addition, memory saturation can easily happen at the receiver because of the large number of packets to be buffered before their delivery. These disadvantages are not supported by all applications. For that, a new solution has been developed, it is called Hybrid Automatic Repeat reQuest (HARQ). Hybrid-ARQ mechanism is a combination of FEC, ARQ and an error detection mechanism such as CRC. It is a powerful technique used to deal with critical transmission channels conditions especially in case of very noisy environments. Principally, there are three different types of HARQ protocols, namely type I HARQ, type II HARQ and type III HARQ.

Figure 3. Selective-Repeat ARQ scheme

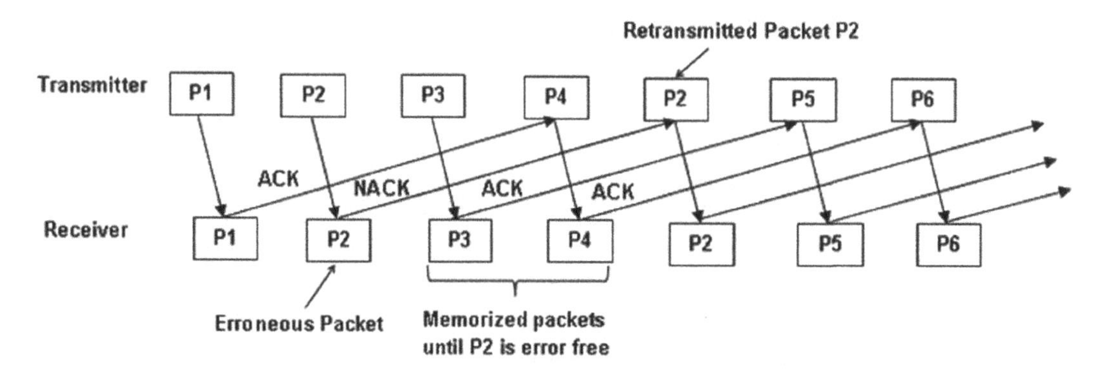

Type I HARQ Protocol

This protocol combines FEC and ARQ in order to improve the throughput. Sending coded packets instead of non-coded packets helps to protect the transmitted data against noise. As a result, the error rate decreases and the throughput increases. To illustrate, refer to the work of Uchoa et al. (2010). It is to note that this type I HARQ discards erroneous packets at the receiver.

Type II HARQ Protocol

The type II HARQ, as illustrated by the authors of Ksairi et al. (2013) and Bosisio et al. (2006), introduces the combination of erroneous packets instead of discarding them, it is also referred to as Chase Combining HARQ (CC-HARQ). For more details, readers can refer to the work of Chase (1985). In this case, when a received packet is detected in error, the receiver sends a NACK to the transmitter. Then, the transmitter resends the same previous packet and both of them are soft combined as shown in Figure 4. Theoretically, this retransmission continues until the packet is decoded successfully. Practically, the process is stopped once a pre-defined maximum number of retransmissions is reached or the time-out has expired. Type II HARQ does not require additional processing at the transmitter. However, the receiver becomes more complex. CC-HARQ takes advantage of the accumulated SNR at the receiver as it will be demonstrated in a next section.

Type III HARQ Protocol

The type III HARQ is based on an Incremental Redundancy method (IR-HARQ). IR-HARQ scheme operates as follows: at the first transmission, the transmitter sends a set of coded bits selected according to a pre-defined puncturing procedure. If the received packet is found to be in error, the receiver sends back a NACK to the transmitter which resends an additional packet with different punctured data and parity bits than the initial one. The previous packet and the new retransmitted pattern are soft combined at the receiver as illustrated in Figure 5. Theoretically, the retransmission continues until the packet is decoded successfully. Practically speaking, the retransmissions are stopped once a pre-determined maximum number of retransmissions is reached or the time-out has expired. In case of IR-HARQ, at each new retransmission, the new transmitted packet is different from the previous ones. In fact, this is the main difference between CC-HARQ and IR-HARQ. As consequence, parity bits are increased

Figure 4. Chase combining approach

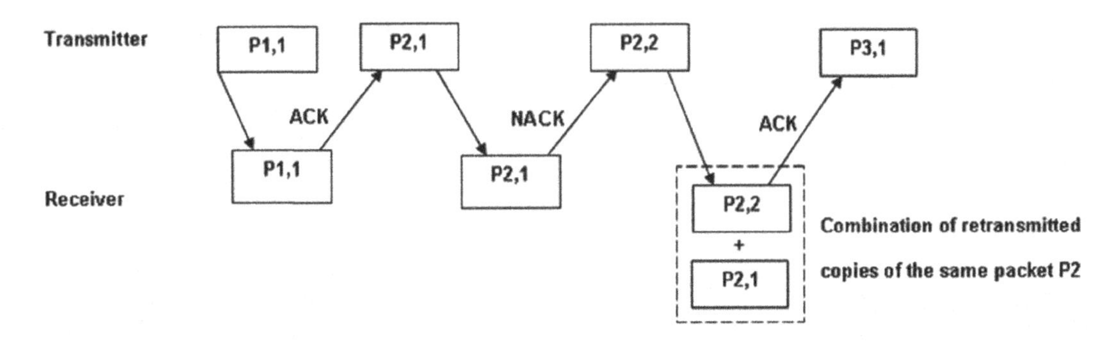

Figure 5. Incremental redundancy approach

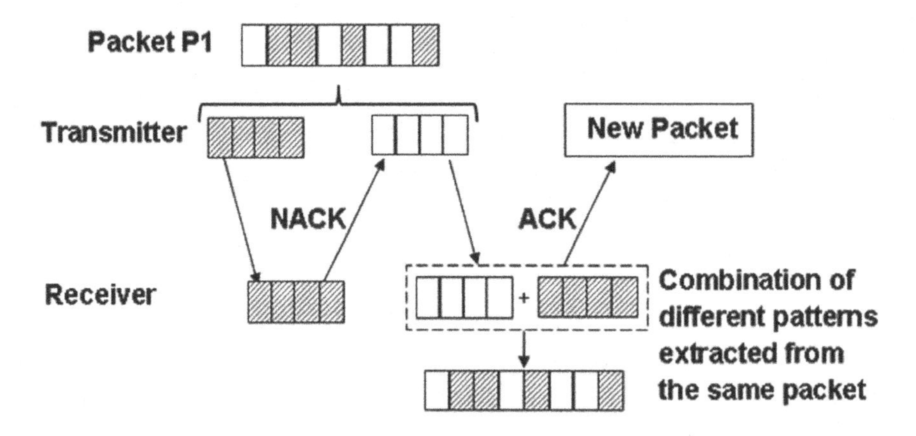

incrementally and the decoding performance is improved for each retransmission. Moreover, IR-HARQ can be classified into two types: full IR-HARQ that retransmits redundant bits only and partial IR-HARQ which retransmits both systematic bits and redundant bits. Readers can also refer to the work of Yafeng et al. (2003) and Boujemaa et al. (2008) to get additional information.

IR-HARQ achieves an interesting gain in terms of coding rates at the expense of additional decoding complexity.

ANALYTICAL ANALYSIS OF TYPE II HARQ PERFORMANCE

Let r_t be the N-length received signal at the tth transmission ($t = 0$ corresponds to the 1st transmission occurring before enabling HARQ processes). Besides, the transmission channel under consideration is a Rayleigh block flat fading channel. As a result, r_t is modeled for the tth transmission as given by Equation 1:

$$r_t = h_t s + w_t \tag{1}$$

where h_t is the Rayleigh channel coefficient, expressed as given by Equation 2.

$$r = \| r \| e^{j\phi} \tag{2}$$

The magnitude $\|r\|$ and the phase ϕ are random variables and statistically independent. The magnitude $\|r\|$ is defined as provided by Equation 3.

$$\| r \| = \sqrt{r_{re}^2 + r_{im}^2} \tag{3}$$

- The h response experienced by different transmissions are considered mutually statistically independent of each other. Also, it is assumed that h is perfectly estimated by the receiver in such way that $\hat{h} = h$. In other words, it is considered that h is simply known at the receiver.
- $w_t \sim N\left(0, N_o\right)$ is the Additive White Gaussian Noise (AWGN) with zero mean $\mu = 0$ and variance $\sigma^2 = N_o$. w of the different transmissions are mutually statistically independent of each other. Also, w and h are mutually statistically independent of each other.
- s is the N-length transmitted signal. Since CC-HARQ is considered, a similar information s is transmitted every retransmission.

Let $\boldsymbol{r}$ be a set of transmissions defined as $\boldsymbol{r} = \left\{\boldsymbol{r}_0, \boldsymbol{r}_1, \boldsymbol{r}_2, \ldots, \boldsymbol{r}_{t-1}, \boldsymbol{r}_t\right\}$. Based on its observation of $\boldsymbol{r}$, the receiver decides that an information $\hat{\boldsymbol{s}}$ was transmitted. So, the probability that the decision made by the receiver is correct is the probability that $\hat{\boldsymbol{s}}$ is really the information that has been transmitted. Let $\boldsymbol{P}\left(\boldsymbol{C}\right)$ be the probability of the correct decision and given by Equation 4:

$$P\left(C\right) = \int P(\hat{s} \mid r) P\left(r\right) dr \tag{4}$$

The main objective of the process is to minimize the error probability. In other words, maximize the probability of the correct decision. Since $P\left(r\right)$ is non-negative for all r, $P(\hat{s} \mid r)$ has to be maximized. According to the Maximum A Posteriori (MAP) probability rule, Equation 5 is obtained. Readers can refer to the work of Proakis and Salehi (2008):

$$\hat{s} = \arg\max_{s} P(s \mid r) \tag{5}$$

According to Bayes's rule:

$$P\left(s \mid r\right) = \frac{P\left(r \mid s\right) P\left(s\right)}{P\left(r\right)} \tag{6}$$

$P\left(\boldsymbol{r}\right)$ is independent of $\boldsymbol{s}$ and remains the same for all $\boldsymbol{s}$. Also, the transmitted signal $\boldsymbol{s}$ is similar for all transmissions. As a result, $\boldsymbol{P}\left(\boldsymbol{r}\right)$ and $\boldsymbol{P}\left(\boldsymbol{s}\right)$ can be neglected and Equation 7 which is the Maximum Likelihood (ML) decision rule, as explained by the authors of Proakis and Salehi (2008), is obtained.

$$\hat{s} = \arg\max_{s} P(r \mid s) \tag{7}$$

Provided that it is assumed that $\boldsymbol{h}$ is known, Equation 7 can be rewritten according to the work of Berrou (2007) as follow:

$$\hat{s} = \arg\max_{s} P(r \mid s, h) \tag{8}$$

In other words, Equation 9 is obtained.

$$\hat{s} = \arg\max_{s} P(r_0, r_1, r_2, \ldots r_t \mid s, h_0, h_1, \ldots, h_t) \tag{9}$$

According to Bayes' rules and provided that $\boldsymbol{w}$ and $\boldsymbol{h}$ are independent of each other. Also, $\boldsymbol{w}$ and $\boldsymbol{h}$ of the different transmissions are independent of each other, the authors can write:

$$P(r_0, r_1, \ldots, r_t \mid s, h_0, h_1, \ldots, h_t) = P(r_0 \mid s, h_0) P(r_1 \mid s, h_1) \ldots\ldots P(r_t \mid s, h_t) \tag{10}$$

By consequence, Equation 11 is obtained.

$$P(r_0, r_1, \ldots, r_t \mid s, h_0, h_1, \ldots, h_t) = \prod_{i=0}^{t} P(r_i \mid s, h_i) \tag{11}$$

Also,

$$P(r_i \mid s, h_i) = \prod_{k=1}^{N} P(r_{i,k} \mid s_k, h_i) \tag{12}$$

$P(r_{i,k} \mid s_k, h_i)$ is the conditional density function of a fading channel with an AWGN and a known h. It can be expressed as given by Equation 13. Readers can refer to the work of Berrou (2007):

$$P\left(r_{i,k} \middle| s_k, h_i\right) = \frac{1}{\sigma\sqrt{2\pi}} exp\left(-\frac{\left(r_{i,k} - h_i s_k\right)^2}{2\sigma^2}\right) \tag{13}$$

So, Equation 12 becomes:

$$P(r_i \mid s, h_i) = \frac{1}{\left(\pi N_o\right)^{\frac{N}{2}}} exp\left(-\frac{\sum_{k=1}^{N} \left|r_{i,k} - h_i s_k\right|^2}{N_0}\right) \tag{14}$$

Since for a given vector x of size n, $\|x\| = \sum_{i=1}^{n} |x_i|$, Equation 14 can be rewritten as follow:

$$P(r_i \mid s, h_i) = \frac{1}{\left(\pi N_o\right)^{\frac{N}{2}}} exp\left(-\frac{\left|r_i - h_i s\right|^2}{N_o}\right) \tag{15}$$

By consequence and provided Equation 11 and Equation 15, $P(r \mid s,h)$ becomes of the following form:

$$P\left(r|s,h\right) \propto \exp\left(-\sum_{i=0}^{t}\frac{\left|r_i - h_i s\right|^2}{N_o}\right) \tag{16}$$

Since if $ln(P(r \mid s,h))$ is maximized, $P(r \mid s,h)$ is also maximized. As a result:

$$ln\left(P\left(r|s,h\right)\right) \propto -\frac{1}{N_o}\sum_{i=0}^{t}\left|r_i - h_i s\right|^2 \tag{17}$$

In that case, ML decision rule reduces to minimize the term $\sum_{i=0}^{t}\left|r_i - h_i s\right|^2$ as given by Equation 18.

$$\hat{s} = \arg\min_s \sum_{i=0}^{t} \| r_i - h_i s \|^2 \tag{18}$$

The authors can write:

$$\| r_i - h_i s \|^2 = \| r_i \|^2 - 2Re\left\{\left\langle r_i, h_i s\right\rangle\right\} + \| h_i s \|^2 \tag{19}$$

where $Re\left(.\right)$ stands for the real part and $\left\langle r_i, h_i s\right\rangle$ is the scalar product of r_i and $h_i s$ (For two complex vectors a and b of size n, $\left\langle a,b\right\rangle = \sum_{i=1}^{n} ab^*$ with $\left(.\right)^*$ refers to the conjugate).

Provided Equation 19 and s is similar for all retransmissions, Equation 20 is obtained.

$$\sum_{i=0}^{t} \| r_i - h_i s \|^2 = \sum_{i=0}^{t} \| r_i \|^2 - 2\sum_{i=0}^{t} Re\left\{\left\langle r_i, h_i s\right\rangle\right\} + \| s \|^2 \sum_{i=0}^{t}\left|h_i\right|^2 \tag{20}$$

Since $\left\|r_i\right\|^2$ is independent of s:

$$\hat{s} = \arg\min_s \left\{-2\sum_{i=0}^{t} Re\left\{\left\langle r_i, h_i s\right\rangle\right\} + \| s \|^2 \sum_{i=0}^{t}\left|h_i\right|^2\right\} \tag{21}$$

The expression given by Equation 21 can be simplified as follows:

$$\hat{s} = \arg\min_{s} \left\{ -\sum_{i=0}^{t} Re\left\{ \left\langle r_i, h_i s \right\rangle \right\} + \frac{\| s \|^2}{2} \sum_{i=0}^{t} \left| h_i \right|^2 \right\} \tag{22}$$

This is equivalent to:

$$\hat{s} = \arg\max_{s} \left\{ \sum_{i=0}^{t} Re\left\{ \left\langle r_i, h_i s \right\rangle \right\} - \frac{\| s \|^2}{2} \sum_{i=0}^{t} \left| h_i \right|^2 \right\} \tag{23}$$

This is also equivalent to:

$$\hat{s} = \arg\max_{s} \left\{ \sum_{i=0}^{t} Re\left\{ \left\langle h_i^* r_i, s \right\rangle \right\} - \frac{\| s \|^2}{2} \sum_{i=0}^{t} \left| h_i \right|^2 \right\} \tag{24}$$

Assuming that h_i with $0 \leq i \leq t$ is known, the observation of the static $\sum_{i=0}^{t} h_i^* r_i$ only is enough to make an optimal decoding decision. More details can be found in the work of Madhow (2008). So, Equation 25 is obtained.

$$\hat{s} = \arg\max_{s} Re \left\{ \sum_{i=0}^{t} \left\langle h_i^* r_i, s \right\rangle \right\} \tag{25}$$

The signal s is independent of h_i and r_i. Thus, Equation 26 is obtained.

$$\hat{s} = \arg\max_{s} Re \left\{ < \sum_{i=0}^{t} h_i^* r_i, s > \right\} \tag{26}$$

It can be observed that the signal denoted previously r, becomes:

$$r = \sum_{i=0}^{t} h_i^* r_i \tag{27}$$

which represents the resulting signal at the receiver after combining all information sequences r_i with $0 \leq i \leq t$ received during t transmissions and equalizing them.

Developing the expression in Equation 27 and provided Equation 1, it can be deduced that the combination of the signals r_i with $0 \leq i \leq t$ is equivalent to a signal, the denoted r, as given by Equation 28.

$$r = \sum_{i=0}^{t} \left|h_i\right|^2 s + \sum_{i=0}^{t} h_i^* w_i \tag{28}$$

For a given individual i^{th} transmission, the energy per symbol is E_s. The related Signal-to-Noise Ratio (SNR), denoted γ_i, is given by Equation 29.

$$\gamma_i = \frac{\left|h_i\right|^2 E_s}{N_o} \tag{29}$$

In the case of t transmissions of the same information s and provided Equation 27, the SNR observed at the tth transmission, denoted γ_t is given by Equation 30:

$$\gamma_t = \frac{\left(\sum_{i=0}^{t} \left|h_i\right|^2\right)^2 E_s}{\sum_{i=0}^{t} \left|h_i\right|^2 N_o} \tag{30}$$

By consequence:

$$\gamma_t = \sum_{i=0}^{t} \frac{\left|h_i\right|^2 E_s}{N_o} \tag{31}$$

From Equation 29 and Equation 31, it can be deduced that:

$$\gamma_t = \sum_{i=0}^{t} \gamma_i \tag{32}$$

So, t individual transmissions of similar information s over a channel with an γ_i where $0 \leq i \leq t$ are equivalent to one transmission of s over a channel with a higher SNR which is given by Equation 32. So, type II HARQ based on chase combining method results in increasing the average SNR of the communication system. Since, the error probability decreases as the SNR increases, it can be deduced that type II HARQ reduces also the error probability. By consequence, the overall reliability of the communication system is enhanced.

HARQ DATA THROUGHPUT

Generally, the throughput is a metric used to measure the transmission efficiency. For an ordinary scheme without HARQ, the throughput, denoted T_h, can be simply given by Equation 33.

$$T_h = \frac{N_{accept}}{N_{trans}} \tag{33}$$

Where N_{accept} is the total number of the frames accepted by the receiver and N_{trans} is the total number of transmitted frames. This is also known as the normalized throughput.

In particular, the throughput of an HARQ scheme is defined as the ratio of a packet's length, denoted L, to the average total number of bits that must be transmitted in order for this packet to be accepted by the receiver, denoted N_{avg}. The total number of bits includes the information bits, the error detection code (CRC) bits and the parity bits.

$$T_h = \frac{L}{N_{avg}} \tag{34}$$

In the case of an HARQ system based on turbo-codes for instance, the throughput is given by Equation 35. Readers can refer to the work of Xiangming Yu et al. (2010).

$$T_h = R.\frac{P}{P + l_{avg}} \tag{35}$$

where R is the coding rate, P is the puncturing period and l_{avg} is the average number of additionally transmitted packets in order to decode correctly the packet initially sent.

The maximum possible value of a system's throughput is 1. This result can be achieved only if all transmitted packets are correctly received at the first attempt. However, in case of an HARQ system, this maximum possible throughput cannot be reached because non-information bits are transmitted as well, such as CRC bits and parity bits.

HARQ SYSTEM GENERAL ARCHITECTURE

Figure 6 shows the architecture of the HARQ system adopted by the 3rd Generation Partnership Long Term Evolution (3GPP LTE). 3GPP LTE is one of the most recent digital cellular mobile technologies. The related specification has been published officially by 3GPP organization in December 2008.

At the NodeB (transmitter), up to two transport blocks are delivered to the physical layer once every TTI (European Telecommunications Standards Institute [ETSI] & Third Generation Partnership Project [3GPP], 2012). Also, assuming that the segmentation has been taken into account according to the required packet size and the maximum code block size (up to 6144bits per code block). In the same way, it is considered that CRC encoding (24bit CRC is adopted) has been performed. The Medium Access Control (MAC), which is a sub-layer of the data link layer, takes also in charge the ARQ retransmission mechanism. As a result, this layer stores the packets transmitted to the receiver until they are positively acknowledged.

At the physical (PHY) layer, the code blocks are channel turbo encoded successively. The resulting coded bits are punctured or repeated by the Rate Matching (RM) block in order to generate the required redundancy version. In case of Incremental Redundancy, the MAC layer controls what redundancy version has to be used for the PHY layer for each TTI (ETSI & 3GPP, 2012), this is performed on the basis of Channel Quality Indicator (CQI) measurement information provided by the User Equipment (UE). The obtained coded sequence is modulated according to the required modulation scheme (QPSK, 16QAM or 64QAM). Then, the modulated symbols are space-time processed with respect to the selected Multiple Input Multiple Output (MIMO) scheme. For each spatial channel, the Orthogonal Frequency Division Multiplexing (OFDM) modulation is performed in such way that an Inverse Fourier Fast Transform (IFFT) is applied and the required cyclic prefix is added.

At the User Equipment (UE) which is the receiver, the reverse processing is performed. Especially, at the PHY layer, if CRC decoding identifies that the turbo decoded sequence is erroneous, an NACK response is sent back to the transmitter. Otherwise, an ACK feedback is sent. It is to note that the HARQ soft combining is handled by the physical layer (ETSI & 3GPP, 2012), and the ARQ mechanism is taken in charge by the Radio Link Control (RLC) sub-layer of the data link layer (ETSI & 3GPP, 2010). In addition, 3GPP LTE uses N-process Stop-and-Wait technique (ETSI & 3GPP, 2010). For more details about the HARQ operation at UE, readers can refer to ETSI & 3GPP (2012).

Figure 6. 3GPP LTE physical layer with HARQ protocols (ETSI & 3GPP, 2012)

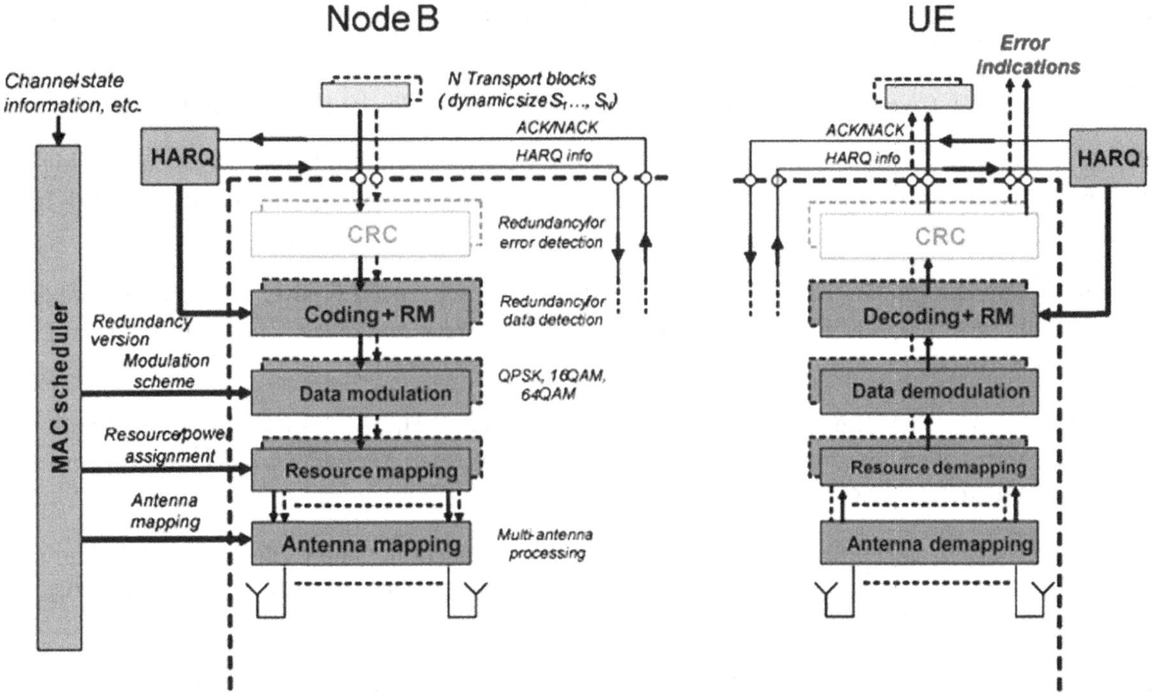

CONCLUSION

HARQ protocols play a key role in enhancing reliability of wireless communication systems. They take advantage of both ARQ mechanisms and FEC techniques. Especially, HARQ protocols are used to deal with critical transmission channels conditions. As indicated in this chapter, there are three main ARQ schemes: Stop-and-Wait, Selective Repeat and Go-Back-N. Selective Repeat technique remains the most complex but it is also the most efficient in terms of latency and throughput. Also, there are three HARQ types, namely type I HARQ, type II HARQ and type III HARQ. The choice of the HARQ combining technique has a great impact on the system's performance. In general, CC-HARQ technique, by combining several received copies of the same originally transmitted packet, is a simple scheme and generates interesting gains in terms of Signal-to-Noise Ratio. Also, IR-HARQ solution, by transmitting different bits patterns each time the retransmission of the packet detected in error is required, achieves beneficial coding gains. Still, the decision of selecting a given combining scheme can be made depending on many factors, such as: throughput and delay constraints, channel conditions, error probability threshold, etc. HARQ protocols have attracted many researchers and have been adopted by several telecommunication systems including the most recent cellular mobile technologies.

REFERENCES

Alsebae, A., Leeson, M., & Green, R. (2013). The throughput benefits of network coding for SR ARQ communication.*Proceedings of the 5th Computer Science and Electronic Engineering Conference*, 45-50. doi:10.1109/CEEC.2013.6659443

Berrou, C. (2007). *Codes et turbocodes*. Springer-Verlag France. doi:10.1007/978-2-287-32740-7

Berrou, C., Glavieux, A., & Thitimajshima, P. (1993). Near Shannon limit error correcting coding and decoding: Turbo-codes.*Proceedings of IEEE International Conference on Communications*, 1064-1070. doi:10.1109/ICC.1993.397441

Bosisio, R., Spagnolini, U., & Bar-Ness, Y. (2006). Multilevel type-II HARQ with adaptive modulation control.*IEEE Wireless Communications and Networking Conference*, 2082-2087. doi:10.1109/WCNC.2006.1696617

Boujemaa, H., Chelly, A., & Siala, M. (2008). Performance of HARQ II and III over multipath fading channels.*Proceedings of the 2nd International Conference on Signals, Circuits and Systems*, 1-5. doi:10.1109/ICSCS.2008.4746946

Chase, D. (1985). Code combining: A maximum-likelihood decoding approach for combining an arbitrary number of noisy packets. *IEEE Transactions on Communications*, *33*(5), 385–393. doi:10.1109/TCOM.1985.1096314

European Telecommunications Standards Institute, & Third Generation Partnership Project. (2010a). *LTE, Evolved Universal Terrestrial Radio Access (E-UTRA) and Evolved Universal Terrestrial Radio Access Network (E-UTRAN), Overall description, Stage 2*. 3GPP TS 36.300 version 8.12.0 Release 8, ETSI TS 136 300 V8.12.0.

European Telecommunications Standards Institute, & Third Generation Partnership Project. (2010b). *LTE, Evolved Universal Terrestrial Radio Access (E-UTRA), Radio Link Control (RLC) protocol specification*. 3GPP TS 36.322 version 8.8.0 Release 8, ETSI TS 136 322 V8.8.0.

European Telecommunications Standards Institute, & Third Generation Partnership Project. (2012a). *LTE, Evolved Universal Terrestrial Radio Access (E-UTRA), Medium Access Control (MAC) protocol specification*. 3GPP TS 36.321 version 8.12.0 Release 8, ETSI TS 136 321 V8.12.0.

European Telecommunications Standards Institute, & Third Generation Partnership Project. (2012b). *LTE, Evolved Universal Terrestrial Radio Access (E-UTRA), Services provided by the physical layer*. 3GPP TS 36.302 version 8.2.1 Release 8, ETSI TS 136 302 V8.2.1.

Holland, I. D., Zepernick, H.-J., & Caldera, M. (2005). Soft combining for hybrid ARQ. *Electronics Letters*, *41*(22), 1230–1231. doi:10.1049/el:20052614

Hong Chen, R. G., Maunder, & Hanzo, L. (2013). A survey and tutorial on low-complexity turbo coding techniques and a holistic hybrid ARQ design example. Proceedings of Fourth Quarter IEEE Communications Surveys and Tutorials, 15(4), 1546-1566.

Institute of Electrical and Electronics Engineers. (2006). *IEEE Std 802.16-2004/Cor1-2005, IEEE Standard for Local and Metropolitan Area Networks Part 16: Air Interface for Fixed and Mobile Broadband Wireless Access Systems Amendment 2: Physical and Medium Access Control Layers for Combined Fixed and Mobile Operation in Licensed Bands And Corrigendum 1*. Author.

Institute of Electrical and Electronics Engineers. (2011). *IEEE Standard for Local and Metropolitan Area Networks Part 16: Air Interface for Broadband Wireless Access Systems Amendment 3: Advanced Air Interface*. Author.

Ksairi, N., Ciblat, P., & Le Martret, C. (2013). Optimal resource allocation for Type-II-HARQ-based OFDMA ad hoc networks.*Proceedings of IEEE Global Conference on Signal and Information Processing*, 379-382. doi:10.1109/GlobalSIP.2013.6736894

Kwatra, P. (2013). ARQ protocol studies in underwater communication networks.*Proceedings of International Conference on Signal Processing and Communication*, 121-126. doi:10.1109/ICSPCom.2013.6719768

Lang, Y., Wuebben, D., Dekorsy, A., & Braun, V. (2013). A turbo-like iterative decoding algorithm for network coded HARQ.*Proceedings of 9th International ITG Conference on Systems, Communication and Coding*, 1-6.

Madhow, U. (2008). *Fundamentals of Digital Communication*. Cambridge University Press. doi:10.1017/CBO9780511807046

Marcille, S., Ciblat, P., & Le Martret, C. J. (2012). Stop-and-Wait Hybrid-ARQ performance at IP level under imperfect feedback.*Proceedings of IEEE Vehicular Technology Conference*, 1-5. doi:10.1109/VTCFall.2012.6399116

Narayanan, K. R., & Stuber, G. L. (1997). A novel ARQ technique using the turbo coding principle. *IEEE Communications Letters*, *1*(2), 49–51. doi:10.1109/4234.559361

Proakis, J. G., & Salehi, M. (2008). *Digital Communications*. Academic Press.

Sheu, S.-T., Kuo, K.-H., Yang, C.-C., & Sheu, Y.-M. (2013). A Go-back-N HARQ time bundling for machine type communication devices in LTE TDD.*Proceedings of IEEE Wireless Communications and Networking Conference*, 280-285.

Shi, Z., Ren, L., & Jin, F. (2002). Design and performance analysis of HARQ for RS-turbo concatenated codes.*Proceedings of IEEE International Conference on Communications, Circuits and Systems and West Sino Expositions*, 56-59.

Third Generation Partnership Project. (2008). *Technical specification group radio access network, Evolved Universal Terrestrial Radio Access (E-UTRA), multiplexing and channel coding* (Release 8). 3GPP TS 36.212 V8.5.0.

Third Generation Partnership Project. (2009). *Technical Specification Group Radio Access Network; Enhanced uplink; Overall description; Stage 2* (Release 7). 3GPP TS 25.319 V7.7.0.

Third Generation Partnership Project. (2010). *Technical Specification Group Radio Access Network, Evolved Universal Terrestrial Radio Access (E-UTRA), Multiplexing and channel coding* (Release 10). 3GPP TS 36.212 V10.0.0.

Uchoa, A. G. D., Demo Souza, R., & Pellenz, M. E. (2010). Type-I HARQ scheme using LDPC codes and partial retransmissions for AWGN and quasi static fading channels.*Proceedings of the 7th International Symposium on Wireless Communication Systems*, 571-575. doi:10.1109/ISWCS.2010.5624326

Wozencraft, J. M., & Horstein, M. (1960). Digitalised communication over two way channels. *The 4th London Symposium of Information Theory*.

Wozencraft, J. M., & Horstein, M. (1961). Coding for two-way channels. Tech. rep. Research Laboratory of Electronics, M.I.T.

Yafeng, W., Lei, Z., & Dacheng, Y. (2003). Performance analysis of type III HARQ with turbo codes. *Proceedings of the 57th IEEE Semiannual Vehicular Technology Conference*, *4*,2740-2744.

Yu, X., Sun, K., & Yuan, D. (2010). Comparative analysis on HARQ with turbo codes in Rician fading channel with low Rician factor.*Proceedings of 12th IEEE International Conference on Communication Technology*, 342-345.

Zhang, M. (2013). Major automatic repeat request protocols generalization and future develop direction. *Proceedings of the 6th International Conference on Information, Management, Innovation Management and Industrial Engineering*, *2*,5-8. doi:10.1109/ICIII.2013.6703196

KEY TERMS AND DEFINITIONS

Acknowledgment: A response sent by the transmitter to the receiver to indicate if the packet is erroneous or correctly received.

Adaptive Modulation and Coding: A technique used to enable a communication system to react dynamically to channel fluctuations by selecting the most adequate modulation order and coding rate.

Automatic Repeat reQuest: An error control mechanism and a feedback-based communication link. It allows the receiver to send data retransmission request to the transmitter.

Forward Error Correction: A technique that refers to channel coding algorithms used to protect data to be transmitted against noise induced by transmission channels.

Hybrid Automatic Repeat reQuest: A protocol used to deal with poor channel conditions by combining Automatic Repeat reQuest and Forward Error Correction codes.

Long Term Evolution: A digital cellular mobile phone standard. It is specification has been officially published in 2008 by the Third Generation Partnership Project organization.

Orthogonal Frequency Division Multiplexing: A multi-carrier transmission technique that divides the available bandwidth into multiple narrower subcarriers.

Soft Combining: A technique used to soft combine several packets in order to improve the reliability of a communication system.

Chapter 12
Security in Mission Critical Communication Systems:
Approach for Intrusion Detection

Karen Medhat
Cairo University, Egypt

Rabie A. Ramadan
Cairo University, Egypt

Ihab Talkhan
Cairo University, Egypt

ABSTRACT

This chapter introduces two different algorithms to detect intrusions in mission critical communication systems to guarantee their security. The first algorithm is a classification algorithm which applies the concept of supervised learning. The second algorithm is a clustering algorithm which applies the concept of unsupervised learning. The algorithms detect intrusions using a set of detection rules that are structured in the form of decision trees. The algorithms are described in details and their results on well-known dataset are introduced. An enhancement for the J48algorithm is also introduced, where the decision tree for the algorithm is changed to a binary tree. The change enhances the complexity to reach a decision. The chapter includes a brief introduction about the security in Mission critical systems and the reason behind securing such systems. It introduces different methodologies that were introduced to detect intrusions in wireless communications.

INTRODUCTION

A mission critical system is essential to the survival of a business or organization. When a mission critical system is attacked or failed, business operations and organizations are significantly impacted. For some governmental organizations and some IT sectors, databases are considered as Mission Critical systems. For the internet applications, servers are considered as Mission Critical systems. For public safety organizations, the systems must be reliable and available around the clock to guarantee instant

DOI: 10.4018/978-1-5225-2113-6.ch012

responses in order to save lives. The security in mission-critical systems and wireless communications has attracted a great attention, especially with its rapid development. Security considerations in mission critical systems and wireless communications act as a challenging research area due to the increase of security-critical applications in which a reliable intrusion detection mechanism is needed. Mission-critical communications are extensively used by public safety responders and organizations where connections and communications have to be done reliably and instantly. Public safety organizations also use the Mission Critical systems to monitor major crime and large scale disasters. The mission Critical systems have three main elements:

1. Interoperability where the communications can be taken instantaneously with different organizations.
2. Critical Networks which offers security to the users of the system.
3. Mission Critical data where the important data needed to secure the network can be easily accessed.

The reliability in Mission Critical Systems is highly needed for the survival of the purpose that they are built for. Securing these systems from attacks increases their reliability greatly. Thus, the security in Mission Critical Systems is one of the important concerns to be addressed in those systems. One of the security issues to be addressed is to detect the intrusions that may attack the devices used in the communications as intrusions can greatly affect the performance of the Mission critical systems or may do unwanted manipulation with the critical data sent over the network. The real-time security monitoring for the Mission-critical systems is highly recommended to protect these systems.

In this chapter, an intrusion detection paradigm is introduced. This paradigm introduces an unsupervised learning algorithm and a supervised learning algorithm to detect intrusions. The algorithms can be embedded in the devices used in the Mission critical systems to detect intrusions. Each one of the algorithms builds a set of the intrusion detection rules. The intrusion detection rules generated from both algorithms are structured in the form of a binary tree which decreases the complexity of reaching a decision. The proposed algorithms provided a high detection accuracy using only 10% of the data for training in addition to less number of features, compared to previous work for intrusion detection, which decreased the complexity and the processing time. An enhancement for J48 classification algorithm is also proposed which decreases the size of the algorithm's decision tree and makes it suitable to be used for intrusion detection in memory constrained devices that are used in mission critical systems.

Background

In order to protect the Mission-Critical systems, there must be a real-time security monitoring systems as these systems can't tolerate any faults. An example of the real-time monitoring services is the one offered by Motorola (MOTOROLA, 2015). Motorola's Security Monitoring service depends on five main concepts. The first is to identify the parts of the system to be monitored. The second one is to protect the system with continuous monitoring of the system's activity. The third one is to detect all the anomalies or the attacks that can threaten the system. The fourth one is to respond by taking corrective actions for the detected threats. The fifth is to recover the system from the attack to a restoration point. Alcatel-Lucent (Alcatel-Lucent, 2013) delivers a Mission-critical communications networks for public safety. They introduced IP/Multiprotocol Label Switching (MPLS)-based communications network for public safety using next-generation products and advanced management tools.

There are different methods to protect the Mission-Critical Communications Networks (Ambady, 2012). Those methods are:

1. Firewalls.
2. Intrusion detection system (IDS).
3. Intrusion prevention systems.
4. demilitarized zone (DMZ).
5. Anti-virus software.
6. Virtual private networks (VPNs), which use transport layer security (TLS) or secure socket layer (SSL) to encrypt transmissions.
7. Authentication.
8. Authorization.
9. Tamper prevention and detection to prevent the unauthorized access.
10. Time-windowed commands to limit the risk of some attacks.
11. Increasing the Entropy of the system, which means increasing the uncertainty of the system's security features.
12. Pass-through devices.
13. Behavior auditing.

One of the main methods to secure the communications systems is to detect intrusions by building a reliable intrusion detection system. The concept of Intrusion Detection was introduced by James Anderson in 1980, as, "Intrusion attempt or threat is the potential possibility of a deliberate unauthorized attempt to access information, manipulate information, or render a system unreliable or unusable" (Anderson, April 1980). Anderson made an investigation about intrusions and intrusion detection, where he discussed the definition of fundamental terms of intrusions and intrusion detection, which are:

- **Risk:** The exposure of information unpredictably.
- **Threat:** The unauthorized access to the data or the network.
- **Attack:** The execution of a plan to perform a threat.
- **Vulnerability:** The flaw in the system or network that makes it vulnerable to attacks.
- **Penetration:** The successful attack.

There are different methodologies to detect intrusions the most commonly used ones are Signature-based intrusion detection and anomaly-based intrusion detection. In signature-based, the patterns of the intrusions are defined in a database. If the system is attacked, the pattern in the received data is compared to the saved ones and an intrusion is reported whenever a match is found. This method is very effective when the attacks are known but is not efficient with unknown attacks. For the unknown attacks, their patterns are not recognized and the attacks pass through the system as if they are normal activity. For the anomaly-based, the normal behavior of the system is defined. If the system finds any activity that differs from the normal behavior, it generates an alarm. This method is effective to detect a new type of attacks to the system, but it may generate false alarms for normal system activity. Monitoring of intrusions can be done at host level or network level or hybrid based (Sandhu, Haider, Naseer, & Ateeb, 2011; Uppal, Javed, & Arshad, 2014). For the host-based intrusion detection systems, the activities on the host are monitored and any suspicious activity is marked as an intrusion. For the network based intrusion detection systems, the activities and the data packets sent on the network are monitored and any suspicious activity or data packet is marked as an intrusion. For the hybrid based approach, both the host based and network based approaches are used in the same system. The Artificial Intelligent techniques are considered the most common techniques used for intrusion detection. The Artificial Neural Network (Al-Subaie & Zulkernine) is used for building a model to recognize patterns in order to recognize the patterns of any abnormal activity in a system. State transitions tables (Ilgun, Kemmerer, Fellow, IEEE, & Porras, 1995) can be used to describe the sequence of activities that are done by an intruder.

The problem of intrusion detection and securing the communications systems has attracted attention recently. There are many researchers that address Intrusion detection in communication systems (Grzech, 2009; Parihar, Rathore, & Burse, April 2014 ; Patel, Qassim, & Wills, 2010; Rassam, Maarof, & Zainal, 2012; Sun, Osborne, Xiao, & Guizani, 2007). In addition to this, many researchers have addressed the security mechanisms that can be used generally in the communications systems (Chen & Gong, 2012 ; Kargl et al., 2008) and those that can be used specifically in mission critical communication systems (Turek, Zerawa, & Anees, 2011).

Ajenjo et al. (Ajenjo & Wietgrefe, 2003) discussed the importance of monitoring of the traffic pattern in the network. The monitoring and the analysis of the traffic were applied on a NATO's system. Kumar et al. in (Kumar & Kumar, 2013) proposed an approach called AMGA2–NB. The approach contains three phases and it uses a genetic algorithm to choose a set of solutions from a pool of proposed solutions to be used for intrusion detection. A set of individual solutions is generated in the first phase of the algorithm from the fitness function. The generated set of solutions is approximated in the second phase to generate an improved chromosome. The output of the first and the second phases acts as an input for the third phase where the prediction of the final ensemble is introduced.

Kruegel et al.(Kruegel, Mutz, Robertson, & Valeur, 2003) proposed an approach to enhance intrusion detection for the systems using anomaly-based algorithms. The authors highlighted some of the reasons behind reporting false alarms from the anomaly-based. One of the main problems for reporting false alarms is the simplicity of combining the model outputs on which the decision is based. The other problem is that the output is not supported with extra information to increase its confidence. The authors based their intrusion detection model on Bayesian network to act as a solution for the mentioned problems. Bayesian network improved the process of combining the model outputs to take a more accurate decision in addition to this it used additional information to strengthen the output credibility.

Krontiris et al. in (Krontiris, Dimitriou, & Freiling, 2007) introduced a decentralized scheme for intrusion detection. Each device in the network has four modules:

1. Local Packet Monitoring module which gathers the data to be sent to the Local Detection module.
2. Local Detection Engine which collects the data sent to it by the Local Packet monitoring module. It analyzes the collected data and stores the specifications that describe the correct operation.
3. Cooperative Detection Engine, if this engine detects an intrusion, it sends the state information of the local detection module to the neighbors and receives information from the same module included in the neighboring devices and then apply a majority vote rule to indicate if there is an intrusion or not.
4. Local Response module takes the appropriate actions to restore the network normal operation and isolate the intruded part when an intrusion is detected.

Silva et al. (Silva et al., 2005) introduced a decentralized intrusion detection algorithm. The authors defined a set of rules to be applied by the algorithm on the collected features for intrusion detection. Each rule can detect a specific type of attack. The proposed algorithm contains three phases; the first phase is data acquisition phase, the second phase is the rule application phase, the third and the last phase is the intrusion detection. The analysis of the acquired data is done in the first phase. In the second phase, the introduced rules are applied on the data and an intrusion is detected if any of the rules fails. In the last phase, an intrusion is reported if the number of detected intrusions increased a certain threshold. The threshold expresses the number of expected attacks in the network.

Decreasing the feature space used by the intrusion detection algorithms is taken in consideration by several researches. For example, Karan et al. (Karan Bajaj, 2013) proposed a technique for feature selection using a combination of feature selection algorithms like Information Gain, Gain Ratio, Correlation Attribute Evaluation. The authors tested the performance of the selected features on different classification algorithms such as J48, Naïve Bayes, NB-Tree, Multi-Layer Perceptron, SVM, and SimpleCart. J48 is a classification algorithm that builds a decision tree using the Entropy. J48 (Bhargava, Sharma, Bhargava, & Mathuria, June 2013) can handle both discrete and continuous data. J48 can simply handle missing attribute values by not including it in the Entropy calculation. Naïve Bayes is a classifier which applies Bayes probability theorem to build conditional probability model from which a classifier is built (Domingos & Pazzani, 1997; Manning, Raghavan, & Schütze, 2009). NB-Tree (Fonseca & Jorge, 2003b) is a decision tree based classifier that uses Naïve Bayes as each node. Multi-Layer Perceptron is a neural network classifier (JK & J., 1999). Support Vector Machines (SVM) also known as Support Vector Networks are sets of supervised learning models that analyze and recognize patterns in the data (Cortes & Vapnik, 1995). SimpleCart is a decision tree based algorithm and builds it based on the Entropy (Kalmegh, 2015).

Ravale et al. (Ravale, Marathe, & Padiya, 2015) introduced an approach to select the significant features according to the type of the attack. The authors used K-means clustering for building the clusters used to take the initial decision. Linear support vector machines were then used to take the most accurate decision regarding reporting an intrusion.

In some of the mentioned research papers, different aspects for the intrusion detection systems were not covered. For example, the approach introduced by Kumar et al. in (Kumar & Kumar, 2013) contains different phases and steps to reach a decision which increases the complexity of the process of taking a decision. The approach introduced by Kruegel et al.(Kruegel et al., 2003) focused only on the anomaly-based algorithms. The accuracy of the decentralized intrusion detection approach introduced by Krontiris et al. (Krontiris et al., 2007) is not mentioned to be a high detection rate and the process for detecting the intrusion is very complex. The accuracy of the approach introduced by Silva et al (Silva et al., 2005) is not mentioned to be high.

In this chapter, high detection accuracy and low complexity intrusion detection algorithms are introduced. One of the algorithms is a supervised learning algorithm and the other is an unsupervised learning algorithm to be used in the devices of the mission-critical systems. The learning algorithms use only 10% of the data for training. An enhancement for the J48 algorithm is also proposed which decreases the size of the decision tree and make it applicable for memory constraint devices.

MAIN FOCUS OF THE CHAPTER

This chapter focuses mainly on introducing high detection accuracy and low complexity intrusion detection algorithms. An enhancement to the J48 algorithm is also introduced to decrease the size of the decision tree and converts it to a binary tree where the decision can be reached easily. The intrusion detection algorithms act as a line of defense to the system where an appropriate action can be taken when the intrusion is reported.

Structure of the Decision Trees

The supervised and the unsupervised algorithms introduced in this chapter produce a set of intrusion detection rules that are structured in the form of a binary tree. Each node in the decision tree expresses a certain feature and a certain value to this feature. When the algorithm is in operation, the measured feature's value (a feature's value from the data sample) is compared with the feature's value in the node of the tree. The comparison of these values determines which branch of the binary tree will be taken next. The nodes of the intrusion detection binary tree are in the form of < feature, value > pair (Bellot & El-Bèze, 2000). The features included in the < feature, value > pairs represents some or all of the features measured by the system. If the number of < feature, value > pairs is relatively large, the Entropy (1) will be measured to select some of the features and to select some of their values as well. The features that have the major effect on dividing the data samples into different classes will be selected.

$$\mathit{Entropy}(p) = -\sum_{j=1}^{n}\left(\frac{|Pj|}{|P|}\log\frac{|Pj|}{|P|}\right) \tag{1}$$

where P_j is the number of samples for class j, P in the denominator is the total number of samples and n is the number of classes. For example, in an intrusion detection system n will be equal to two as there will be only two classes. One of the two classes is for the data samples that describe the normal activity of the system, and the other class for the data samples that describe the abnormal activities (attacks) of the system.

The Supervised Learning Algorithm

The intrusion detection tree of the supervised learning is built using the Ginni Index (2). After the < feature, value > pairs are extracted from the training dataset, the Ginni Index is calculated at each level of the decision tree to choose the pair to be added at that level.

$$\mathit{Gini\ Index} = 1 - \sum_{j=1}^{n}\left(\frac{|Pj|}{|P|}\right)^2 \tag{2}$$

The intrusion detection problem is mainly considered as a classification problem. For the illustration of the algorithm, it will be applied on the simple Weather dataset shown in Table 1. This dataset is one of the sample datasets for classification problems included in WEKA data mining tool (Hall et al., 2009). Each record in the dataset represents a data sample. The dataset has 4 features:

1. Outlook
2. Temperature
3. Humidity
4. Windy

According to the values of the features, the data sample is classified as one of the two existing classes. The two classes are the two values (YES, NO) in the PLAY column. Temperature and humidity are continuous features as they take numerical values. Outlook and Windy are discrete features as they take non-numerical values.

The extracted < feature, value > pairs for this dataset are:

- < Outlook, Sunny >
- < Outlook, overcast >
- < Outlook, rainy >
- < Temp, <= 75 >
- < Humidity, <= 75 >
- < Windy, false >

For Outlook, its values can take any of the three non-numerical values (Sunny, Overcast, and Rainy). Thus, all the values will be taken in separate < feature, value > pairs. For Temperature and Humidity, the numerical value that can divide the data samples into two classes (or clusters) evenly will be selected in

Table 1. Weather dataset

OUTLOOK	TEMPERATURE	HUMIDITY	WINDY	PLAY
Sunny	85	85	FALSE	NO
Sunny	80	90	TRUE	NO
Overcast	83	78	FALSE	YES
Rain	70	96	FALSE	YES
Rain	68	80	FALSE	YES
Rain	65	70	TRUE	NO
Overcast	64	65	TRUE	YES
Sunny	72	95	FALSE	NO
Sunny	69	70	FALSE	YES
Rain	75	80	FALSE	YES
Sunny	75	70	TRUE	YES
Overcast	72	90	TRUE	YES
Overcast	81	75	FALSE	YES
Rain	71	80	TRUE	NO

Figure 1. Division of data samples of the Weather dataset for <outlook, Sunny> pair

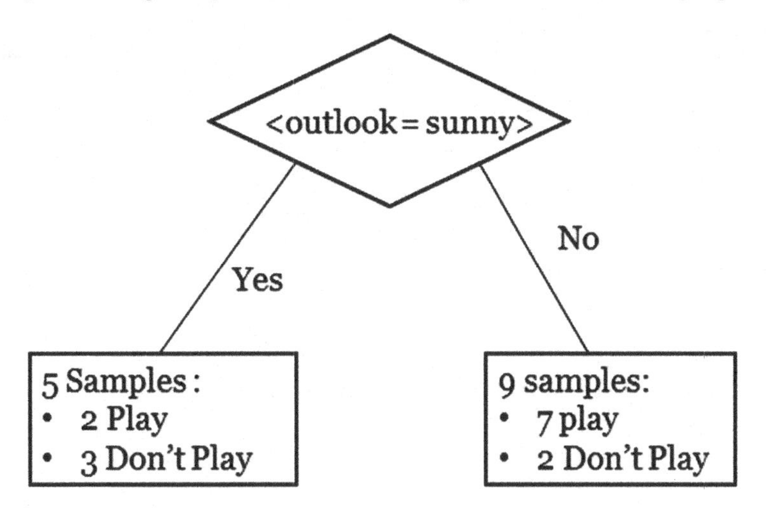

the <feature, value> pair. The selected value was found to be 75 for Temperature and 75 for Humidity. As Windy is a discrete feature of binary values, any of the two values will be selected in the <feature, value> pair. The value of the Ginni Index is calculated to each pair of the <feature, value> pairs. For the first pair, < Outlook, Sunny >, the dataset will be divided into two sets as shown in Figure 1.

The data samples having outlook = sunny are five data samples from the dataset and are included in the YES set. Two out of the five data samples have value play = yes and three out of the five data samples have value play = no. The data samples having other values for outlook are nine data samples from the dataset and are included in the NO set. Seven out of the nine data samples have value play = yes and two out of the nine data samples have value play= no. Then the value of the Ginni Index will be calculated for the < Outlook, Sunny > pair as the following:

For the YES set: $G^{yes} = 1 - (\left(\frac{2}{5}\right)^2 * \left(\frac{3}{5}\right)^2) = 0.48$

For the NO set: $G^{no} = 1 - (\left(\frac{2}{9}\right)^2 * \left(\frac{7}{9}\right)^2) = 0.346$

The total value of the Ginni Index for the < Outlook, Sunny > Pair:

$$G^{Total} = (\frac{5}{14})\left(0.48\right) + \left(\frac{9}{14}\right)\left(0.346\right) = 0.39365$$

The same calculations were applied to all the < feature, value > pairs to give the results shown in Table 2.

The largest value of the Ginni Index is the one for the < Outlook, overcast > pair. Thus, none of the < feature, value > pairs having the Outlook feature will be selected at the current level of the tree. The smallest value of the Ginni Index is the one for < Windy, false > pair. Thus, < Windy, false > pair will be selected at the current level of the tree which is level zero (root node) as shown in Figure 2.

Table 2. Ginni Index values for the WEATHER's dataset < feature, value >pairs

Feature Value Pairs	Total Value of Ginni Index
Outlook = Sunny	$G^{Total} = 0.39365$
Outlook = overcast	$G^{Total} = 0.5$
Outlook = rainy	$G^{Total} = 0.457$
Temp < = 75	$G^{Total} = 0.4428$
Humidity	$G^{Total} = 0.43157$
Windy = false	$G^{Total} = 0.4285$

Figure 2. Level 0 of the Decision Tree for the supervised learning algorithm

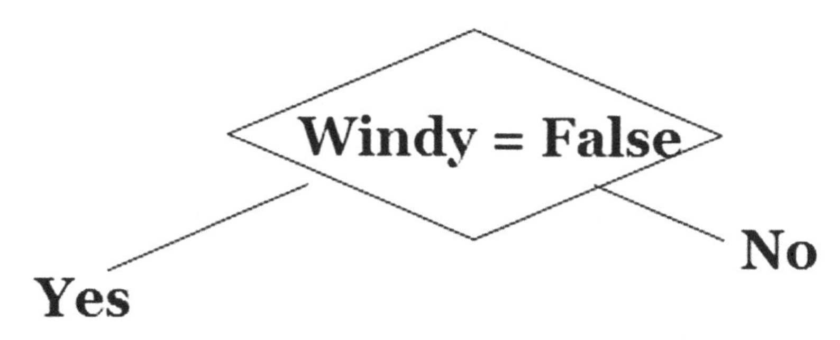

The value of the Ginni Index is calculated for the data samples in the YES set, and for the data samples in the No set for the < Windy, false> pair. The process is repeated till reaching the leaf nodes. At the leaf nodes, each data sample in the dataset is assigned to one class whether play = yes or play = no. The levels of the decision tree are shown in the Figures 3, Figure 4, Figure 5, and Figure 6.

Figure 3. Level 1 of the Decision Tree for WEATHER dataset using the supervised learning algorithm

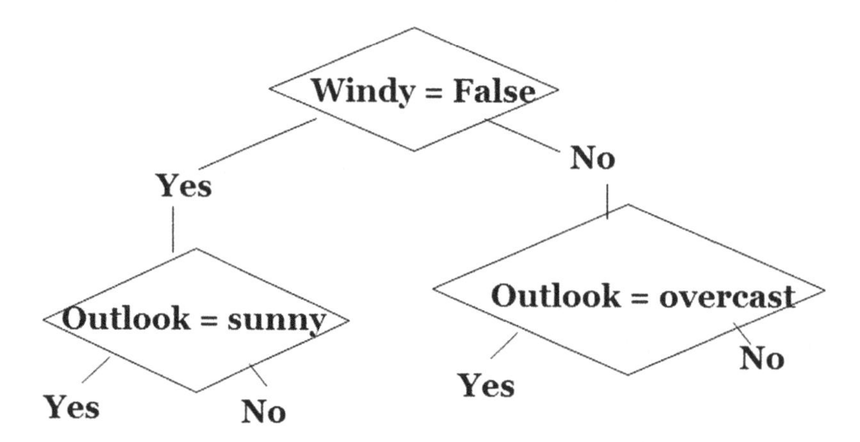

Figure 4. Level 2 of the Decision Tree for WEATHER dataset using the supervised learning algorithm

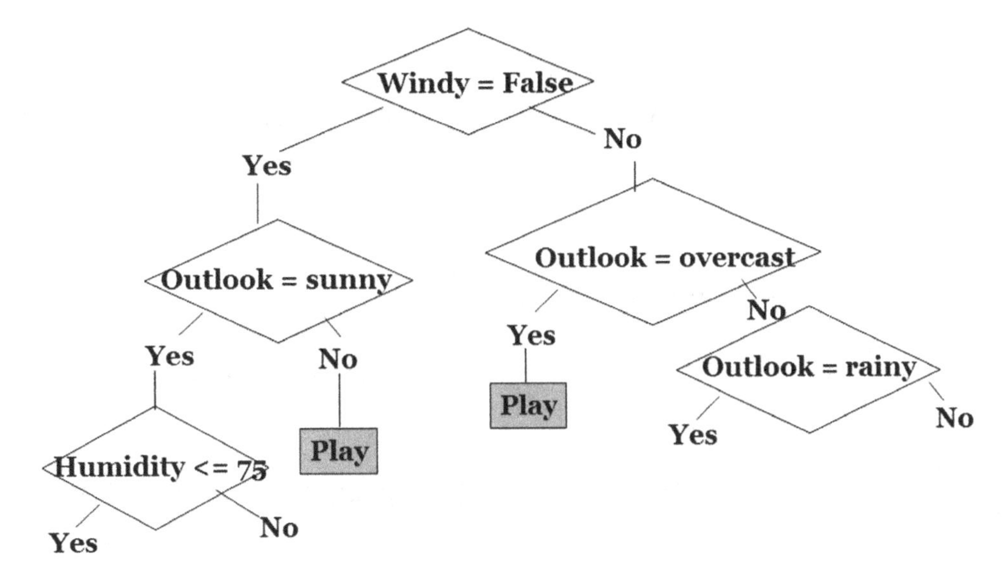

Figure 5. Level 3 of the Decision Tree for WEATHER dataset using the supervised learning algorithm

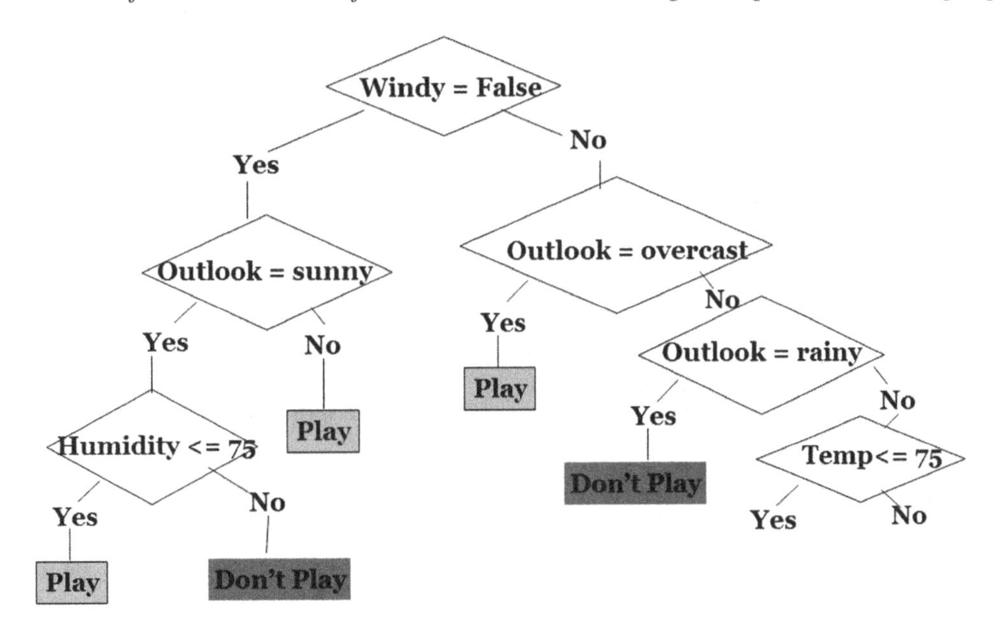

The Unsupervised Learning Algorithm

The intrusion detection tree of the unsupervised learning is built using the Optimal Grouping (3). After the < feature, value > pairs are extracted from the training dataset, the value of Optimal Grouping (Wu, Liu, Zhou, & Zhan, 2012) is calculated at each level of the decision tree to choose the pair to be added at that level. The optimum grouping value is the summation of all taxon (4) values for the sets of each < feature, value > pair

Figure 6. Level 4 of the Decision Tree for WEATHER dataset using the supervised learning algorithm (The final decision tree)

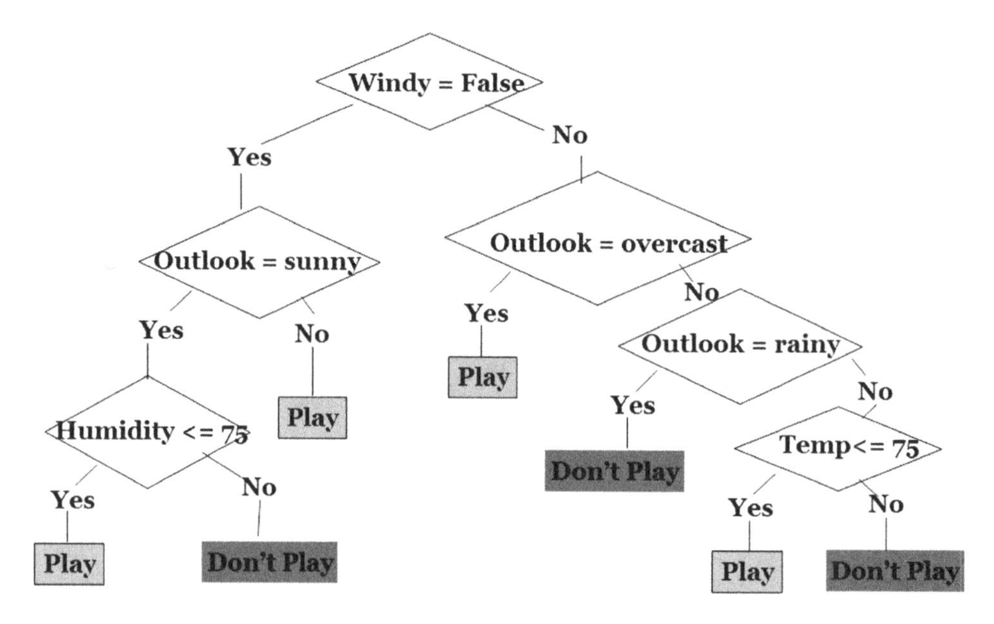

$$g = \sum_{i=1}^{L} »^{i} \tag{3}$$

$$»^{i} = \prod_{j=1}^{n} \frac{\left|V_j^i\right|}{\left|D_j\right|}, \tag{4}$$

where $\left|V_j^i\right|$ is the length of an interval (in case of the continuous features) or number of values of appropriate subset $\mathbf{V}_j^i$ (in case of the discrete features); $\left|D_j\right|$ is the length of an interval between the minimal and maximal values of continues feature X for all objects from the initial dataset (for the continuous features) or the general number of values of discrete feature X for all objects from the initial dataset (for the discrete features).

For the illustration of the algorithm, it will be applied on the simple Weather dataset shown in Table 1. The < feature, value > pairs are as introduced before to be:

- < Outlook, Sunny >
- < Outlook, overcast >
- < Outlook, rainy >
- < Temp, <= 75 >
- < Humidity, <= 75 >
- < Windy, false >

Table 3. Data samples having Outlook = Sunny

OUTLOOK	TEMPERATURE	HUMIDITY	WINDY	PLAY
Sunny	85	85	FALSE	NO
Sunny	80	90	TRUE	NO
Sunny	72	95	FALSE	NO
Sunny	69	70	FALSE	YES
Sunny	75	70	TRUE	YES

For the first pair, < Outlook, Sunny >, the dataset will be divided into two sets as shown in Figure 1. The data samples having outlook = sunny are five data samples from the dataset (shown in Table 3) and are included in the YES set.

The value of Taxon $»^{i} = \prod_{j=1}^{n} \frac{|V_j^i|}{|D_j|}$ will be calculated as below for the YES set:

$$»^{yes} = \frac{1}{3} * \frac{(85-69)}{85-64} * \frac{(95-70)}{96-65} * \frac{2}{2} = 0.2048$$

In the above, the outlook feature has only one value in the YES set (sunny) out of the three values that it can take. For the temperature, the maximum value is 85 and the minimum value is 69 in the data samples included in the YES set. The numerator expresses the difference between these two values (85-69). The denominator expresses the difference between the maximum and the minimum values of the temperature for all the data samples in the dataset. For the humidity, the maximum value is 95 and the minimum value is 70 in the data samples included in the YES set. The numerator expresses the difference between these two values (95-70). The denominator expresses the difference between the maximum and the minimum values of the humidity for all the data samples in the dataset. For Windy, the data samples in the YES set have the two values of this feature. The same calculations were applied for the data samples in the NO set to get the result below:

$$»^{No} = 0.5059$$

The value of optimal grouping was then calculated from the summation of the Taxon values for both the YES and NO sets:

$$g = \sum_{i=1}^{L} »^{i} = »^{yes} + »^{No} = 0.2048 + 0.5059$$

$$= 0.7107$$

The same calculations were applied to all the < feature, value > pairs to give the results shown in Table 4.

Table 4. Optimal grouping values for the WEATHER's dataset < feature, value >pairs

Feature Value Pairs	Value of Optimum Grouping
Outlook = Sunny	g = 0.7107
Outlook = overcast	g = 0.7757
Outlook = rainy	g = 0.7783
Temp < = 75	g = 0.6006
Humidity	g = 0.6789
Windy = false	g = 0.65898

The smallest value of the Optimal Grouping is the one for the < Temp, 75 > pair. Thus, it will be selected at the current level (level 0) of the decision tree as shown in Figure 7.

The levels of the decision tree are shown in the Figure 8Figure 9, Figure 10, and Figure 11.

The decision tree shown in Figure 10 has seven clusters of one, three, four, two, one, one, two data samples from the left to the right respectively. The dataset has 14 data samples as shown in Table 1, nine of them are classified as "play", and five of them are classified as "don't play". Thus, "play" class appears more in the dataset than "don't play" class. According to the dataset analysis, the clusters having more data samples will be labelled with the class label that appears more in the dataset "play". Thus, the four clusters having three, four, two, two data samples will labelled as "play" and the other clusters will be labelled as "don't play" to get the latest decision tree shown in Figure 11.

Figure 7. Level 0 of the Decision Tree for WEATHER dataset using the unsupervised learning algorithm

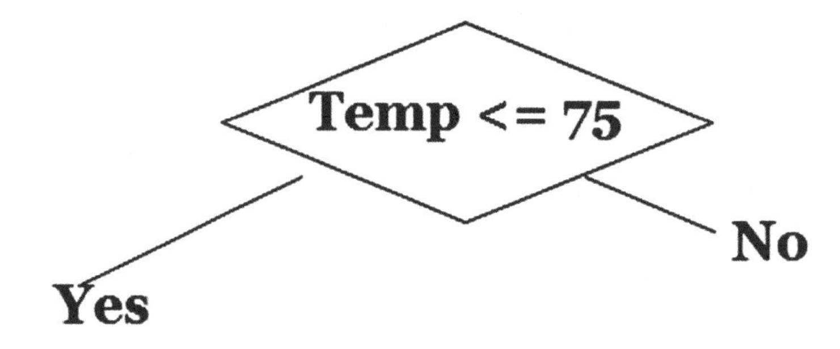

Figure 8. Level 1 of the Decision Tree for WEATHER dataset using the unsupervised learning algorithm

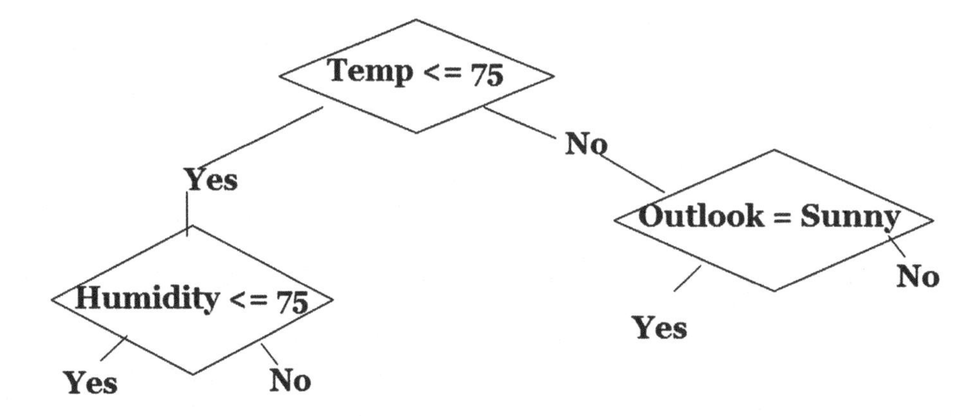

Figure 9. Level 2 of the Decision Tree for WEATHER dataset using the unsupervised learning algorithm

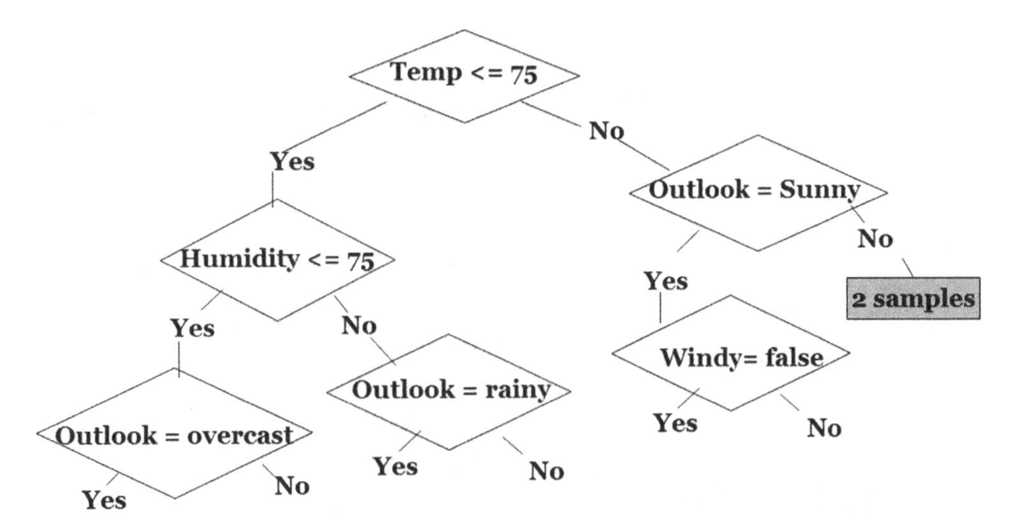

Figure 10. Level 3 of the Decision Tree for WEATHER dataset using the unsupervised learning algorithm

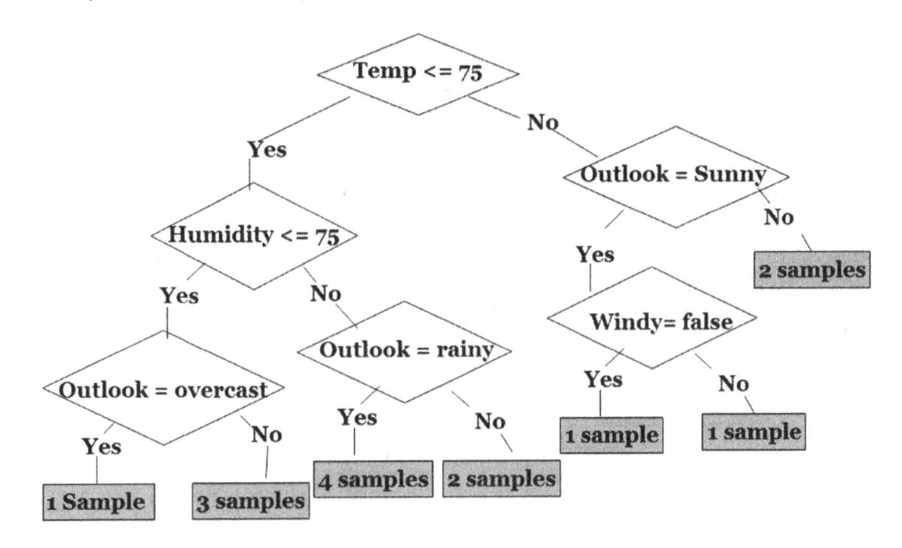

The J48 Algorithm's Enhancement

The enhancement for the J48 decision tree algorithm is to prune its branches that have no data samples from the training dataset for the large datasets and to convert the decision tree of the algorithm to a binary tree. The enhancement was applied to the KDD dataset (Mahbod Tavallaee, 2009) and the size of the decision tree was greatly decreased. The size of the decision tree was reduced from 392 nodes to 145 nodes and from 331 leaves to 146 leaves. Part of the decision tree for the KDD dataset is shown in Figure 12 and the enhancement for this part of the decision tree is shown in Figure 13. See also Appendix.

Figure 11. Final Decision Tree for WEATHER dataset using the unsupervised learning algorithm

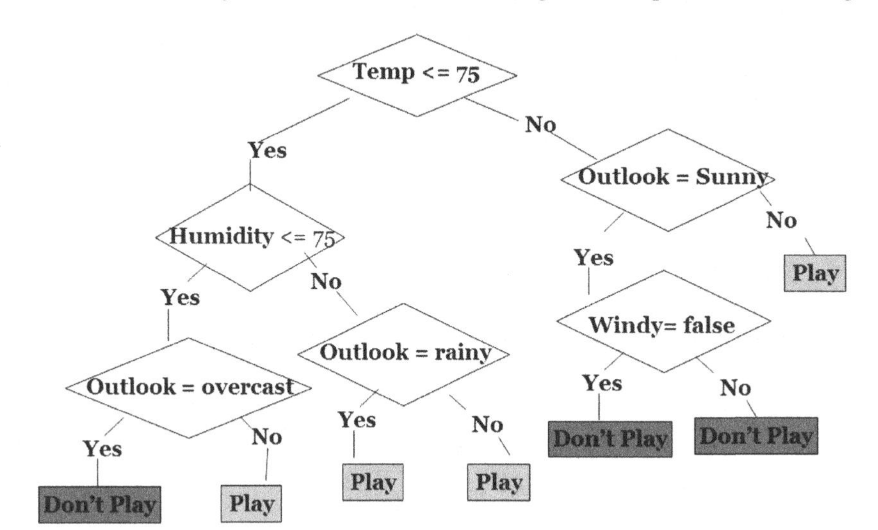

Figure 12. Part of J48 Decision tree for KDD dataset

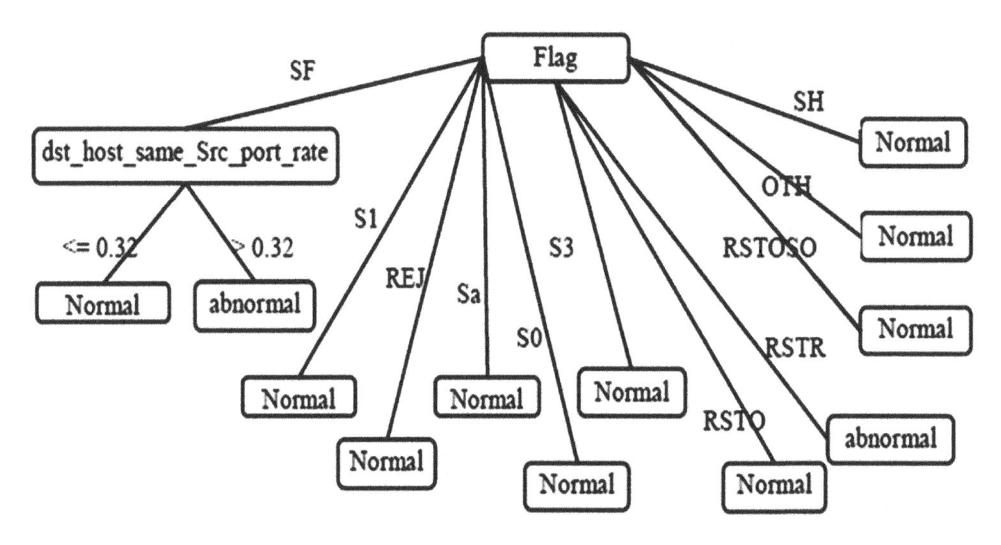

Figure 13. Enhancement for part of J48 Decision tree for KDD dataset shown in Figure 12

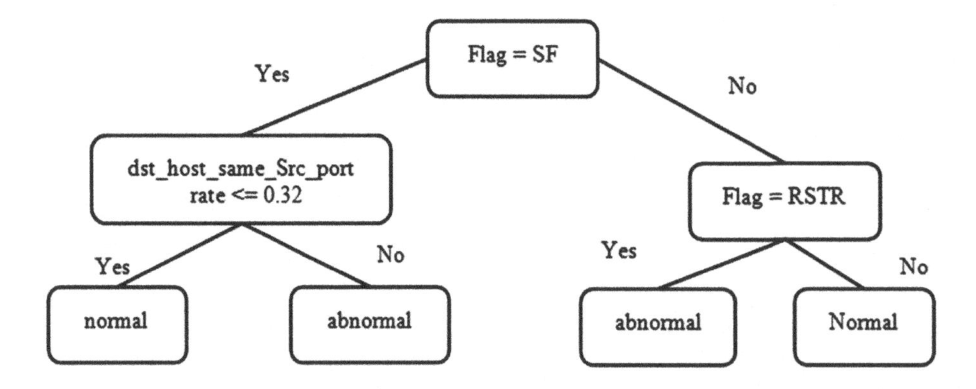

Detection Accuracy of the Algorithms on KDD Dataset

For the application of the intrusion detection algorithms, 494,021 data samples of the KDD Cup '99 dataset were selected. Only 10% of the data was used as a training set which is about 49,402 data samples out of 494,021 total data samples.

The detection accuracy was calculated on the selected dataset and a high detection accuracy was reached given that only 10% of the data was used for training. The detection accuracy results are introduced in Table 5. The features space was reduced by selecting the most significant features. The dataset consists of 41 features, only 15 features were selected by the unsupervised learning algorithm, 14 features were selected by the supervised learning algorithm, and 22 features were selected after applying the enhancements for the J48 decision tree.

FUTURE RESEARCH DIRECTIONS

Future work may involve some modifications on the algorithms for higher detection accuracy. Data aggregation for reporting the intrusions detected may be studied to decrease energy consumption and communication overhead across the network. The actions taken on detecting an intrusion may be analyzed based on the application of the Mission Critical System.

CONCLUSION

In this chapter, high detection accuracy intrusion detection algorithms for Mission critical systems were proposed. One of the algorithms used supervised learning and the other used unsupervised learning. The aim of the learning methods used by the algorithms is to build decision trees which will be used to take the decision of intrusion. The learning algorithms used only 10% of the data for training and gave high detection accuracy on the used dataset using a reduced number of features. The proposed algorithms involved a reduction in the decision complexity.

Table 5. Detection accuracy of the proposed algorithms

Intrusion Detection Algorithm	Detection Accuracy
Unsupervised Learning Algorithm	84.15%
Supervised Learning Algorithm	95.08%
The J48 Enhancement	98.89%

REFERENCES

Ajenjo, A. D., & Wietgrefe, H. (2003). Minimal-Intrusion Traffic Monitoring And Analysis In Mission-Critical Communication Networks. *Journal of Systemics, Cybernetics and Informatics, 1*(5).

Al-Subaie, M., & Zulkernine, M. (2006). *Efficacy of Hidden Markov Models Over Neural Networks in Anomaly Intrusion Detection*. Paper presented at the Computer Software and Applications Conference, 2006. COMPSAC '06. 30th Annual International, Chicago, IL. doi:10.1109/COMPSAC.2006.40

Alcatel-Lucent. (2013). *Mission-critical Communications Networks for Public Safety Highly reliable converged IP/MPLS-based backhaul Application Note*. Retrieved from http://www.tmcnet.com/tmc/white-papers/documents/whitepapers/2013/9270-mission-critical-communications-networks-public-safet.pdf

Ambady, B. (2012). A Guide to Security Methods for Mission-Critical Communications Networks. *Mission Critical Communications has more details about security options for utilities and smart-grid applications*. Retrieved from http://www.radioresourcemag.com/Features/FeaturesDetails/FID/341

Anderson, J. P. (April1980). *Computer Security Threat Monitoring and Surveillance* (J. P. A. Co, Ed.). Fort Washington, PA: James P Anderson Company.

Bellot, P., & El-Bèze, M. (2000). *Clustering by means of Unsupervised Decision Trees or Hierarchical and K-means-like Algorithm*. Paper presented at the RIAO'2000, Paris, France.

Bhargava, N., Sharma, G., Bhargava, R., & Mathuria, M. (2013, June). Decision Tree Analysis on J48 Algorithm for Data Mining. *International Journal of Advanced Research in Computer Science and Software Engineering, 3*(6), 1114–1119.

Chen, L., & Gong, G. (2012). *Communication System Security*. Academic Press.

Cortes, C., & Vapnik, V. (1995). Support-Vector Networks. *Machine Learning, 20*(3), 273–297. doi:10.1007/BF00994018

Domingos, P., & Pazzani, M. (1997). On the Optimality of the Simple Bayesian Classifier under Zero-One Loss. *Machine Learning, 29*(2-3), 103–130. doi:10.1023/A:1007413511361

Fonseca, M. J., & Jorge, J. A. (2003b). *NB-Tree: An Indexing Structure for Content-Based Retrieval in Large Databases*. INESC-ID.

Grzech, A. (2009). *Intelligent Distributed Intrusion Detection Systems of Computer Communication Systems*. Paper presented at the Asian Conference on, Intelligent Information and Database Systems, Dong Hoi City, Vietnam. doi:10.1109/ACIIDS.2009.87

Hall, M., Frank, E., Holmes, G., Pfahringer, B., Reutemann, P., & Witten, I. H. (2009). The WEKA Data Mining Software: An Update. *SIGKDD Explorations, 11*(1), 10. doi:10.1145/1656274.1656278

Ilgun, K., Kemmerer, R. A.,, & Porras, P. A. (1995). State Transition Analysis: A Rule-Based Intrusion Detection Approach. *IEEE Transactions on Software Engineering*.

Kalmegh, S. (2015). Analysis of WEKA Data Mining Algorithm REPTree, SimpleCart and RandomTree for Classification of Indian News. *International Journal of Innovative Science, Engineering & Technology, 2*(2).

Karan Bajaj, A. A. (2013). *Dimension Reduction in Intrusion Detection Features Using Discriminative Machine Learning Approach. International Journal of Computer Science Issues, 10*(4), 324–328.

Kargl, F., Papadimitratos, P., Buttyan, L., Müter, M., Schoch, E., Wiedersheim, B., . . . Hubaux, J.-P. (2008). Secure Vehicular Communication Systems: Implementation, Performance, and Research Challenges. Communications Magazine, IEEE, 110 - 118.

Krontiris, I., Dimitriou, T., & Freiling, F. C. (2007). *Towards Intrusion Detection in Wireless Sensor Networks*. Paper presented at the 13th European Wireless Conference, Paris, France.

Kruegel, C., Mutz, D., Robertson, W., & Valeur, F. (2003). *Bayesian Event Classification for Intrusion Detection*. Paper presented at the 19th Annual Computer Security Applications Conference, Las Vegas, NV.

Kumar, G., & Kumar, K. (2013). Design of an Evolutionary Approach for Intrusion Detection. *The Scientific World Journal, 2013*, 1–14. PMID:24376390

Mahbod Tavallaee, Lu, & Ghorbani. (2009). *A Detailed Analysis of the KDD CUP 99 Data Set*. Paper presented at the CISDA'09 Second IEEE international conference on Computational intelligence for security and defense applications.

Manning, C. D., Raghavan, P., & Schütze, H. (2009). Naive Bayes text classification. In C. D. Manning, P. Raghavan, & H. Schütze (Eds.), *Introduction to Information Retrieval. Cambridge University Press.*

Motorola. (2015). *Proactively Protect Mission-Critical Infrastructure From Cyber Attacks And Security Threats*. Retrieved from https://www.motorolasolutions.com/content/dam/msi/docs/services/network-infrastructure-management/monitoring/security-monitoring-data-sheet.pdf

Parihar, J. S., Rathore, J. S., & Burse, K. (April 2014). *Agent Based Intrusion Detection System to Find Layers Attacks*. Paper presented at the 2014 Fourth International Conference on Communication Systems and Network Technologies, Bhopal, India doi:10.1109/CSNT.2014.144

Patel, A., Qassim, Q., & Wills, C. (2010). A survey of intrusion detection and prevention systems. *Information Management & Computer Security, 18*(4), 277–290. doi:10.1108/09685221011079199

Rassam, M. A., Maarof, M. A., & Zainal, A. (2012). A Survey of Intrusion Detection Schemes in Wireless Sensor Networks. *American Journal of Applied Sciences, 9*(10), 1636–1652. doi:10.3844/ajassp.2012.1636.1652

Ravale, U., Marathe, N., & Padiya, P. (2015). *Feature Selection Based Hybrid Anomaly Intrusion Detection System Using K-Means and RBF Kernel Function*. Paper presented at the International Conference on Advanced Computing Technologies and Applications (ICACTA-2015), Mumbai, India. doi:10.1016/j.procs.2015.03.174

Sandhu, U. A., Haider, S., Naseer, S., & Ateeb, O. U. (2011). *A Survey of Intrusion Detection & Prevention Techniques* Paper presented at the 2011 International Conference on Information Communication and Management, Singapore.

Silva, A. P. R. d., Martins, M. H., Rocha, B. P., Loureiro, A. A., Ruiz, L. B., & Wong, H. C. (2005). *Decentralized intrusion detection in wireless sensor networks*. Paper presented at the 1st ACM international workshop on Quality of service & security in wireless and mobile networks. doi:10.1145/1089761.1089765

Sun, B., Osborne, L., Xiao, Y., & Guizani, S. (2007). Intrusion detection techniques in mobile ad hoc and wireless sensor networks. *IEEE Wireless Communications*, *14*(5), 56–63. doi:10.1109/MWC.2007.4396943

Turek, T., Zerawa, S.-A., & Anees, T. (2011). *Towards safety and security critical communication systems based on SOA paradigm*. Paper presented at the 2011 IEEE 20th International Symposium on Industrial Electronics (ISIE), Gdansk, Poland. doi:10.1109/ISIE.2011.5984337

Uppal, H. A. M., Javed, M., & Arshad, M. J. (2014). An Overview of Intrusion Detection System (IDS) along with its Commonly Used Techniques and Classifications. *International Journal of Computer Science and Telecommunications*, *5*(2), 20–24.

Wang, Y. (2009). *Statistical Techniques for Network Security: Modern Statistically-Based Intrusion Detection and Protection*. Hershey, PA: IGI Global. doi:10.4018/978-1-59904-708-9

Wu, J., Liu, S., Zhou, Z., & Zhan, M. (2012). Toward Intelligent Intrusion Prediction forWireless Sensor Networks Using Three-Layer Brain-Like Learning. *International Journal of Distributed Sensor Networks*, *2012*, 1–14. doi:10.1155/2012/243841

ADDITIONAL READING

El-Taj, H., Najjar, F., Alsenawi, H., & Najjar, M. (2012). Intrusion Detection and Prevention Response based on Signature-Based and Anomaly-Based: Investigation Study *(IJCSIS). International Journal of Computer Science and Information Security*, *10*(6).

Li, X., & Claramunt, C. (2006). A Spatial Entropy-Based Decision Tree for Classification of Geographical Information. *Transactions in GIS*, *10*(3), 451–467. doi:10.1111/j.1467-9671.2006.01006.x

Mukhopadhyay, I., Chakraborty, M., & Chakrabarti, S. (2011). A Comparative Study of Related Technologies of Intrusion Detection & Prevention Systems *Journal of Information Security, 2011*(2), 28-38

Sharma, A. K., & Sahni, S. (2011). A Comparative Study of Classification Algorithms for Spam Email Data Analysis. *International Journal on Computer Science and Engineering*, *3*(5), 1890–1895.

Wu, S. X., & Banzhaf, W. (2010). The use of computational intelligence in intrusion detection systems. *Applied Soft Computing*, *10*(1), 1–35. doi:10.1016/j.asoc.2009.06.019

KEY TERMS AND DEFINITIONS

Cluster: An expression can be called on a group of people, devices or data that has something in common or are positioned closely. For example, a computer cluster is a group of computers connected together and can be viewed to the external world as a single system. In data mining, a cluster is a group of data that are similar to each other in some features and is different to those assigned to other clusters according to these features.

Continuous Feature: A feature that takes numerical values.

Discrete Feature: A feature that takes non-numerical values.

Entropy: A measure of uncertainty in a group of information. The value of the entropy increases when the information is random and decreases otherwise.

Feature: A certain characteristic of something that is unique to it and distinguishes it from similar items.

Ginni Index: A measure of dispersion of inequality. When it approaches zero, it expresses equality. When it approaches one, it expresses maximal inequality among values.

Optimal Grouping: The optimal value of summation of the values of Taxa among all the classes of the dataset.

Supervised Learning: A machine learning technique in which the implemented learning algorithms are trained on labeled data.

Taxa: Plural of Taxon

Taxon: A value that represents the differences between the features' values of the data samples in a class.

Unsupervised Learning: A machine learning mechanism where the learning algorithms are trained on unlabeled data.

APPENDIX

KDD Dataset

KDD dataset was collected by MIT Lincoln Laboratory, under Defense Advanced Research Projects Agency (DARPA ITO) and Air Force Research Laboratory (AFRL/SNHS) sponsorship for evaluation of computer network intrusion detection systems. KDD Cup '99 dataset is the data set used for The Third International Knowledge Discovery and Data Mining Tools Competition. The dataset consists of 41 features as shown in Table 6. The defined features were selected to help in distinguishing normal connections from attacks. The dataset was collected from a simulation of Air Force environment which was attacked by several types of simulated attacks. The simulated attacks were of four main types:

1. Denial-of-service
2. unauthorized access from a remote machine
3. unauthorized access to local root privileges
4. probing

Table 6. KDD dataset features (Mahbod Tavallaee, 2009; Wang, 2009)

Feature Number	Feature Name	Description	Type
1	duration	length (number of seconds) of the connection	continuous
2	protocol_type	type of the protocol, e.g. tcp, udp, etc.	discrete
3	service	network service on the destination, e.g., http, telnet, etc.	discrete
4	src_bytes	number of data bytes from source to destination	continuous
5	dst_bytes	number of data bytes from destination to source	continuous
6	flag	normal or error status of the connection	discrete
7	land	1 if connection is from/to the same host/port; 0 otherwise	discrete
8	wrong_fragment	number of ``wrong'' fragments	continuous
9	urgent	number of urgent packets	continuous
10	hot	number of ``hot'' indicators	continuous
11	num_failed_logins	number of failed login attempts	continuous
12	logged_in	1 if successfully logged in; 0 otherwise	discrete
13	num_compromised	number of ``compromised'' conditions	continuous
14	root_shell	1 if root shell is obtained; 0 otherwise	discrete
15	su_attempted	1 if ``su root'' command attempted; 0 otherwise	discrete
16	num_root	number of ``root'' accesses	continuous
17	num_file_creations	number of file creation operations	continuous
18	num_shells	number of shell prompts	continuous
19	num_access_files	number of operations on access control files	continuous

continued on following page

Table 6. Continued

Feature Number	Feature Name	Description	Type
20	num_outbound_cmds	number of outbound commands in an ftp session	continuous
21	is_hot_login	1 if the login belongs to the ``hot'' list; 0 otherwise	discrete
22	is_guest_login	1 if the login is a "guest" login; 0 otherwise	discrete
23	count	number of connections to the same host as the current connection in the past two seconds	continuous
24	serror_rate	% of connections that have ``SYN'' errors	continuous
25	rerror_rate	% of connections that have ``REJ'' errors	continuous
26	same_srv_rate	% of connections to the same service	continuous
27	diff_srv_rate	% of connections to different services	continuous
28	srv_count	number of connections to the same service as the current connection in the past two seconds	continuous
29	srv_serror_rate	% of connections that have ``SYN'' errors	continuous
30	srv_rerror_rate	% of connections that have ``REJ'' errors	continuous
31	srv_diff_host_rate	% of connections to different hosts	continuous
32	dst_host_count	Count of connections having the same destination host	continuous
33	dst_host_srv_count	% of connections having the same destination host and using the same service	continuous
34	dst_host_same_srv_rate	% of connections having the same destination host and using the same service	continuous
35	dst_host_diff_srv_rate	% of different services on the current host	continuous
36	dst_host_same_src_port_rate	% of connections to the current host having the same src port	continuous
37	dst_host_srv_diff_host_rate	% of connections to the same service coming from different hosts	continuous
38	dst_host_serror_rate	% of connections to the current host that have an S0 error	continuous
39	dst_host_srv_serror_rate	% of connections to the current host and specified service that have an S0 error	continuous
40	dst_host_rerror_rate	% of connections to the current host that have an RST error	continuous
41	dst_host_srv_rerror_rate	% of connections to the current host and specified service that have an RST error	continuous

Compilation of References

3GPP TS 23.228. (2005). *Service Requirements for the Internet Protocol (IP) Multimedia Core Network Subsystem (IMS), Stage 1*. 3GPP.

3GPP TS. (2014). *Multimedia Broadcast/Multicast Service (MBMS); Architecture and functional description. 3GPP TS 23.246 V12.1.0*. Author.

3rd Generation Partnership Project. (2012). *Technical Specification Group Services and System Aspects, Performance Management (PM); Performance measurements; IP Multimedia Subsystems (IMS) (Release 11)*. 3GPP TS 32.409 (v 11.4.0). 3GPP.

3rd Generation Partnership Project. (2013). *Technical Specification Group Services and System Aspects. Feasibility study on IP Multimedia Subsystems (IMS) evolution (Release 11)*. 3GPP TR23.812 (v12.1.0). 3GPP.

Abdelhalim, A., Ahmed, T., Walid-Khaled, H., & Matsuoka, S. (2012) Using Bit-torrent and SVC for efficient video sharing and streaming. *IEEE Symp on Comput. Commun, 12*, 537–543.

Abdul-Hameed, O. (2008). *Quality of Service for Multimedia Applications over Wireless Networks* (Doctoral Thesis). University of Surrey, Guildford, UK.

Abolhasan, M., Hagelstein, B., & Wang, J. C.-P. (2009). Real-world Performance of Current Proactive Multi-hop Mesh Protocols. *Communications, 2009. APCC 2009, 15th Asia-Pacific Conference on*. doi:10.1109/APCC.2009.5375690

Adeyemi-Ejeye, A. O., & Walker, S. D. (2013). *Ultra-high definition Wireless Video transmission using H. 264 over 802.11 n WLAN: Challenges and performance evaluation*. Paper presented at the Telecommunications (ConTEL), 2013 12th International Conference on.

Adeyemi-Ejeye, A. O. (2015). *Ultra-High Defnition Wireless Video Transmission (PhD Thesis)*. University of Essex.

Adeyemi-Ejeye, A. O., & Walker, S. (2014). 4kUHD H264 wireless live video streaming using CUDA. *Journal of Electrical and Computer Engineering*.

Agarwal, S., Singh, J. P., Mavlankar, A., Baccichet, P., & Girod, B. (2008). Performance and Quality-of-Service Analysis of a Live P2P Video Multicast Session on the Internet. *Quality of Service, 2008. IWQOS 2008. 16th International Workshop on*, 11-19. doi:10.1109/IWQOS.2008.7

Agbinya, J. I. (2010). *IP Communications and Services for NGN*. CRC Press.

Aguiar, E., Riker, A., Cerqueira, E., Abelém, A., Mu, M., Braun, T., & Zeadally, S. et al. (2014). A real-time video quality estimator for emerging wireless multimedia systems. *Wireless Networks, 20*(7), 1759–1776. doi:10.1007/s11276-014-0709-y

Aguiar, E., Riker, A., Mu, M., Zeadally, S., Cerqueira, E., & Abelem, A. (2012). Real-time QoE prediction for multimedia applications in wireless mesh networks. *Proc.IEEE Consumer Communications and Networking Conference*, 592–596.

Ahluwalia, P., & Varshney, U. (2007). Managing end-to-end quality of service in multiple heterogeneous wireless networks. International Journal of Network Management,John Wiley & Sons. *Inc.*, *17*(3), 243–260. doi:10.1002/nem.639

Ahmed, T., Buridant, G., & Mehaoua, A. (2001). Encapsulation and marking of MPEG-4 video over IP differentiated services. *ISCC*, *1*, 346–352.

Ahmed, T., Mehaoua, A., & Buridant, G. (2001). Implementing MPEG-4 video on demand over IP differentiated services. *GLOBECOM*, *1*, 2489–2493.

Ahmed, T., Nafaa, A., & Mehaoua, A. (2003) An Object-Based MPEG-4 Multimedia Content Classification Model for IP QOS Differentiation. *IEEE Symposium on Computers and Communications*. doi:10.1109/ISCC.2003.1214260

Aimar, L., Merritt, L., Petit, E., Chen, M., Clay, J., Rullgrd, M., . . . Izvorski, A. (2005*). x264-a free h264/AVC encoder.* Retrieved from http://www. videolan. org/developers/x264. html

Ajenjo, A. D., & Wietgrefe, H. (2003). Minimal-Intrusion Traffic Monitoring And Analysis In Mission-Critical Communication Networks. *Journal of Systemics, Cybernetics and Informatics, 1*(5).

Akbari & Movaghar. (2010). A hybrid mesh-tree peer-to-peer overlay structure for layered video streaming. *5th International Symposium on Telecommunications,* 706 – 710.

Albero T., et al. (2012). *AODV Performance Evaluation and Proposal of Parameters Modification for Multimedia Traffic on Wireless Ad Hoc Networks*. Academic Press.

Alcatel-Lucent. (2013). *Mission-critical Communications Networks for Public Safety Highly reliable converged IP/MPLS-based backhaul Application Note*. Retrieved from http://www.tmcnet.com/tmc/whitepapers/documents/whitepapers/2013/9270-mission-critical-communications-networks-public-safet.pdf

Alcatel-Lucent. (2013). *Mission-critical Communications Networks for Public Safety*. Retrieved from http://www.tmcnet.com/tmc/whitepapers/documents/whitepapers/2013/9270-mission-critical-communications-networks-public-safet.pdf

Ali, M., Liang, L., Sun, Z., & Cruickshank, H. (2011). Optimisation of SIP Session Setup for VoIP over DVB-RCS Satellite Networks. *International Journal of Satellite Communications Policy and Management*, *1*(1), 55–76. doi:10.1504/IJSCPM.2011.039741

Al-Kanj, L., & Dawy, Z. (2012). Energy-aware resource allocation in OFDMA wireless multicasting networks in*Proc. IEEE 19th International Conference on Telecommunications*, 1–5. doi:10.1109/ICTEL.2012.6221318

Allani, M., Garbinato, B., & Pietzuch, P. (2012). Chams: Churn-aware overlay construction for media streaming. *Peer-to-Peer Networking and Applications*, *5*(4), 412–427. doi:10.1007/s12083-012-0161-7

Alreshoodi, M., & Woods, J. (2013). An empirical study based on a fuzzy logic system to assess the QoS/QoE correlation for layered video streaming. *Proc. IEEE International Conference on Computational Intelligence and Virtual Environments for Measurement Systems and Applications,*180–184. doi:10.1109/CIVEMSA.2013.6617417

Alreshoodi, M., & Woods, J. (2013). QoE prediction model based on fuzzy logic system for different video contents. *Proc. IEEE European Modelling Symposium*, 635–639. doi:10.1109/EMS.2013.106

Alreshoodi, M., & Woods, J. (2013). Survey on QoE\QoS correlation models for multimedia services. *International Journal of Distributed and Parallel Systems*, *4*(3), 53–72. doi:10.5121/ijdps.2013.4305

Alsebae, A., Leeson, M., & Green, R. (2013). The throughput benefits of network coding for SR ARQ communication. *Proceedings of the 5th Computer Science and Electronic Engineering Conference*, 45-50. doi:10.1109/CEEC.2013.6659443

Alshamrani, A., Xuemin, S., & Liang-Liang, X. (2011). QoS Provisioning for Heterogeneous Services in Cooperative Cognitive Radio Networks. Selected Areas in Communications. *IEEE Journal on*, *29*(4), 819–830. doi:10.1109/JSAC.2011.110413

Al-Subaie, M., & Zulkernine, M. (2006). *Efficacy of Hidden Markov Models Over Neural Networks in Anomaly Intrusion Detection.* Paper presented at the Computer Software and Applications Conference, 2006. COMPSAC '06. 30th Annual International, Chicago, IL. doi:10.1109/COMPSAC.2006.40

Alwis, C., Arachchi, H., Silva, V., Fernando, A., & Kondoz, A. (2012). Robust video communication using random linear network coding with pre-coding and interleaving. *Proc.IEEE International Conference on Image Processing*, 2269–2272.

Ambady, B. (2012). A Guide to Security Methods for Mission-Critical Communications Networks. *Mission Critical Communications has more details about security options for utilities and smart-grid applications.* Retrieved from http://www.radioresourcemag.com/Features/FeaturesDetails/FID/341

Amnai, M., Fakhri, Y., & Abouchabaka, J. (2011). *QoS Routing and Performance Evaluation for Mobile Ad Hoc Networks Using OLSR Protocol.* arXiv preprint arXiv: 1107.3656

Anderson, J. P. (April1980). *Computer Security Threat Monitoring and Surveillance* (J. P. A. Co, Ed.). Fort Washington, PA: James P Anderson Company.

Angelini, C. (2013). Next-Gen Video Encoding: x265 Tackles HEVC/H. 265. *Tom's Hardware Online Magazine.*

Anh, T. N., Baochun, L., & Frank, E. (2010). Chameleon: adaptive peer-to peer streaming with network coding.*Proceedings of IEEE INFOCOM*, 1–9.

ANSI Standard. (2003). *Digital transport of one-way video signals - parameters for objective performance assessment,* ANSI Standard ATIS-0100801.03.2003(R2013).

ANSI Standard. (2003). *Digital transport of one-way video signals - parameters for objective performance assessment.* ANSI Standard ATIS-0100801.03.2003(R2013).

Argyropoulos, S., Raake, A., Garcia, M.-N., & List, P. (2011). No-reference bit stream model for video quality assessment of H.264/AVC video based on packet loss visibility. *Proc. IEEE International Conference on Acoustics, Speech and Signal Processing,* 1169–1172. doi:10.1109/ICASSP.2011.5946617

Aroussi, S., & Mellouk, A. (2014). Survey on machine learning-based QoE-QoS correlation models. *Computing, Management and Telecommunications (ComManTel), 2014 International Conference on*, 200–204.

Babar, U., & Malliah. (2011). *Call Setup Delay Analysis of H. 323 and SIP.* Academic Press.

Bah, S., Glitho, R., & Dssouli, R. (2012). A SIP Servlets-based Framework for Service Provisioning in Stand-Alone MANETs. *Journal of Network and Computer Applications.*

Bakanoğlu, K., Mingquan, W., Hang, L., & Saurabh, M. (2010). Adaptive resource allocation in multicast OFDMA systems. *Proc. IEEE Wireless Communications and Networking Conference*, 1–6. doi:10.1109/WCNC.2010.5506213

Balachandran, K., Budka, K. C., Chu, T. P., Doumi, T. L., & Kang, J. H. (2006, January). Mobile Responder Communication Networks for Public Safety. *IEEE Communication Magazine.*

Balachandran, K., Budka, K.C., Chu, T.P., Doumi, T.L., & Kang, J.H. (2006, January). Mobile Responder Communication Networks for Public Safety. *IEEE Communication Magazine.*

Balachandran, K., Budka, K.C., Chu, T.P., Doumi, T.L., Kang, J.H., & Whinnery, R. (2005). *Converged Wireless Network Architecture for Homeland Security.* Military Communications Conference, IEEE MILCOM2005, Atlantic City, NJ.

Balachandran, K., Budka, K. C., Chu, T. P., Doumi, T. L., & Kang, J. H. (2006). Mobile responder communication networks for public safety. *IEEE Communications Magazine*, *44*(1), 56–64. doi:10.1109/MCOM.2006.1580933

Baldini, G. (2012). The Evolution of Public Safety Communications in Europe: the results from the FP7 HELP project. *ETSI Reconfigurable Radio Systems Workshop*.

Baldini, G., Karanasios, S., Allen, D., & Vergari, F. (2014). Survey of Wireless Communication Technologies for Public Safety. *IEEE Communications Surveys and Tutorials*, *16*(2), 619–641. doi:10.1109/SURV.2013.082713.00034

Banerjee, D., Saha, S., Sen, S., & Dasgupt, P. (2005). Reciprocal Resource Sharing in P2P Environments. *Proceedings of ACM AAMAS05*.

Banerjee, S., Bhattacharjee, B., & Kommareddy, C. (2002). Scalable application layer multicast.*Proceedings of the 2002 Conference on Applications, Technologies, Architectures, and Protocols for Computer Communications*, 205–217.

Bankoski, J., Bultje, R. S., Grange, A., Gu, Q., Han, J., Koleszar, J., . . . Xu, Y. (2013). *Towards a next generation open-source video codec*. Paper presented at the IS&T/SPIE Electronic Imaging. doi:10.1117/12.2009777

Barachi, M. E., Glitho, R., & Dssouli, R. (2008). Charging for multi-grade services in the IP Multimedia Subsystem. *Proc.International Conference on Next Generation Mobile Applications, Services and Technologies*, 10–17.

Batteram, H., Damm, G., Mukhopadhyay, A., Philippart, L., Odysseos, R., & Urrutia-Valdes, C. (2010). Delivering quality of experience in multimedia networks. *Bell Labs Technical Journal*, *15*(1), 75–193.

Bellini, A., Leone, A., Rovatti, R., Franchi, E., & Manaresi, N. (1996). Analog fuzzy implementation of a perceptual classifier for videophone sequences. *IEEE Transactions on Consumer Electronics*, *42*(3), 787–794. doi:10.1109/30.536186

Bellot, P., & El-Bèze, M. (2000). *Clustering by means of Unsupervised Decision Trees or Hierarchical and K-means-like Algorithm*. Paper presented at the RIAO'2000, Paris, France.

Berrou, C. (2007). *Codes et turbocodes*. Springer-Verlag France. doi:10.1007/978-2-287-32740-7

Berrou, C., Glavieux, A., & Thitimajshima, P. (1993). Near Shannon limit error correcting coding and decoding: Turbo-codes.*Proceedings of IEEE International Conference on Communications*, 1064-1070. doi:10.1109/ICC.1993.397441

Bhargava, N., Sharma, G., Bhargava, R., & Mathuria, M. (2013, June). Decision Tree Analysis on J48 Algorithm for Data Mining. *International Journal of Advanced Research in Computer Science and Software Engineering*, *3*(6), 1114–1119.

Birrer, S., & Bustamante, F. (2005). Resilient peer-to-peer multicast without the cost. *Proceedings 12th Annual Multimedia Computing and Networking Conf.*, 113–120.

Bisio, I., Delucchi, S., Lavagetto, F., & Marchese, M. (2014). *Transmission rate allocation over satellite networks with quality of experience - based performance metrics in*. Livorno: Advanced Satellite Multimedia Systems Conference and Signal Processing for Space Communications Workshop. doi:10.1109/ASMS-SPSC.2014.6934576

Bittorrent. (2015) Available at http://www.bittorrent.com/

Blom, R., de Bruin, P., Eman, J., & Folke, M. (2008). Public Safety Communication Using Commercial Cellular Technology. *The Second International Conference on Next Generation Mobile Applications, Services and Technologies*. doi:10.1109/NGMAST.2008.78

Bosisio, R., Spagnolini, U., & Bar-Ness, Y. (2006). Multilevel type-II HARQ with adaptive modulation control.*IEEE Wireless Communications and Networking Conference*, 2082-2087. doi:10.1109/WCNC.2006.1696617

Boujemaa, H., Chelly, A., & Siala, M. (2008). Performance of HARQ II and III over multipath fading channels.*Proceedings of the 2nd International Conference on Signals, Circuits and Systems*, 1-5. doi:10.1109/ICSCS.2008.4746946

Boumezzough, M., Idboufker, N., & Ouahman, A. (2013). Evaluation of SIP Call Setup Delay for VoIP in IMS. In *Advanced Infocomm Technology* (pp. 16–24). Springer Berlin Heidelberg. doi:10.1007/978-3-642-38227-7_6

Bradai, A., Ahmed, T., Boutaba, R., & Ahmed, R. (2014). Efficient content delivery scheme for layered video streaming in large-scale networks. *Journal of Network and Computer Applications*, *45*, 1–14. doi:10.1016/j.jnca.2014.07.004

Brajdic, A., Suznjevic, M., & Matijaševic, M. (2009). Measurement of SIP Signaling Performance for Advanced Multimedia Services. *Proceedings of the 10th International Conference on Telecommunications*, 381-388.

Bross, B., Han, W., Sullivan, G., Ohm, J., & Wiegand, T. (2013). *High efficiency video coding (HEVC) text specification draft 10 (JCTVCL1003)*. Paper presented at the JCT-VC Meeting (Joint Collaborative Team of ISO/IEC MPEG & ITU-T VCEG).

Bross, B., Han, W.-J., Ohm, J.-R., Sullivan, G. J., & Wiegand, T. (2012). *High efficiency video coding (HEVC) text specification draft 8*. JCTVC-J1003.

Brunnström, K., Sedano, I., Wang, K., Barkowsky, M., Kihl, M., Andrén, B., & Aurelius, A. et al. (2012). 2D no-reference video quality model development and 3D video transmission quality. *Proc. International Workshop on Video Processing and Quality Metrics for Consumer Electronics*, 1–6.

Budka, K. C., Chu, T., Doumi, T. L., Brouwer, W., Lamoureux, P., & Palamara, M. E. (2011). Public safety mission critical voice services over LTE. *Bell Labs Technical Journal*, *16*(3), 133–149. doi:10.1002/bltj.20526

Butyka, Z., Jursonovics, T., & Imre, S. (2008). New fair QoS-based charging solution for mobile multimedia streams. *International Journal of Virtual Technology and Multimedia*, *1*(1), 3–22. doi:10.1504/IJVTM.2008.017107

Castro, M., Druschel, P., Kermarrec, A. M., Nandi, A., Rowstron, A., & Singh, A. (2003). Split-stream: High-bandwidth multicast in cooperative environments. *SIGOPS Operating System Review*, *37*(5), 298–313. doi:10.1145/1165389.945474

Chakareski, J., Han, S., & Girod, B. (2003). *Layered coding vs. multiple descriptions for video streaming over multiple paths* (A. C. M. Multimedia, Ed.). doi:10.1145/957013.957100

Chalmers, R. C., Krishnamurthi, G., & Almeroth, K. C. (2006). Enabling intelligent handovers in heterogeneous wireless networks. *Mobile Networks and Applications, Springer-Verlag*, *11*(2), 215–227. doi:10.1007/s11036-006-4474-8

Chang, D.-Y. (1996). Applications of the extent analysis method on fuzzy AHP. *European Journal of Operational Research*, *95*(3), 649–655. doi:10.1016/0377-2217(95)00300-2

Chang, L., Sung, C., Chiu, S., & Lin, Y. (2010). Design and Realization of Ad Hoc VoIP with Embedded p-SIP Server. *Journal of Systems and Software*, *83*(12), 2536–2555. doi:10.1016/j.jss.2010.07.053

Changqiao, X., Tianjiao, L., Jianfeng, G., Hongke, Z., & Muntean, G. M. (2013). CMT-QA: Quality-Aware Adaptive Concurrent Multipath Data Transfer in Heterogeneous Wireless Networks. Mobile Computing. *IEEE Transactions on*, *12*(11), 2193–2205. doi:10.1109/TMC.2012.189

Chase, D. (1985). Code combining: A maximum-likelihood decoding approach for combining an arbitrary number of noisy packets. *IEEE Transactions on Communications*, *33*(5), 385–393. doi:10.1109/TCOM.1985.1096314

Chen, J. (2010). *Resource Allocation for Delay Constrained Wireless Communications (*Doctoral Thesis). University College London, London, UK.

Chen, L., & Gong, G. (2012). *Communication System Security*. Academic Press.

Chen, Y.-K., Tian, X., Ge, S., & Girkar, M. (2004). *Towards efficient multi-level threading of H. 264 encoder on Intel hyper-threading architectures.* Paper presented at the Parallel and Distributed Processing Symposium.

Chen, X., Hwang, J., Lee, C., & Chen, S. (2015). A near optimal QoE-driven power allocation scheme for scalable video transmissions over MIMO systems. *IEEE Journal of Selected Topics in Signal Processing*, *9*(1), 76–88. doi:10.1109/JSTSP.2014.2336603

Chen, X., Hwang, J., Wang, C., & Lee, C. (2014). A near optimal QoE-driven power allocation scheme for SVC-based video transmissions over MIMO systems. *Proc. IEEE International Conference on Communications*, 1675–1680. doi:10.1109/ICC.2014.6883563

Chen, Y., Yang, Y., & Hwang, R. (2007). SIP-based MIP6-MANET: Design and Implementation of Mobile IPv6 and SIP-based Mobile Ad Hoc Networks. *Computer Communications*, *29*(8), 1226–1240. doi:10.1016/j.comcom.2005.08.015

Chiang, W., Dai, H., & Luo, C. (2012). Cross-layer Handover for SIP Applications Based on Media-Independent Pre-Authentication with Redirect Tunneling. *Digital Information and Communication Technology and it's Applications (DICTAP), 2012Second International Conference on*, 348-353. doi:10.1109/DICTAP.2012.6215373

Chikkerur, S., Sundaram, V., Reisslein, M., & Karam, L. J. (2011). Objective video quality assessment methods: A classification, review, and performance comparison. *IEEE Transactions on Broadcasting*, *57*(2), 165–182. doi:10.1109/TBC.2011.2104671

Chu, Y., Chuang, J., & Zhang, H. (2004). A Case for Taxation in Peer-to-Peer Streaming Broadcast. Proc of ACM SIGCOMM '04 Workshops.

Chuah, S., Chen, Z., & Tan, Y. (2010). An optimized resource allocation algorithm for scalable video delivery over wireless multicast links. *International Packet Video Workshop, Hong Kong*, 41–47. doi:10.1109/PV.2010.5706818

Chuah, S., Chen, Z., & Tan, Y. (2012). Energy-efficient resource allocation and scheduling for multicast of scalable video over wireless networks. *IEEE Transactions on Multimedia*, *14*(4), 1324–1336. doi:10.1109/TMM.2012.2193560

Chuanxiong, G., Zihua, G., Qian, Z., & Zhu, W. (2004). A seamless and proactive end-to-end mobility solution for roaming across heterogeneous wireless networks. Selected Areas in Communications. *IEEE Journal on*, *22*(5), 834–848. doi:10.1109/JSAC.2004.826921

Cisco White Paper, F. L. G. D. (2014). *Cisco Visual Networking Index: Forecast and Methodology.* 2013–2018 Cisco, USA, White Paper FLGD 11684.

CISCO. (2009). *Voice over IP – Per Call Bandwidth Consumption.* Available: http://www.cisco.com/en/US/tech/tk652/tk698/technologies_tech_note09Goswami,a0080094ae2.shtml

Clift, L., Adeyemi-Ejeye, A. O., Koczian, G., Walker, S. D., & Clarke, A. (2014). Delivering Live 4K Broadcasting Using Today's Technology.*International Broadcast Convention (IBC 2014).*

Cohen, B. (2003). Incentives build robustness in bit torrent. *Proceedings ACM SIGCOMM Workshop Economics of Peer-to-Peer Systems*, 1–5.

Corson, S., & Macker. (1999). *Mobile Ad Hoc Networking (MANET): Routing Protocol Performance Issues and Evaluation Considerations.* RFC 2501.

Cortes, C., & Vapnik, V. (1995). Support-Vector Networks. *Machine Learning*, *20*(3), 273–297. doi:10.1007/BF00994018

Cortes, M., Ensor, J. R., & Esteban, J. O. (2004). On SIP performance. *Bell Labs Technical Journal*, *9*(3), 155–172. doi:10.1002/bltj.20048

Creswell, J. W. (2003). *Research Design: Qualitative, Quantitative, and Mixed Methods Approaches*. Research Design Qualitative Quantitative and Mixed Methods Approaches.

Das, S. K. (2006). *Feasibility study of IP Multimedia Subsystems (IMS) based Push To Talk over Cellular for Public Safety and Security Communications* (Master's Thesis). Department of Electrical and Communication Engineering. Helsinki University of Technology (HUT).

De Rango, F., Tropea, M., Fazio, P., & Marano, S. (2006). Overview on VoIP: Subjective and Objective Measurement Methods. *International Journal of Computer Science and Network Security*, *6*(1), 140–153.

Diaz-Sanchez, Proserpio, Marin-Lopez, Mendoza, & Weik. (2010). *A General IMS Registration Protocol for Wireless Network Interworking*. Academic Press.

Dingledine, R., Freedman, M., & Molnar, D. (2001). Free Haven. In Peer-to-Peer: Harnessing the Power of Disruptive Technologies. O'Reilly & Associates.

Dixit, S., & Wu, T. (Eds.). (2004). *Content networking in the mobile Internet*. Hoboken, NJ: John Wiley & Sons. doi:10.1002/047147827X

Domingos, P., & Pazzani, M. (1997). On the Optimality of the Simple Bayesian Classifier under Zero-One Loss. *Machine Learning*, *29*(2-3), 103–130. doi:10.1023/A:1007413511361

Dong, M., & Maode, M. (2012). A QoS Oriented Vertical Handoff Scheme for WiMAX/WLAN Overlay Networks. Parallel and Distributed Systems. *IEEE Transactions on*, *23*(4), 598–606. doi:10.1109/TPDS.2011.216

Doumi, T., Dolan, M. F., Tatesh, S., Casati, A., Tsirtsis, G., Anchan, K., & Flore, D. (2013). LTE for public safety networks. *IEEE Communications Magazine*, *51*(2), 106–112. doi:10.1109/MCOM.2013.6461193

Douros, V. G., & Polyzos, G. C. (2011). Review of some fundamental approaches for power control in wireless networks. *Computer Communications*, *34*(13), 1580–1592. doi:10.1016/j.comcom.2011.03.001

Eden, A. (2007). No-reference estimation of the coding PSNR for H.264-coded sequences. *IEEE Transactions on Consumer Electronics*, *53*(2), 667–674. doi:10.1109/TCE.2007.381744

Ejeye, A. O., & Walker, S. D. (2012). *Uncompressed Quad-1080p Wireless Video Streaming*. Paper presented at the 4th Computer science and Electronic Engineering Conference.

El-Gamal, A., & Cover, T. M. (1982). Achievable rates for multiple descriptions. *IEEE Transactions on Information Theory*, *28*(06), 851–857. doi:10.1109/TIT.1982.1056588

Elliott, E. (1963). Estimates of error rates for codes on burst-noise channels. *The Bell System Technical Journal*, *42*(5), 1977–1997. doi:10.1002/j.1538-7305.1963.tb00955.x

ETSI, European Telecommunication Standard Institute, IMS Network Testing (INT). (2012). *IMS/NGN Performance Benchmark*. ETSI TS 186 008. Author.

European Telecommunications Standards Institute, & Third Generation Partnership Project. (2010a). *LTE, Evolved Universal Terrestrial Radio Access (E-UTRA) and Evolved Universal Terrestrial Radio Access Network (E-UTRAN), Overall description, Stage 2*. 3GPP TS 36.300 version 8.12.0 Release 8, ETSI TS 136 300 V8.12.0.

European Telecommunications Standards Institute, & Third Generation Partnership Project. (2010b). *LTE, Evolved Universal Terrestrial Radio Access (E-UTRA), Radio Link Control (RLC) protocol specification*. 3GPP TS 36.322 version 8.8.0 Release 8, ETSI TS 136 322 V8.8.0.

European Telecommunications Standards Institute, & Third Generation Partnership Project. (2012a). *LTE, Evolved Universal Terrestrial Radio Access (E-UTRA), Medium Access Control (MAC) protocol specification.* 3GPP TS 36.321 version 8.12.0 Release 8, ETSI TS 136 321 V8.12.0.

European Telecommunications Standards Institute, & Third Generation Partnership Project. (2012b). *LTE, Evolved Universal Terrestrial Radio Access (E-UTRA), Services provided by the physical layer.* 3GPP TS 36.302 version 8.2.1 Release 8, ETSI TS 136 302 V8.2.1.

Europian Telecommunication Standardization Institution. (n.d.). Retrieved from http://www.tandcca.com/about/page/12024

Eyers, T., & Schulzrinne, H. (2000). Predicting Internet Telephony Call Setup Delay. *Proc. 1st IP-Telephony Workshop.* 2000.

Fabini, H. Kuthan, & Wiedermann. (2013). Mobile SIP: An Empirical Study on SIP Retransmission Timers in HSPA 3G Networks. In Advances in Communication Networking (pp. 78-89). Springer Berlin Heidelberg.

Fabini, J., Hirschbichler, M., Kuthan, J., & Wiedermann, W. (2013). Mobile SIP: An Empirical Study on SIP Retransmission Timers in HSPA 3G Networks. In Advances in Communication Networking, (pp. 78-89). Springer Berlin Heidelberg. doi:10.1007/978-3-642-40552-5_8

Fangmin, X., Luyong, Z., & Zheng, Z. (2007). Interworking of Wimax and 3GPP networks based on IMS. *Communications Magazine, IEEE, 45*(3), 144–150. doi:10.1109/MCOM.2007.344596

Farahbaksh, R., Varposhti, M., & Movahhedinia, N. (2007). Transmission Delay reduction in IMS by Re-registration Procedure modification. *The Second International Conference on Next Generation Mobile Applications, Services and Technologies.*

Farid, F., Shahrestani, S., & Ruan, C. (2013). *QoS analysis and evaluations: Improving cellular-based distance education.* Paper presented at the Local Computer Networks Workshops (LCN Workshops), 2013 IEEE 38th Conference on.

Farnaz Farid, S. S. (2012). *Quality of service requirements in wireless and cellular networks: application-based analysis.* Paper presented at the International Business Information Management Association Conference, Bercelona, Spain.

Fathi, Chakraborty, & Prasad. (2009). *Voice over IP in Wireless Heterogeneous Networks: Signaling, Mobility, and Security.* Springer Science Business Media B.V.

Fathi, H., Chakraborty, S., & Prasadm, R. (2006). Optimization of SIP Session Setup Delay for VoIP in 3G Wireless Networks. *IEEE Transactions on Mobile Computing, 5*(9), 1121-1132. doi:10.1109/TMC.2006.135

Fehn, C. (2004). *Depth-image-based rendering (DIBR), compression, and transmission for a new approach on 3D-TV.* Academic Press.

Fei, Z., Xing, C., & Li, N. (2015). QoE-driven resource allocation for mobile IP services in wireless network *Science China. Information Sciences, 58*(1), 1–10.

Ferrús, R., Sallent, O., Baldini, G., & Goratti, L. (2013). LTE: The technology driver for future public safety communications. *IEEE Communications Magazine, 51*(10), 154–161. doi:10.1109/MCOM.2013.6619579

Fesci-Sayit, M., Turhan Tunali, E., & Murat Tekalp, A. (2009) Bandwidth-aware multiple multicast tree formation for P2P scalable video streaming using hierarchical clusters. ICIP'09. IEEE, 945–948 doi:10.1109/ICIP.2009.5414021

Fesci-Sayit, M., Turhan Tunali, E., & Murat Tekalp, A. (2012). Resilient peer-to peer streaming of scalable video over hierarchical multicast trees with backup parent pools. *Signal Processing Image Communication, 27*(2), 113–125. doi:10.1016/j.image.2011.11.004

Fokus, F. (2004). *Open IMS core*. Fraunhofer Institute FOKUS Available: http://www.openimscore.org

Fonseca, M. J., & Jorge, J. A. (2003b). *NB-Tree: An Indexing Structure for Content-Based Retrieval in Large Databases*. INESC-ID.

Forge, S., Horvitz, R., & Blackman, C. (2014). *Is Commercial Cellular Suitable for Mission Critical Broadband? Academic Press.*

Forum, D. S. L. (2006). *Triple-play services quality of experience (QoE) requirements*. DSL Forum, Technical Report TR-126.

Freenet. (2015). Available at http://freenet.sourceforge.net/

Fujii, T., Kitamura, M., Murooka, T., Shirai, D., & Takahara, A. (2009, 2009). *4K& 2K multi-resolution video communication with 60 fps over IP networks using JPEG2000*. Paper presented at the Intelligent Signal Processing and Communication Systems, 2009. ISPACS 2009. International Symposium on.

Gandhi, S., Chaubey, N., Shah, P., & Sadhwani, M. (2012). Performance Evaluation of DSR, OLSR and ZRP Protocols in MANETs. *Computer Communication and Informatics (ICCCI),2012International Conference on*, 1–5. doi:10.1109/ICCCI.2012.6158841

Ganguly, S., & Bhatnagar, S. (2008). VoIP: Wireless, P2P and New Enterprise Voice Over IP. Chichester, UK: Wiley. doi:10.1002/9780470997925

Gannoune & Robert. (2003). *A Survey of QoS Techniques and Enhancements for IEEE 802.11b Wireless LAN's*. Technical report, EIVD-Swisscom. IEEE.

Gao, Q., Fei, N., Xing, C., & Kuang, J. (2015). Balancing QoE and fairness of heterogeneous traffics based on Equivalent Rate Scaling. *China Communications*, *12*(1), 136–144.

Garmonov, A. V., Seok Ho, C., Do Hyon, Y., Ki Lae, H., Yun Sang, P., Savinkov, A. Y., & Kondakov, M. S. (2008). QoS-Oriented Intersystem Handover Between IEEE 802.11b and Overlay Networks. Vehicular Technology. *IEEE Transactions on*, *57*(2), 1142–1154. doi:10.1109/TVT.2007.906347

Ge, S., Tian, X., & Chen, Y.-K. (2003). *Efficient multithreading implementation of H. 264 encoder on Intel hyper-threading architectures*. Paper presented at the Information, Communications and Signal Processing, 2003 and Fourth Pacific Rim Conference on Multimedia.

German, S., Markus, W., & Herwig, U. (2006). Search Methods in P2P Networks: A Survey. Lecture Notes in Computer Science, 3473, 59-68.

Ghalut, T., & Larijani, H. (2014). Non-intrusive method for video quality prediction over LTE using random neural networks (RNN). *International Symposium on Communication Systems, Networks Digital Signal Processing*, 519–524. doi:10.1109/CSNDSP.2014.6923884

Ghosh, S. (2013). *Comparative Study of QoS Parameters of SIP Protocol in 802.11a and 802.11b Network*. arXiv preprint arXiv: 1311.3184

Gizelis, C., & Vergados, D. (2011). A survey of pricing schemes in wireless networks. *IEEE Communications Surveys and Tutorials*, *13*(1), 126–145. doi:10.1109/SURV.2011.060710.00028

Goldmann, L., & Ebrahimi, T. (2011). Towards reliable and reproducible 3D video quality assessment. *Proc. SPIE*, 8043, 302–307. doi:10.1117/12.887037

Goswami, Joardar, & Das. (2014). Reactive and Proactive Routing Protocols Performance Metric Comparison in Mobile Ad hoc Networks NS 2. *Memory, 3*(1).

Goudarzi, P. (2013). On the differentiated QoE enforcement between competing scalable video flows over wireless networks. *Wireless Communications and Mobile Computing, 13*(7), 633–649.

Grasic & Kos. (2012). Structuring the RCS Services on the IMS Application Layer Part I: Description and Comparison Parameters. *Elektrotehniski vestnik, 79*(3).

Grecos, C., & Wang, Q. (2011). Advances in video networking: Standards and applications. *International Journal of Pervasive Computing and Communications, 7*(1), 22–43. doi:10.1108/17427371111123676

Grgic, T., & Matijasevic, M. (2013). Performance Metrics for Context-based Charging in 3GPP Online Charging System. *Telecommunications (ConTEL),201312th International Conference on*, 171-178.

Grzech, A. (2009). *Intelligent Distributed Intrusion Detection Systems of Computer Communication Systems*. Paper presented at the Asian Conference on, Intelligent Information and Database Systems, Dong Hoi City, Vietnam. doi:10.1109/ACIIDS.2009.87

GSM. (2000). *Full Rate Speech; Transcoding (GSM 06.10)*. ETSI std. EN 300 961 v. 8.1.1.

Guha, S., Daswani, N., & Jainet, R. (2006). An experimental study of the Skype peer-to-peer VoIP system. *Proceedings of the 5th International Workshop on Peer-to-Peer Systems.*

Guo, Y., Liang, C., & Liu, Y. (2008). Adaptive Queue-based Chunk Scheduling for P2P Live Streaming. *Proceedings of 7th International IFIP-TCP Networking conference on Adhoc and sensor networks, wireless networks, next generation internet*, 433-444.

Gupta, Sadawarti, & Verma. (2011). Review of Various Routing Protocols for MANETs. *International Journal of Information and Electronics Engineering, 1*(3), 99.

Gurbani, K., Jagadeesan, L., & Mendiratta, V. (2005). Characterizing Session Initiation Protocol (SIP) Network Performance and Reliability. In Service Availability, (pp. 196-211). Springer Berlin Heidelberg. doi:10.1007/11560333_16

Gutkowski, P., & Kaczmarek, S. (2012). The Model of end-to-end Call Setup Time Calculation for Session Initiation Protocol. *Bulletin of the Polish Academy of Sciences: Technical Sciences, 60*(1), 95–101.

H.264/AVC JM. (2014). *H.264/AVC JM Reference Software*. Available: http://iphome.hhi.de/suehring/tml/

Haider, F., Wang, C., Haas, H., Hepsaydir, E., & Ge, X. (2012). Energy-efficient subcarrier-and-bit allocation in multiuser OFDMA systems. *Proc. IEEE 75th Vehicular Technology Conference,*1–5. doi:10.1109/VETECS.2012.6240331

Halák, J., Krsek, M., Ubik, S., Žejdl, P., & Nevřela, F. (2011). Real-time long-distance transfer of uncompressed 4K video for remote collaboration. *Future Generation Computer Systems, 27*(7), 886–892. doi:10.1016/j.future.2010.11.014

Hall, M., Frank, E., Holmes, G., Pfahringer, B., Reutemann, P., & Witten, I. H. (2009). The WEKA Data Mining Software: An Update. *SIGKDD Explorations, 11*(1), 10. doi:10.1145/1656274.1656278

Hamam, A., Eid, M., El Saddik, A., & Georganas, N. (2008). A fuzzy logic system for evaluating quality of experience of haptic-based applications. In M. Ferre (Ed.), *Haptics: Perception, Devices and Scenarios* (Vol. 5024, pp. 129–138). Springer Berlin Heidelberg. doi:10.1007/978-3-540-69057-3_14

Han, S., Joo, H., Lee, D., & Song, H. (2011, May). An end-to-end virtual path construction system for stable live video streaming over heterogeneous wireless networks. *IEEE Journal on Selected Areas in Communications, 29*(5), 1032–1041. doi:10.1109/JSAC.2011.110513

Hao, H., Yang, G., & Yong, L. (2010). Mesh-based peer-to-peer layered video streaming with taxation.*Proceedings of ACM Workshop on Network and Operating System Support for Digital Audio and Video NOSSDAV'10*, 27–32.

Happenhofer, M., Egger, C., & Reichl, P. (2010). Quality of Signaling: A New Concept for Evaluating the Performance of non-INVITE SIP Transactions. *Proceedings of the 22nd International Teletraffic Congress (ITC),* 1-8.

Hareesh, K., & Manjaiah, D. H. (2011). Peer to Peer live streaming and video on demand design issues and its challenges. *International Journal of Peer to Peer Networks*, *02*(04), 1–11.

Hasib, A., & Fapojuwo, A. (2008). Analysis of Common Radio Resource Management Scheme for End-to-End QoS Support in Multiservice Heterogeneous Wireless Networks. Vehicular Technology. *IEEE Transactions on*, *57*(4), 2426–2439. doi:10.1109/TVT.2007.912326

Hasslinger, G., & Hohlfeld, O. (2008). The Gilbert-Elliott model for packet loss in real time services on the internet. *Proc. GI/ITG Conference - Measuring, Modelling and Evaluation of Computer and Communication Systems,* 1–15.

Haykin, S. (Ed.). (2003). *Communication Systems* (4th ed.). Wiley.

Haykin, S., & Moher, M. (2005). *Modern wireless communications.* Pearson/Prentice Hall.

He, L., & Liu, G. (2014). Quality-driven cross-layer design for H.264/AVC video transmission over OFDMA system. *IEEE Transactions on Wireless Communications*, *13*(12), 6768–6782. doi:10.1109/TWC.2014.2364603

Hewage, C., & Martini, M. (2011). Reduced-reference quality assessment for 3D video compression and transmission. *IEEE Transactions on Consumer Electronics*, *57*(3), 1185–1193. doi:10.1109/TCE.2011.6018873

Hoeher, T., Petraschek, M., Tomic, S., & Hirschbichler, M. (2007). Evaluating Performance Characteristics of SIP over IPv6. *Journal of Networks*, *2*(4), 40–50. doi:10.4304/jnw.2.4.40-50

Holland, I. D., Zepernick, H.-J., & Caldera, M. (2005). Soft combining for hybrid ARQ. *Electronics Letters*, *41*(22), 1230–1231. doi:10.1049/el:20052614

Hong Chen, R. G., Maunder, & Hanzo, L. (2013). A survey and tutorial on low-complexity turbo coding techniques and a holistic hybrid ARQ design example. Proceedings of Fourth Quarter IEEE Communications Surveys and Tutorials, 15(4), 1546-1566.

Hong Yi, C., Ya Yueh, S., & Yuan, W. L (2012). Cloud PP: A novel cloud-based P2P live video streaming platform with SVC technology. *8th International Conference on Computer Technology and Information Management*, 64–68

Hoong, P., & Matsuo, H. (2008). Palms: A reliable and incentive-based P2P live media streaming system. Lecture Notes in Electrical Engineering, 4, 51–66. doi:10.1007/978-0-387-74938-9_5

Hosek, J. (2006). *Performance Analysis: Impact of Signalling Load over IMS Core on KPIs. Recent Advances in Circuits, Systems*. Telecommunications and Control.

Hosek, J., & Molnar, K. (2011). Investigation on OLSR Routing Protocol Efficiency. *Proceedings of the 2011 international conference on Computers, digital communications and computing.*

Hossain, E. (2008). *Heterogeneous Wireless Access Networks: Architectures and Protocols.* Springer.

Hsu, R., & Chen, A. (2015). *Ensure Nonstop IP Surveillance with an Optimized Industrial Ethernet Network.* Retrieved from http://www.remotemagazine.com/main/articles/ensure-nonstop-ip-surveillance-with-an-optimized-industrial-ethernet-network/

Hsu, J. (2004). Performance of Mobile Ad Hoc Networking Routing Protocols in Large Scale Scenarios. *Military Communications Conference, 2004. MILCOM 2004. 2004 IEEE*, 1. doi:10.1109/MILCOM.2004.1493241

Hu, J. (2013). *Resource Allocation and Optimization Techniques in Wireless Relay Networks* (Doctoral Thesis). Loughborough University, Loughborough, UK.

Hu, Z., Zhu, G., Xia, Y., & Liu, G. (2004). Multiuser subcarrier and bit allocation for MIMO-OFDM systems with perfect and partial channel information (Vol. 2). Academic Press.

Huang, C., & Leung, C. (2012). BitQoS-aware resource allocation for multi-user mixed-traffic OFDM systems. *IEEE Transactions on Vehicular Technology*, *61*(5), 2067–2082. doi:10.1109/TVT.2012.2189030

Hu, H., Guo, Y., & Liu, Y. (2011). Peer to Peer Streaming of Layered Video: Efficiency, Fairness and Incentive. *IEEE Transactions on Circuits and Systems for Video Technology*, *21*(08), 1013–1026. doi:10.1109/TCSVT.2011.2129290

Husic, Hidic, Hadzialic, & Barakovic. (2014). *Simulation-based Optimization of Signaling Procedures in IP Multimedia Subsystem*. Academic Press.

Ibarrola, E., Saiz, E., Zabala, L., Cristobo, L., & Xiao, J. (2014). A new global quality of service model: QoXphere. *IEEE Communications Magazine*, *52*(1), 193–199. doi:10.1109/MCOM.2014.6710083

IEEE Standard 802.16-2012. (2012). *IEEE Standard for Air Interface for Broadband Wireless Access Systems*. IEEE, New York, Standard 802.16-2012.

Ilgun, K., Kemmerer, R. A.,, & Porras, P. A. (1995). State Transition Analysis: A Rule-Based Intrusion Detection Approach. *IEEE Transactions on Software Engineering*.

Institute of Electrical and Electronics Engineers. (2006). *IEEE Std 802.16-2004/Cor1-2005, IEEE Standard for Local and Metropolitan Area Networks Part 16: Air Interface for Fixed and Mobile Broadband Wireless Access Systems Amendment 2: Physical and Medium Access Control Layers for Combined Fixed and Mobile Operation in Licensed Bands And Corrigendum 1*. Author.

Institute of Electrical and Electronics Engineers. (2011). *IEEE Standard for Local and Metropolitan Area Networks Part 16: Air Interface for Broadband Wireless Access Systems Amendment 3: Advanced Air Interface*. Author.

IPWireless. (2012). *LTE addressing the needs of the Public Safety Community*. 3GPP RAN Workshop on Rel-12 and Onward RWS-120030.

Issa, O., Li, W., & Liu, H. (2010). Performance evaluation of TV over broadband wireless access networks. *IEEE Transactions on Broadcasting*, *56*(2), 201–210. doi:10.1109/TBC.2010.2046979

ITU. (2012). *Parameter values for ultra-high definition television systems for production and international programme exchange*. Geneva: International Telecommunication Union.

ITU-R BT.1683. (2004). *Objective perceptual video quality measurement techniques for standard definition digital broadcast television in the presence of a full reference*. ITU-R Recommendation BT.1683.

ITU-T & ISO/IEC. (2003). *Recommendation and Final Draft International Standard of Joint Specification. ITU-T Rec. H. 264/IEC 14496-10 AVC*. Author.

ITU-T FG IPTV. (2007). *Definition of Quality of Experience (QoE)*. International Telecommunication Union, Geneva, Switzerland, Technical Report ITU-T FG IPTV.

ITU-T G. 1000. (2001). *Communications quality of service: a framework and definitions (2001)*. International Telecommunication Union, Technical Report ITU-T G.1000.

ITU-T G. 1010. (2001). *End-user multimedia QoS categories.* International Telecommunication Union, Technical Report ITU-T G.1010.

ITU-T G.1011. (2010). *Multimedia Quality of Service and performance – Generic and user-related aspects - Reference guide to quality of experience assessment methodologies.* ITU, Switzerland, Recommendation ITU-T G.1011.

ITU-T H.264. (2014). *H.264: Advanced video coding for generic audiovisual services.* ITU, Switzerland, Recommendation ITU-T H.264.

ITU-T J. 143. (2000). *User requirements for objective perceptual video quality measurements in digital cable television.* International Telecommunication Union, Geneva, Switzerland, Technical Report ITU-T J.143.

ITU-T J.144. (2004). *Objective perceptual video quality measurement techniques for digital cable television in the presence of a full reference.* ITU-T Recommendation J.144.

ITU-T Rec. G.107. (2003). *The E Model, A Computational Model for Use in Transmission Planning.* ITU-T.

ITU-T Rec. G.114. (2003). *One-way Transmission Time.* ITU-T.

ITU-T Rec. G.723.1. (2006). *Dual Rate Speech Coder for Multimedia Communications Transmitting at 5.3 and 6.3 kbit/s.* ITU-T.

ITU-T Rec. G.728. (2012). *Coding of Speech at 16 Kbit/s Using Low-delay Code Excited Linear Prediction.* ITU-T.

ITU-T Recommendation P.800. (1996). *Methods for Subjective Determination of Transmission Quality.* ITU-T.

ITU-T TR Q-series supplements 51 signaling requirements for IP-QoS. (2004). ITU-T.

ITU-T. (2004). *TR Q-series Supplements 51 Signaling Requirements for IP-QoS.* ITU-T.

Jagadish, H. V., Ooi, B. C., & Vu, Q. H. (2005). BATON: a balanced tree structure for peer to-peer networks.*Proceedings of the International Conference on Very Large Databases*, 661–672

Jaganathan, & Eranti, J. (2011). High quality of service on video streaming in P2P networks using FST-MDC. *International Journal of Multimedia and Its Applications, 3*(2), 33-43.

Jain, Somasundaram, Wang, Baras, & Roy-Chowdhury. (2010). *Study of OLSR for Real-time Media Streaming over 802.11 Wireless Network in Software Emulation Environment.* Academic Press.

Jammeh, E., Mkwawa, I., Khan, A., Goudarzi, M., Sun, L., & Ifeachor, E. (2012). Quality of experience (QoE) driven adaptation scheme for voice/video over IP. *Telecommunication Systems, 49*(1), 99–111. doi:10.1007/s11235-010-9356-5

Javed, Q., & Prakash, R. (2012). Improving the Performance of Hybrid Wireless Mesh Protocol. *Mobile Ad hoc and Sensor Networks (MSN),2012Eighth International Conference on.* doi:10.1109/MSN.2012.35

Jayakumar, G., & Ganapathi, G. (2008). Reference Point Group Mobility and Random Waypoint Models in Performance Evaluation of MANET Routing Protocols. *Journal of Computer Systems, Networks, and Communications*, 13.

Jayant, N. S. (1981). Sub-sampling of a DPCM speech channel to provide two self-contained half-rate channels. *The Bell System Technical Journal, 60*(04), 501–509. doi:10.1002/j.1538-7305.1981.tb03069.x

Joskowicz, J., & Ardao, J. C. L. (2011). Combining the effects of frame rate, bit rate, display size and video content in a parametric video quality model. *Proceedings of the 6th Latin America Networking Conference*, 4–11. doi:10.1145/2078216.2078218

Joskowicz, J., & Sotelo, R. (2014). A model for video quality assessment considering packet loss for broadcast digital television coded in H.264. *International Journal of Digital Multimedia Broadcasting*, *2014*, 11. doi:10.1155/2014/242531

Joskowicz, J., Sotelo, R., & Ardao, J. (2013). Towards a general parametric model for perceptual video quality estimation. *IEEE Transactions on Broadcasting*, *59*(4), 569–579. doi:10.1109/TBC.2013.2277951

Juncheng, M., Falei, L., Shanshe, W., & Siwei, M. (2014). *Flexible CTU-level parallel motion estimation by CPU and GPU pipeline for HEVC.* Paper presented at the Visual Communications and Image Processing Conference.

Kalmegh, S. (2015). Analysis of WEKA Data Mining Algorithm REPTree, SimpleCart and RandomTree for Classification of Indian News. *International Journal of Innovative Science, Engineering & Technology, 2*(2).

Kan, G. (2001). Gnutella. In *Peer-to-Peer: Harnessing the Power of Disruptive Technologies* (pp. 94–122). O'Reilly & Associates.

Karan Bajaj, A. A. (2013). *Dimension Reduction in Intrusion Detection Features Using Discriminative Machine Learning Approach. International Journal of Computer Science Issues, 10*(4), 324–328.

Karapantazis, S., & Stylianos, F. P. (2009). VoIP: A Comprehensive Survey on A Promising Technology. *Computer Networks*, *53*(12), 2050–2090. doi:10.1016/j.comnet.2009.03.010

Kargl, F., Papadimitratos, P., Buttyan, L., Müter, M., Schoch, E., Wiedersheim, B., . . . Hubaux, J.-P. (2008). Secure Vehicular Communication Systems: Implementation, Performance, and Research Challenges. Communications Magazine, IEEE, 110 - 118.

Karimi, O. B., & Fathy, M. (2010). Adaptive end-to-end QoS for multimedia over heterogeneous wireless networks. *Computers & Electrical Engineering*, *36*(1), 45–55. doi:10.1016/j.compeleceng.2009.04.006

Kataoka, N. (2009). *4K Uncompressed Streaming over Colored Optical Packet Switching Network.* Paper presented at the OptoElectronics and Communications Conference.

Katoozian, M., Navaie, K., & Yanikomeroglu, H. (2009). Utility-based adaptive radio resource allocation in OFDM wireless networks with traffic prioritization. *IEEE Transactions on Wireless Communications*, *8*(1), 66–71. doi:10.1109/TWC.2009.080033

Katsaggelos, A., Zhai, F., Eisenberg, Y., & Berry, R. (2005). Energy-efficient wireless video coding and delivery. IEEE Wireless Communications, 12(4), 24–30.

Kaur & Singh. (2012). Performance Comparison of OLSR, GRP and TORA Using OPNET. *International Journal of Advanced Research in Computer Science and Software Engineering*, *2*(10), 1-7.

Kazemitabar, H., Ahmed Ali, S., Nisar, K., Md Said, A., & Hasbullah, H. (2010). A Survey on Voice Over IP Over Wireless LANs. *World Academy of Science, Engineering and Technology*, *71*, 352–358.

Kelly, F., Maulloo, A., & Tan, D. (1998). Rate control for communication networks: Shadow prices, proportional fairness and stability. *The Journal of the Operational Research Society*, *49*(3), 237–252. doi:10.1057/palgrave.jors.2600523

Khan, N. (2014). *Quality-Driven Multi-User Resource Allocation and Scheduling Over LTE for Delay Sensitive Multimedia Applications* (Doctoral Thesis). Kingston University, Kingston, UK.

Khan, A., Sun, L., & Ifeachor, E. (2010). Learning models for video quality prediction over wireless local area network and universal mobile telecommunication system networks. *IET Communications*, *4*(12), 1389–1403. doi:10.1049/iet-com.2009.0649

Khan, A., Sun, L., & Ifeachor, E. (2012). QoE prediction model and its application in video quality adaptation over UMTS networks. *IEEE Transactions on Multimedia, 14*(2), 431–442. doi:10.1109/TMM.2011.2176324

Khan, A., Sun, L., Ifeachor, E., Fajardo, J., Liberal, F., & Koumaras, H. (2015). Video quality prediction models based on video content dynamics for H.264 video over UMTS networks. *International Journal of Digital Multimedia Broadcasting, 2010*, 1–17. doi:10.1155/2010/608138

Khan, I., & Qayyum, A. (2009). Performance Evaluation of AODV and OLSR in Highly Fading Vehicular Ad Hoc Network Environments. *Multitopic Conference, 2009. INMIC 2009. IEEE 13th International*, 1–5. doi:10.1109/INMIC.2009.5383121

Khan, S., Schroeder, D., El Essaili, A., & Steinbach, E. (2014). Energy-efficient and QoE-driven adaptive HTTP streaming over LTE. *Proc.IEEE Wireless Communications and Networking Conference*, 2354–2359.

Khirman, S., & Henriksen, P. (2002). Relationship between quality-of-service and quality-of-experience for public internet service. *Proc. Passive and Active Network Measurement Workshop.*

Kianoosh, M., & Mohamed, H. (2013). Capacity management of seed servers in peer to peer streaming systems with scalable video streams. *IEEE Transactions on Multimedia, 15*(1), 181–194. doi:10.1109/TMM.2012.2225042

Kim, H., Duong, D., Yun, J., Jang, Y., & Ko, S. (2006). Power-aware rate-control for QoS provisioning over CDMA networks. *Digest of Technical Papers International Conference on Consumer Electronics*, 139–140.

Kim, J., Kwon, T., & Cho, D.-H. (2007). Resource allocation scheme for minimizing power consumption in OFDM multicast systems. *Communications Letters, IEEE, 11*(6), 486–488.

Kim, H., & Choi, S. (2010). A study on a QoS/QoE correlation model for QoE evaluation on IPTV service. *Proc. International Conference on Advanced Communication Technology,*2, 1377–1382.

Kim, H.-J., Lee, D. H., Lee, J. M., Lee, K.-H., Lyu, W., & Choi, S.-G. (2008). The QoE evaluation method through the QoS-QoE correlation model. *Proc.International Conference on Networked Computing and Advanced Information Management*, 2, 719–725.

Kivanc, D., Li, G., & Liu, H. (2003). Computationally efficient bandwidth allocation and power control for OFDMA. *IEEE Transactions on Wireless Communications, 2*(6), 1150–1158. doi:10.1109/TWC.2003.819016

Klingberg, T., & Manfredi, R. (2002). *Gnutella 0.6.* Technical report. Available at: http://rfcgnutella.sourceforge.net/src/rfc-0_6-draft.html

Kofler, I., Seidl, J., Timmerer, C., Hellwagner, H., Djama, I., & Ahmed, T. (2008). Using MPEG-21 for cross-layer multimedia content adaptation. *Signal Image Video Process., 2*(4), 355–370. doi:10.1007/s11760-008-0088-x

Kondoz, A. M. (2004). *Digital Speech* (2nd ed.). John Wiley & Sons, Ltd. doi:10.1002/0470870109

Krongold, B., Ramchandran, K., & Jones, D. (2000). Computationally efficient optimal power allocation algorithms for multicarrier communication systems. *IEEE Transactions on Communications, 48*(1), 23–27. doi:10.1109/26.818869

Krontiris, I., Dimitriou, T., & Freiling, F. C. (2007). *Towards Intrusion Detection in Wireless Sensor Networks.* Paper presented at the 13th European Wireless Conference, Paris, France.

Kruegel, C., Mutz, D., Robertson, W., & Valeur, F. (2003). *Bayesian Event Classification for Intrusion Detection.* Paper presented at the 19th Annual Computer Security Applications Conference, Las Vegas, NV.

Ksairi, N., Ciblat, P., & Le Martret, C. (2013). Optimal resource allocation for Type-II-HARQ-based OFDMA ad hoc networks.*Proceedings of IEEE Global Conference on Signal and Information Processing*, 379-382. doi:10.1109/GlobalSIP.2013.6736894

Kulin, M., Kazaz, T., & Mrdovic, S. (2012). SIP Server Security with TLS: Relative Performance Evaluation. *Telecommunications (BIHTEL),2012IX International Symposium on*, 1-6. doi:10.1109/BIHTEL.2012.6412062

Kumar, G., & Kumar, K. (2013). Design of an Evolutionary Approach for Intrusion Detection. *The Scientific World Journal, 2013*, 1–14. PMID:24376390

Kuo, J. L., Shih, C. H., Ho, C. Y., & Chen, Y. C. (2014). *Advanced bootstrap and adjusted bandwidth for content distribution in peer-to-peer live streaming*. Peer-to-Peer Networking and Applications.

Kwatra, P. (2013). ARQ protocol studies in underwater communication networks.*Proceedings of International Conference on Signal Processing and Communication*, 121-126. doi:10.1109/ICSPCom.2013.6719768

Kwon, O. C., & Song, H. (2013). Adaptive tree-based P2P video streaming multicast system under high peer-churn rate. *Journal of Visual Communication and Image Representation, 24*(3), 203–216. doi:10.1016/j.jvcir.2012.11.004

Labib, Y., Sherbini, A. E., & Sabri, A. (2009). A Clustered P2P Proxy-Assisted Architecture for On Demand Media Streaming. *Computer Technology and Development, ICCTD '09. International Conference on, 2*, 95-100.

Lahbabi, Y., Ibn Elhaj, E. H., & Hammouch, A. (2014) Quality adaptation using scalable video coding (SVC) in peer-to-peer (P2P) video-on demand (VoD) streaming. *International conference on Multimedia Computing and Systems ICMCS'14*, 1140–1146.

Lande, S., Helonde, S., Pande, R., & Pathak, H. (2011). *Adaptive Subcarrier and Bit Allocation for Downlink OFDMA System with Proportional Fairness. International Journal of Wireless & Mobile Networks*, *3*(5), 125–140.

Lang, Y., Wuebben, D., Dekorsy, A., & Braun, V. (2013). A turbo-like iterative decoding algorithm for network coded HARQ.*Proceedings of 9th International ITG Conference on Systems, Communication and Coding*, 1-6.

Le Callet, P., Möller, S., & Perkis, A. (2013). Qualinet White Paper on Definitions of Quality of Experience. *European Network on Quality of Experience in Multimedia Systems and Services (COST Action IC 1003).*

Lee, G. R., & Wen, J. (2007). The performance of subcarrier allocation schemes combined with error control codings in OFDM systems. *IEEE Transactions on Consumer Electronics*, *53*(3), 852–856. doi:10.1109/TCE.2007.4341556

Lee, S., Jung, K., & Sim, D. (2010). Real-time objective quality assessment based on coding parameters extracted from H.264/AVC bitstream. *IEEE Transactions on Consumer Electronics*, *56*(2), 1071–1078. doi:10.1109/TCE.2010.5506041

Lee, S., Koo, J., & Chung, K. (2010). Content-aware rate control scheme to improve the energy efficiency for mobile IPTV. *Proc. IEEE International Conference on Consumer Electronics*, 445–446.

Lei, Y., Xiang, W., Chunping, H., Jianjun, L., & Yonghong, H. (2013). Trade-off optimization for scalable video coding streaming in relay-based OFDMA networks. *China Communications*, *10*(5), 99–113.

Leszek, C. (2006). Scalable Video Coding for Flexible Multimedia Services. *IEEE Transactions on Circuits and Systems for Video Technology*, *18*(7).

Levine, D. A., Akyildiz, I. F., & Naghshineh, M. (1997). A resource estimation and call admission algorithm for wireless multimedia networks using the shadow cluster concept. *Networking, IEEE/ACM Transactions on*, *5*(1), 1-12. doi: 10.1109/90.554717

Li, B., Li, S., Xing, C., Fei, Z., & Kuang, J. (2012). A QoE-based OFDM resource allocation scheme for energy efficiency and quality guarantee in multiuser-multiservice system. *Proc. IEEE Globecom Workshops*, 1293–1297. doi:10.1109/GLOCOMW.2012.6477768

Liew, C., Worrall, S., Goldshtein, M., Navarro, A., & Mota, M. (2006). *WIMAX modelling*. Project Deliverable SUIT_208 D2.3.

Li, F., Liu, G., & He, L. (2010). A low complexity algorithm of packet scheduling and resource allocation for wireless VoD systems. *IEEE Transactions on Consumer Electronics*, *56*(2), 1057–1062. doi:10.1109/TCE.2010.5506039

Li, M., Chen, Z., Tan, P., Sun, S., & Tan, Y. (2015). QoE-aware video streaming for SVC over multiuser MIMO–OFDM systems. *Journal of Visual Communication and Image Representation*, *26*, 24–36. doi:10.1016/j.jvcir.2014.10.011

Li, M., Chen, Z., & Tan, Y. (2012). QoE-aware resource allocation for scalable video transmission over multiuser MIMO-OFDM systems. *Proc. IEEE Visual Communications and Image Processing*, 1–6. doi:10.1109/VCIP.2012.6410733

Lin, X., Ma, H., Luo, L., & Chen, Y. (2012). No-reference video quality assessment in the compressed domain. *IEEE Transactions on Consumer Electronics*, *58*(2), 505–512. doi:10.1109/TCE.2012.6227454

Li, P., Chang, Y., Feng, N., & Yang, F. (2011). A cross-layer algorithm of packet scheduling and resource allocation for multi-user wireless video transmission. *IEEE Transactions on Consumer Electronics*, *57*(3), 1128–1134. doi:10.1109/TCE.2011.6018865

Li, P., Wang, Y., Zhang, W., & Huang, Y. (2014). QoE-oriented two-stage resource allocation in femtocell networks. *Proc. IEEE Vehicular Technology Conference*, 1–5. doi:10.1109/VTCFall.2014.6966143

Liu, Wen, Yeung, & Lei. (2012). Request-peer selection for load-balancing in P2P live streaming systems. *IEEE Conference on Wireless Communications and Networking*, 3227–3232.

Liu, Z., Shen, Y., Panwar, S. S., Ross, K. W., & Wang, Y. (2007). P2P video live streaming with MDC: Providing incentives for redistribution. ICME. IEEE, 48–51.

Liu, H., Ho, H., & Chou, C. (2014). Content-aware spectrum and power allocation for video multicast in two-tier femtocell networks. *Proc. IEEE Wireless Communications and Networking Conference*, 3213–3217. doi:10.1109/WCNC.2014.6953056

Liu, Y., Guo, Y., & Liang, C. (2008). A survey on peer-to-peer video streaming systems. *Journal of Peer-to-Peer Networking and Applications*, *1*(1), 18–28. doi:10.1007/s12083-007-0006-y

Liu, Y., Li, C., & Yang, Z. (2015). Tradeoff between energy and user experience for multimedia cloud computing. *Computers & Electrical Engineering*, *47*, 161–172. doi:10.1016/j.compeleceng.2015.04.016

Liu, Z., Shen, Y., Ross, K. W., Panwar, S. S., & Wang, Y. (2009). Layer P2P: Using layered video Chunks in P2P live streaming. *IEEE Transactions on Multimedia*, *11*(7), 1340–1352. doi:10.1109/TMM.2009.2030656

Luan, H., Kwong, K. W., Huang, Z., & Tsang, D. H. K. (2008). P2P live streaming towards best quality video. 5th IEEE Consumer Communications and Networking Conference (IEEE CCNC), 458-460.

Luby, M. (2012). *Broadcast Delivery of Multimedia Content to Mobile Users*. Qualcomm, Technical Report.

Luo, H., & Shyu, M.-L. (2011). Quality of service provision in mobile multimedia - a survey. *Human-centric Computing and Information Sciences*, *1*(1), 1–15. doi:10.1186/2192-1962-1-5

Lu, Q., & Koutsakis, P. (2011). Adaptive Bandwidth Reservation and Scheduling for Efficient Wireless Telemedicine Traffic Transmission. Vehicular Technology. *IEEE Transactions on*, *60*(2), 632–643. doi:10.1109/TVT.2010.2095472

Macker, J., & Corson, M. (1998). Mobile Ad Hoc Networking and The IETF. *Mobile Computing and Communications Review*, *2*(1), 9–14. doi:10.1145/584007.584015

Madhow, U. (2008). *Fundamentals of Digital Communication*. Cambridge University Press. doi:10.1017/CBO9780511807046

Magharei, N., Rejaie, R., & Guo, Y. (2007). Mesh or multiple-tree: A comparative study of live P2P streaming approaches. *Proceedings of IEEE INFOCOM*.

Magi. (2015). Available at http://www.endeavors.com

Mahadevan, V., Chaczko, Z., & Braun, R. (2004). Mastering the mystery through "SAIQ" metrics of user experience in telecollaboration business systems. *Proc. IADIS International Conference on WWW/Internet,* 1029–1034.

Mahbod Tavallaee, Lu, & Ghorbani. (2009). *A Detailed Analysis of the KDD CUP 99 Data Set*. Paper presented at the CISDA'09 Second IEEE international conference on Computational intelligence for security and defense applications.

Malas, D., & Morton, A. (2011). *Basic Telephony SIP end-to-end Performance Metrics*. Technical Report RFC 6076. Internet Engineering Task Force (IETF). Retrieved from http://tools.ietf.org/html/rfc6076

Malas, D., & Morton, A. (2011). *Basic Telephony SIP End-to-End Performance Metrics*. Technical Report RFC 6076. Internet Engineering Task Force (IETF). Retrieved from http://tools.ietf.org/html/rfc6076

Malekmohamadi, H., Fernando, A., Danish, E., & Kondoz, A. (2014). Subjective quality estimation based on neural networks for stereoscopic videos. *Proc. IEEE International Conference on Consumer Electronics*, 107–108. doi:10.1109/ICCE.2014.6775929

Manning, C. D., Raghavan, P., & Schütze, H. (2009). Naive Bayes text classification. In C. D. Manning, P. Raghavan, & H. Schütze (Eds.), *Introduction to Information Retrieval. Cambridge University Press.*

Marcille, S., Ciblat, P., & Le Martret, C. J. (2012). Stop-and-Wait Hybrid-ARQ performance at IP level under imperfect feedback.*Proceedings of IEEE Vehicular Technology Conference*, 1-5. doi:10.1109/VTCFall.2012.6399116

Martini, M. (2013). Cross-layer design for quality-driven multi-user multimedia transmission in mobile networks. *IEEE COMSOC MMTC E-Letter*, *8*(2), 18–20.

Matsuura, N. (2014). *Wireless LAN device and controlling method thereof*. Google Patents.

Ma, W., Zhang, H., Zheng, W., Lu, Z., & Wen, X. (2012). MOS-driven energy efficient power allocation for wireless video communications. *Proc. IEEE Globecom Workshops*, 52–56. doi:10.1109/GLOCOMW.2012.6477543

McCann, K., Bross, B., Kim, I.-K., Sekiguchi, S.-I., & Han, W.-J. (2011). *HM5: High Efficiency Video Coding (HEVC) Test Model 5 Encoder Description*. Paper presented at the JCTVC-G1102, JCT-VC Meeting, Geneva, Switzerland.

McGee, A. R., Coutière, M., & Palamara, M. E. (2012). Public safety network security considerations. *Bell Labs Technical Journal*, *17*(3), 79–86. doi:10.1002/bltj.21559

Medjiah, S., Ahmed, T., & Boutaba, R. (2014). Avoiding quality bottlenecks in P2P adaptive streaming. *IEEE Journal on Selected Areas in Communications*, *32*(4), 734–745. doi:10.1109/JSAC.2014.140406

Merritt, L., & Vanam, R. (2007). *Improved rate control and motion estimation for H. 264 encoder*. Paper presented at the Image Processing, 2007. ICIP 2007. IEEE International Conference on. doi:10.1109/ICIP.2007.4379827

Metha, P., & Udani, S. (2001, October-November). Voice over IP. *IEEE Potentials*, *20*(4), 36–40. doi:10.1109/45.969596

Minoli, D. (2011). *Voice over IPv6: Architectures for Next Generation VoIP Networks*. Newnes.

Mir, Musa, Gao, & Shivakumar. (2012). Performance Analysis of IMS Signaling Multimedia Networks. *Information Engineering, 1*(1).

Mohamed, H., & Cheng, H.H. (2008). Rate-distortion optimized streaming of fine grained scalable video sequences. *ACM Transactions on Multimedia Computing, Communications and Applications, 4*(1), 2:1–2:28.

Mol, J. D., Epema, D. H. P., & Sips, H. J. (2007). The orchard algorithm: Building multicast trees for P2P video multicasting without free-riding. *IEEE Transactions on Multimedia, 9*(8), 1593–1604. doi:10.1109/TMM.2007.907450

Moller, S., Engelbrecht, K.-P., Kühnel, C., Wechsung, I., & Weiss, B. (2009). A taxonomy of quality of service and quality of experience of multimodal human-machine interaction. *Proc. International Workshop on Quality of Multimedia Experience*, 7–12. doi:10.1109/QOMEX.2009.5246986

Montazeri, A., Akbari, B., & Ghanbari, M. (2012). An incentive scheduling mechanism for peer-to-peer video streaming. *Peer-to-Peer Networking and Applications, 5*(3), 257–278. doi:10.1007/s12083-011-0121-7

Moorsel, A. (2001). *Metrics for the internet age: quality of experience and quality of business.* Hewlett Packard (HP), HP Labs Technical Report HPL-2001-179.

Moorsel, A. (2001). Metrics for the internet age: quality of experience and quality of business. *Proc. International Workshop on Performability Modeling of Computer and Communication Systems*, 34, 26–31.

Motorola. (2015). *Proactively Protect Mission-Critical Infrastructure From Cyber Attacks And Security Threats.* Retrieved from https://www.motorolasolutions.com/content/dam/msi/docs/services/network-infrastructure-management/monitoring/security-monitoring-data-sheet.pdf

Mourtaji, I., Bouhorma, M., Benahmed, M., & Bouhdir, A. (2014). Proposition of a new approach to adapt SIP protocol to Ad hoc Networks. *International Journal of Software Engineering and Its Applications, 8*(7), 133–148.

Muge, S., Sercan, D., Yagiz, K., & Turhan, T. (2015). Adaptive, incentive and scalable dynamic tree overlay for P2P live video streaming. *Springer Journal of Peer to Peer networking and Applications.* DOI 10.1007/s12083-015-0390-7

Munir, A. (2008). Analysis of SIP-based IMS Session Establishment Signaling for WiMax-3G Networks. *Networking and Services, 2008. ICNS 2008. Fourth International Conference on*, 282-287. doi:10.1109/ICNS.2008.7

Munir, A., & Gordon-Ross, A. (2010, May). SIP-Based IMS Signaling Analysis for WiMAX-3G Interworking Architectures. *IEEE Transactions on Mobile Computing, 9*(5), 733–750. doi:10.1109/TMC.2010.16

Mushtaq, M., Augustin, B., & Mellouk, A. (2012). Empirical study based on machine learning approach to assess the QoS/QoE correlation. *Proc. European Conference on Networks and Optical Communications,*1–7. doi:10.1109/NOC.2012.6249939

Muthusamy, S. (2011). *Increasing broadcast and multicast service capacity and quality using LTE and MBMS.* Teleca, White paper.

Nahum, Tracey, & Wright. (2007). Evaluating SIP Server Performance. ACM SIGMETRICS Performance Evaluation Review, 35(1), 349-350.

Nakasu, E. (2012). Super Hi-Vision on the Horizon: A Future TV System That Conveys an Enhanced Sense of Reality and Presence. *Consumer Electronics Magazine, IEEE, 1*(2), 36–42. doi:10.1109/MCE.2011.2179821

Napster. (2015). Available at http://www.napster.com/

Narayan, S. (2010). VoIP Network Performance Evaluation of Operating Systems with IPv4 and IPv6 Network Implementations. *Computer Science and Information Technology (ICCSIT),20103rd IEEE International Conference on.* doi:10.1109/ICCSIT.2010.5564004

Narayanan, K. R., & Stuber, G. L. (1997). A novel ARQ technique using the turbo coding principle. *IEEE Communications Letters*, *1*(2), 49–51. doi:10.1109/4234.559361

Nee, R., & Prasad, R. (2000). OFDM for wireless multimedia communications. Boston: Artech House.

Ng, D., Lo, E., & Schober, R. (2013). Energy-efficient resource allocation in OFDMA systems with hybrid energy harvesting base station. *IEEE Transactions on Wireless Communications*, *12*(7), 3412–3427. doi:10.1109/TWC.2013.052813.121589

Ngo, D., Tellambura, C., & Nguyen, H. (2009). Efficient resource allocation for OFDMA multicast systems with fairness consideration. *Radio and Wireless Symposium*, 392–395.

Ngo, D., Tellambura, C., & Nguyen, H. (2009). Efficient resource allocation for OFDMA multicast systems with spectrum-sharing control *Vehicular Technology. IEEE Transactions on*, *58*(9), 4878–4889.

Nightingale, J., Wang, Q., Grecos, C., & Goma, S. (2014). The impact of network impairment on quality of experience (QoE) in H.265/HEVC video streaming. *IEEE Transactions on Consumer Electronics*, *60*(2), 242–250. doi:10.1109/TCE.2014.6852000

Niyato, D., & Hossain, E. (2008). A Noncooperative Game-Theoretic Framework for Radio Resource Management in 4G Heterogeneous Wireless Access Networks. Mobile Computing. *IEEE Transactions on*, *7*(3), 332–345. doi:10.1109/TMC.2007.70727

Noh, J., & Girod, B. (2012). Time-shifted streaming in a tree-based peer-to-peer system. *Journal of Communication*, *7*(3), 202–212.

Nokia. (2004). *Quality of Experience (QoE) of mobile services: Can it be measured and improved?* Nokia, White paper.

NTIA - US Department of Commerce. (2013). *Video Quality Metric (CVQM) Software*. Available: http://www.its.bldrdoc.gov/resources/video-quality-research/guides-and-tutorials/cvqm-overview.aspx

Nvidia, C. (2011). *Compute unified device architecture programming guide*. Academic Press.

Nvidia. (2015). *JETSON TK1*. Author.

Oh, S., Hoogs, A., Perera, A., Cuntoor, N., Chen, C.-C., Lee, J. T., . . . Davis, L. (2011). *A large-scale benchmark dataset for event recognition in surveillance video.* Paper presented at the Computer Vision and Pattern Recognition (CVPR), 2011 IEEE Conference on. doi:10.1109/CVPR.2011.5995586

Ohm, J., Sullivan, G. J., Schwarz, H., Tan, T. K., & Wiegand, T. (2014). Comparison of the Coding Efficiency of Video Coding Standards 2014; Including High Efficiency Video Coding (HEVC). *Circuits and Systems for Video Technology. IEEE Transactions on*, *12*(12), 1669–1684.

On SIP Performance. (2004). *Bell Labs Technical Journal.*

Orosz, P., Skopkó, T., Nagy, Z., Varga, P., & Gyimóthi, L. (2014). A case study on correlating video QoS and QoE. *Proc. IEEE Network Operations and Management Symposium*, 1–5. doi:10.1109/NOMS.2014.6838399

Pack, & Lee. (2008). Call Setup Latency Analysis in SIP-based Voice over WLANs. *IEEE Communications Letters, 12* (2), 103-105. doi:10.1109/LCOMM.2008.071230

Pack, S., Jeong, P., & Kim, Y. (2010). Proactive Route Optimization in SIP Mobility Support Protocol. *Consumer Communications and Networking Conference (CCNC), 20107th IEEE*, 1–2. doi:10.1109/CCNC.2010.5421774

Padmanabhan, V. N., Wang, H. J., & Chou, P. A. (2003). Resilient Peer-to-Peer Streaming.*International conference on network protocols (IEEE ICNP 03)*, 16-27.

Palta & Goyal. (2012). Comparison of OLSR and TORA Routing Protocols Using OPNET Modeler. *International Journal of Engineering Research and Technology, 1*(5).

Pandey & Swaroop. (2011). A Comprehensive Performance Analysis of Proactive, Reactive and Hybrid MANETs Routing Protocols. *International Journal of Computer Science Issues, 8*(6).

Papavassiliou, S., & Tassiulas, L. (1996). Joint optimal channel base station and power assignment for wireless access. *IEEE/ACM Transactions on Networking*, *4*(6), 857–872. doi:10.1109/90.556343

Papoutsis, V., & Kotsopoulos, S. (2011). Chunk-based resource allocation in multicast OFDMA systems with average BER constraint. *Communications Letters, IEEE*, *15*(5), 551–553.

Parihar, J. S., Rathore, J. S., & Burse, K. (April 2014). *Agent Based Intrusion Detection System to Find Layers Attacks*. Paper presented at the 2014 Fourth International Conference on Communication Systems and Network Technologies, Bhopal, India doi:10.1109/CSNT.2014.144

Park, J., & van der. Schaar, M. (2010). A game theoretic analysis of incentives in content production and sharing over P2P networks. *IEEE Journal of Selected Topics in Signal Processing*, *4*(4), 704–717. doi:10.1109/JSTSP.2010.2048609

Patel, A., Qassim, Q., & Wills, C. (2010). A survey of intrusion detection and prevention systems. *Information Management & Computer Security*, *18*(4), 277–290. doi:10.1108/09685221011079199

Patil, V. (2012). *Qualitative and Quantitative Performance Evaluation of Ad hoc on Demand Routing Protocol in MANET*. Academic Press.

Paudel, I., Pokhrel, J., Wehbi, B., Cavalli, A., & Jouaber, B. (2014). Estimation of video QoE from MAC parameters in wireless network: a random neural network approach. *Proc. International Symposium on Communications and Information Technologies*, 51–55. doi:10.1109/ISCIT.2014.7011868

Pearson, D. E. (1975). *Transmission and display of pictorial information*. Pentech Press Limited.

Perera, R., Fernando, A., Mallikarachchi, T., Arachchi, H., & Pourazad, M. (2014). QoE aware resource allocation for video communications over LTE based mobile networks. *Proc. International Conference on Heterogeneous Networking for Quality, Reliability, Security and Robustness*, 63–69. doi:10.1109/QSHINE.2014.6928661

Pinson, M. H., & Wolf, S. (2004). A new standardized method for objectively measuring video quality. *IEEE Transactions on Broadcasting*, *50*(3), 312–322. doi:10.1109/TBC.2004.834028

Pirhadi, M., Hemami, S., & Tabrizipoor, A. (2009). Call Set-up Time Modeling for SIP-based Stateless and Stateful Calls in Next Generation Networks. *Advanced Communication Technology, 2009. ICACT 2009. 11th International Conference on, 2,* 1299-1304.

Pokhrel, J., Wehbi, B., Morais, A., Cavalli, A., & Allilaire, E. (2013). Estimation of QoE of video traffic using a fuzzy expert system.*Proc. IEEE Consumer Communications and Networking Conference*, 224–229. doi:10.1109/CCNC.2013.6488450

Poynton, C. (2002). Chroma subsampling notation. In Digital Video and HDTV: Algorithms and Interfaces. Morgan Kaufmann.

PPLive. (2015). Available at http://www.pplive.com/

PPStream. (2015). Available at http://www.ppstream.com

Proakis, J. G., & Salehi, M. (2008). *Digital Communications*. Academic Press.

Project 25. (1995). *Project 25 System and Standard Definition, TIA, TSB102-A*. Retrieved from http://www.project25.org

Protocol, U. D. (1980). RFC 768 J. Postel ISI 28 August 1980. *Isi*.

Public Safety mobile broadband and spectrum needs. (2010). *Report for the TETRA Association*.

Radicke, S., Hahn, J., Grecos, C., & Wang, Q. (2013). *Highly-parallel HVEC motion estimation with CUDA*. Paper presented at the Visual Information Processing (EUVIP), 2013 4th European Workshop on.

Raheel, M. S., Raad, R., & Ritz, C. (2014). Efficient utilization of peer's upload capacity in P2P networks using SVC. *IEEECommunications and Information Technologies (ISCIT), 2014 14th International Symposium on*, 66-70. doi:10.1109/ISCIT.2014.7011871

Raheel, Raad, & Ritz. (2015, August). Achieving maximum utilization of peer's upload capacity in P2P networks using SVC. *Springer Peer-to-Peer Networking and Applications*, 1-21.

Ramirez, R. (2014). *Low Complexity Radio Resource Management for Energy Efficient Wireless Networks* (Doctoral Thesis). The University of Edinburgh, Edinburgh, UK.

Rankin, Costaiche, & Zeto. (2013). *Validating VoLTE – A Definitive Guide to Successful Deployments*. IXIA.

Rashad, S., Kantardzic, M., & Kumar, A. (2006). User mobility oriented predictive call admission control and resource reservation for next-generation mobile networks. *Journal of Parallel and Distributed Computing*, *66*(7), 971–988. doi:10.1016/j.jpdc.2006.03.007

Rassam, M. A., Maarof, M. A., & Zainal, A. (2012). A Survey of Intrusion Detection Schemes in Wireless Sensor Networks. *American Journal of Applied Sciences*, *9*(10), 1636–1652. doi:10.3844/ajassp.2012.1636.1652

Ratnasamy, Francis, Handley, Karp, & Shenker. (2001). *A scalable content-addressable network*. Technical report, ACM SIGCOMM'01.

Ravale, U., Marathe, N., & Padiya, P. (2015). *Feature Selection Based Hybrid Anomaly Intrusion Detection System Using K-Means and RBF Kernel Function*. Paper presented at the International Conference on Advanced Computing Technologies and Applications (ICACTA-2015), Mumbai, India. doi:10.1016/j.procs.2015.03.174

Rec, I. (2007). T. 800| ISO/IEC 15444-3: 2007. *Information technology -- JPEG 2000 image coding system -- Part 3: Motion JPEG 2000*.

Reddy, Y. (2007). Genetic algorithm approach for adaptive subcarrier, bit, and power allocation. *Proc.IEEE International Conference on Networking, Sensing and Control*, 14–19. doi:10.1109/ICNSC.2007.372925

Reeves, C., & Bednar, D. (1994). Defining quality: Alternatives and implications. *Academy of Management Review*, *19*(3), 419–445.

Reiter, U. (2009). Perceived quality in consumer electronics - from quality of service to quality of experience. *Proc. IEEE International Symposium on Consumer Electronics*, 958–961. doi:10.1109/ISCE.2009.5156963

Ren, D., Li, Y. T. H., & Chan, S.-H. G. (2009). Fast-Mesh: A Low-Delay High-Bandwidth Mesh for Peer-to-Peer Live Streaming. *IEEE Transactions on Multimedia*, *11*(8), 1446–1456. doi:10.1109/TMM.2009.2032677

Rezac, F., Voznak, M., Partila, P., & Tomala, K. (2013). Interactive Video Audio System and Its Performance Evaluation. *Telecommunications and Signal Processing (TSP),201336th International Conference on*, 43-46. doi:10.1109/TSP.2013.6613888

Reza, R., & Antonio, O. (2003). PALS: Peer-to-peer adaptive layered streaming.*Proceedings of ACM International Workshop on Network and Operating System Support for Digital Audio and Video NOSSDAV'03*, 153–161.

Riabov, A., Liu, Z., & Zhang, L. (2004). Overlay multicast trees of minimal delay. *Distributed Computing Systems, 2004. Proceedings. 24th International Conference on*, 654-661. doi:10.1109/ICDCS.2004.1281633

Rodríguez, D., Rosa, R., & Bressan, G. (2013). A billing system model for voice call service in cellular networks based on voice quality. *Proc. IEEE International Symposium on Consumer Electronics*, 89–90. doi:10.1109/ISCE.2013.6570267

Rohling, H., & Gruneid, R. (1997). Performance comparison of different multiple access schemes for the downlink of an OFDM communication system. *Proc. IEEE 47th Vehicular Technology Conference*, 3, 1365–1369. doi:10.1109/VETEC.1997.605406

Rosenberg, J., Schulzrinne, H., Camarillo, G., Johnston, A., Peterson, J., Sparks, R., … Schooler. (2002). *SIP: Session Initiation Protocol*. RFC 3261.

Rosenberg, J., Schulzrinne, H., Camarillo, G., Johnston, A., Peterson, J., Sparks, R., Handley, M., & Schooler. (2002). *SIP: Session Initiation Protocol*. RFC 3261.

Rozhon, J., & Voznak, M. (2011). SIP Registration Burst Load Test. In Digital Information Processing and Communications, (pp. 329-336). Springer Berlin Heidelberg. doi:10.1007/978-3-642-22410-2_29

Rozhon, Voznak, Tomala, & Vychodil. (2012). Updated Approach to SIP Benchmarking. *Telecommunications and Signal Processing (TSP),201235th International Conference on*, 251-254.

Sacchi, C., Granelli, F., & Schlegel, C. (2011). A QoE-oriented strategy for OFDMA radio resource allocation based on min-MOS maximization. *IEEE Communications Letters*, *15*(5), 494–496. doi:10.1109/LCOMM.2011.031411.101672

Sadr, S., Anpalagan, A., & Raahemifar, K. (2009). Radio resource allocation algorithms for the downlink of multiuser OFDM communication systems. *IEEE Communications Surveys Tutorials*, *11*(3), 92–106.

Saki, H., & Shikh-Bahaei, M. (2015). Cross-layer resource allocation for video streaming over OFDMA cognitive radio networks. *IEEE Transactions on Multimedia*, *17*(3), 333–345. doi:10.1109/TMM.2015.2389032

Samara, C., Karapistoli, E., & Economides, A. (2012). Performance Comparison of MANET Routing Protocols Based on Real-life Scenarios. *Ultra Modern Telecommunications and Control Systems and Workshops (ICUMT),20124th International Congress on*. doi:10.1109/ICUMT.2012.6459784

Sandhu, U. A., Haider, S., Naseer, S., & Ateeb, O. U. (2011). *A Survey of Intrusion Detection & Prevention Techniques* Paper presented at the 2011 International Conference on Information Communication and Management, Singapore.

Saponara, S., Denolf, K., Lafruit, G., Blanch, C., & Bormans, J. (2004). Performance and complexity co-evaluation of the advanced video coding standard for cost-effective multimedia communications. *EURASIP Journal on Applied Signal Processing*, *2004*(2), 220–235. doi:10.1155/S111086570431019X

Satsiou, A., & Tassiulas, L. (2010). Reputation-Based Resource Allocation in P2P Systems of Rational Users. *IEEE Transactions on Parallel and Distributed Systems*, *21*(4), 476–479. doi:10.1109/TPDS.2009.80

Schulzrinne, Casner, Frederick, & Jacobson. (2003). *RTP: A Transport Protocol for Realtime Applications*. RFC 3550.

Schulzrinne, H., Casner, S., Frederick, R., & Jacobson, V. (2003). *Real-time transport protocol*. RFC1899.

Seyfabad, M. S., & Akbari, B. (2014). CAC-live: Centralized assisted cloud P2P live streaming. *Electrical Engineering (ICEE), 2014 22nd Iranian Conference on*, 908-913.

Shabnam, M., & Mohamed, H. (2010). Live peer to peer streaming with scalable video coding and network coding. *Proceedings of the first annual ACM SIGMM conference on Multimedia systems MMSys'10*, 123-132.

Shay, H., & Danny, D. (2010). *On the Role of Helper Peers inP2P Networks. Parallel and Distributed Computing* (A. Ros, Ed.). InTech. doi:10.5772/9450

Shen, J., Yi, N., Wu, B., Jiang, W., & Xiang, H. (2009). A greedy-based resource allocation algorithm for multicast and unicast services in OFDM system. *InternationalConference on Wireless Communications Signal Processing,*1–5.

Shen, X., Yu, H., Buford, J., & Akon, M. (2010). *Handbook of Peer-to-Peer Networking*. Springer Publishing Company. doi:10.1007/978-0-387-09751-0

Shen, Z., Andrews, J., & Evans, B. (2003). Optimal power allocation in multiuser OFDM systems. *Proc. IEEE Global Telecommunications Conference*, 1, 337–341.

Sheu, S.-T., Kuo, K.-H., Yang, C.-C., & Sheu, Y.-M. (2013). A Go-back-N HARQ time bundling for machine type communication devices in LTE TDD.*Proceedings of IEEE Wireless Communications and Networking Conference*, 280-285.

Shi, Z., Zou, H., Rank, M., Chen, L., Hirche, S., & Muller, H. J. (2010). Effects of packet loss and latency on the temporal discrimination of visual-haptic events. *IEEE Transactions on Haptics*, *3*(1), 28–36.

Shimizu, T., Shirai, D., Takahashi, H., Murooka, T., Obana, K., Tonomura, Y., & Ohta, N. et al. (2006). International real-time streaming of 4K digital cinema. *Future Generation Computer Systems*, *22*(8), 929–939. doi:10.1016/j.future.2006.04.001

Shirai, D., Kawano, T., & Fujii, T. (2007). *6 Gbit/s uncompressed 4K video stream switching on a 10 Gbit/s network.* Paper presented at the Intelligent Signal Processing and Communication Systems, 2007. ISPACS 2007. International Symposium on.

Shirai, D., Shimizu, K., Sameshima, Y., & Takahashi, H. (2007). 6-Gbit/s Uncompressed 4K Video IP Stream Transmission and OXC Stream Switching Trial Using JGN II. *NTT Technical Review*, *5*(1).

Shirai, D., Yamaguchi, T., Shimizu, T., Murooka, T., & Fujii, T. (2006, 2006). *4K SHD real-time video streaming system with JPEG 2000 parallel codec.* Paper presented at the Circuits and Systems, 2006. APCCAS 2006. IEEE Asia Pacific Conference on.

Shirai, D., Kawano, T., Fujii, T., Kaneko, K., Ohta, N., Ono, S., & Ogoshi, T. et al. (2009). Real time switching and streaming transmission of uncompressed 4K motion pictures. *Future Generation Computer Systems*, *25*(2), 192–197. doi:10.1016/j.future.2008.07.003

Shi, Z., Ren, L., & Jin, F. (2002). Design and performance analysis of HARQ for RS-turbo concatenated codes.*Proceedings of IEEE International Conference on Communications, Circuits and Systems and West Sino Expositions*, 56-59.

Shrestha, A., & Tekiner, F. (2009). On MANET Routing Protocols for Mobility and Scalability. *International Conference on Parallel and Distributed Computing, Applications and Technologies*, 451-456. doi:10.1109/PDCAT.2009.88

Shun-Ren, Y., & Wen-Tsuen, C. (2008). SIP Multicast-Based Mobile Quality-of-Service Support over Heterogeneous IP Multimedia Subsystems. Mobile Computing. *IEEE Transactions on*, *7*(11), 1297–1310. doi:10.1109/TMC.2008.53

Signalling flows for the IP multimedia call control based Session Initiation Protocol (SIP) and Session Description Protocol; Stage3 (Release 5). (2006). 3GPP TS 24.228 v5.15.0. 3GPP.

Silva, A. P. R. d., Martins, M. H., Rocha, B. P., Loureiro, A. A., Ruiz, L. B., & Wong, H. C. (2005). *Decentralized intrusion detection in wireless sensor networks*. Paper presented at the 1st ACM international workshop on Quality of service & security in wireless and mobile networks. doi:10.1145/1089761.1089765

Simic, M. B. (2012). Feasibility of long term evolution (LTE) as technology for public safety. In *2012 20th Telecommunications Forum (TELFOR)*, 158–161. http://doi.org/ doi:<ALIGNMENT.qj></ALIGNMENT>10.1109/TELFOR.2012.6419172

Sinnreich, H., & Johnston, A. B. (2012). *Internet Communications Using SIP: Delivering VoIP and Multimedia Services with Session Initiation Protocol* (Vol. 27).Wiley.

Skordoulis, D., Ni, Q., Chen, H.-H., Stephens, A. P., Liu, C., & Jamalipour, A. (2008). IEEE 802.11 n MAC Frame Aggregation Mechanisms for Next-Generation High-Throughput WLANs. *Wireless Communications, IEEE, 15*(1), 40–47. doi:10.1109/MWC.2008.4454703

So-In, C., Jain, R., and Tamimi, A.-K. (2010). Capacity evaluation for IEEE 802.16e mobile WiMAX. *Journal of Computer Systems, Networks, and Communications*, 1:1–1:12.

Sondi, P., & Gantsou, D. (2009). Voice communication over Mobile Ad Hoc Networks: Evaluation of A QoS Extension of OLSR Using OPNET. *Asian Internet Engineering Conference*, 61–68. doi:10.1145/1711113.1711125

Song, G., & Li, Y. (2003). Adaptive subcarrier and power allocation in OFDM based on maximizing utility. *Proc.IEEE Semiannual Vehicular Technology Conference*, 2, 905–909.

Song, W., Tjondronegoro, D., & Docherty, M. (2011). Saving bitrate vs. pleasing users: where is the break-even point in mobile video quality?*Proc. ACM International Conference on Multimedia*, 403–412. doi:10.1145/2072298.2072351

SopCast. (2015). Available at http://www.sopcast.com/

Stoica, I., Morris, R., Liben, N.D., Karger, D.R., Kaashoek, M.F., Dabek, F., & Balakrishnan, H. (2003). Chord: A Scalable Peer-to-Peer Lookup Protocol for Internet Applications. *IEEE/ACM Transaction of Networking, 11*(1), 17–32.

Sudipta, S., Shao, L., Minghua, C., Mung, C., Jin, L., & Philip, A. C. (2011). Peer to peer streaming capacity. *IEEE Transactions on Information Theory, 57*(08), 5072–5088. doi:10.1109/TIT.2011.2145630

Su, H., Wen, M., Wu, N., Ren, J., & Zhang, C. (2014). Efficient parallel video processing techniques on GPU: From framework to implementation. *The Scientific World Journal.* PMID:24757432

Sühring, K., Heising, G., & Marpe, D. (2009). *H. 264/AVC reference software*. Academic Press.

Sullivan, G. J., & Ohm, J.-R. (2010). *Recent developments in standardization of high efficiency video coding (HEVC).* Paper presented at the SPIE Optical Engineering+ Applications.

Sullivan, G. J., Ohm, J., Han, W.-J., & Wiegand, T. (2012). Overview of the High Efficiency Video Coding (HEVC) Standard. *Circuits and Systems for Video Technology. IEEE Transactions on, 22*(12), 1649–1668.

Sun, P.-T. (2002). *Similarity of Discrete Gilbert-Elliot and Polya Channel Models to Continuous Rayleigh Fading Channel Model* (Doctoral Thesis). National Chiao Tung University, Taiwan.

Sun, B., Osborne, L., Xiao, Y., & Guizani, S. (2007). Intrusion detection techniques in mobile ad hoc and wireless sensor networks. *IEEE Wireless Communications, 14*(5), 56–63. doi:10.1109/MWC.2007.4396943

Sundaram, Palani, & Babu. (2013). Performance Evaluation of AODV, DSR and OLSR Mobile Ad Hoc Network Routing Protocols using OPNET Simulator. *International Journal of Computer Science & Communication Networks, 3*, 54-63.

Sung, Bishop, & Rao. (2006). Enabling contribution awareness in an overlay broadcasting system. *Proceedings of ACM SIGCOMM.*

Sun, Z. (2005). *Satellite Networking: Principles and Protocols.* John Wiley & Sons. doi:10.1002/047087029X

Suzuki, T., Kutsuna, T., & Tasaka, S. (2008). QoE estimation from MAC-level QoS in audio-video transmission with IEEE 802.11e EDCA. *Proc. IEEE International Symposium on Personal, Indoor and Mobile Radio Communications,* 1–6. doi:10.1109/PIMRC.2008.4699471

SYNESIS. (2010). *Megapixel video analytics in difficult surveillance conditions.* Retrieved from http://en.synesis.ru/en/surveillance/contents/megapixel-analytics

Takahashi, A., Hands, D., & Barriac, V. (2008). Standardization activities in the ITU for a QoE assessment of IPTV. *Communications Magazine, IEEE, 46*(2), 78–84.

Tang, J. (2013). *Mathematical optimization techniques for resource allocation and spatial multiplexing in spectrum sharing networks* (Doctoral Thesis). Loughborough University, Loughborough, UK.

Tang, J., Davids, C., & Cheng, Y. (2008). A Study of an Open Source IP Multimedia Subsystem Test Bed. *Proceedings of the 5th International ICST Conference on Heterogeneous Networking for Quality, Reliability, Security and Robustness.* doi:10.4108/ICST.QSHINE2008.3952

Tang, Y., Luo, J. G., Zhang, Q., Zhang, M., & Yang, S. Q. (2007). Deploying P2P networks for large-scale live video-streaming service. *IEEE Communications Magazine, 45*(6), 100–106. doi:10.1109/MCOM.2007.374426

Tebbani, B., Haddadou, K., & Pujolle, G. (2009). A Session-based Management Architecture for QoS Assurance to VoIP Applications on Wireless Access Networks. *Consumer Communications and Networking Conference, 2009. CCNC 2009. 6th IEEE,* 1–5. doi:10.1109/CCNC.2009.4784757

Technical Specification Group Services and System Aspects. (2006). *IP Multimedia Subsystem (IMS), Stage 2, TS 23.228, 3rd Generation Partnership Project TS 23.228 v8.2.0.* Author.

Tektronix. (2008). *A Guide to IPTV The Technologies, the Challenges and How to Test IPTV.* Tektronix, Inc.

Tektronix. (2009). *A Guide to MPEG Fundamentals and Protocol Analysis.* Tektronix, Inc., Technical Report.

TeleManagement Forum. (2005). SLA Management Handbook, Concepts and Principles. TeleManagement Forum.

TeleManagement Forum. (2009). Best Practice: Video over IP SLA Management. *TeleManagement Forum.*

Terrestrial Trunked Radio (TETRA). (2003). *Voice plus Data (V+D); Part 2: Air Interface (AI), ETSI, EN 300 392-2 v2.3.10.* Author.

TETRA and Critical Communication Association. (2010). Retrieved from http://www.tandcca.com/

TETRA MoU Association. (2004). *Push To Talk over Cellular (PoC) and Professional Mobile Radio (PMR).* TETRA.

Third Generation Partnership Project, Organization, 3GPP. (2008). *3rd Generation Partnership Project;Technical Specification Group Radio Access Network.* Evolved Universal Terrestrial Radio Access(EUTRA).

Third Generation Partnership Project. (2008). *Technical specification group radio access network, Evolved Universal Terrestrial Radio Access (E-UTRA), multiplexing and channel coding* (Release 8). 3GPP TS 36.212 V8.5.0.

Third Generation Partnership Project. (2009). *Technical Specification Group Radio Access Network; Enhanced uplink; Overall description; Stage 2* (Release 7). 3GPP TS 25.319 V7.7.0.

Third Generation Partnership Project. (2010). *Technical Specification Group Radio Access Network, Evolved Universal Terrestrial Radio Access (E-UTRA), Multiplexing and channel coding* (Release 10). 3GPP TS 36.212 V10.0.0.

Thisen, D. J. M., Espinosa, C., & Herpertz, R. (2009). Evaluating the Performance of an IMS/NGN Deployment. *Proceedings of the 2nd Workshop on Services, Platforms, Innovations and Research for new Infrastructures in Telecommunications*, 2561-2573.

Thompson, J., Ge, X.-L., Wu, H.-C., Irmer, R., Jiang, H., Fettweis, G., & Alamouti, S. (2014). 5G wireless communication systems: Prospects and challenges[guest editorial]. *Communications Magazine, IEEE, 52*(2), 62–64. doi:10.1109/MCOM.2014.6736744

Thompson, M. S., MacKenzie, A. B., & DaSilva, L. A. (2011). A Method of Proactive MANET Routing Protocol Evaluation Applied to The OLSR Protocol. *Proceedings of the 6th ACM international workshop on Wireless network testbeds, experimental evaluation and characterization*, 27–34. doi:10.1145/2030718.2030726

Tillo, T., Grangetto, M., & Olmo, G. (2008). Redundant slice optimal allocation for H.264 multiple description coding. *IEEE Transactions on Circuits and Systems for Video Technology, 18*(01), 59–70. doi:10.1109/TCSVT.2007.913751

TKN. (2012). *MPEG-4 and H.263 Video Traces for Network Performance Evaluation*. Retrieved from http://trace.eas.asu.edu/TRACE/ltvt.html

Tran, D. A., Hua, K. A., & Do, T. T. (2004). A peer-to-peer architecture for media streaming. *IEEE Journal on Selected Areas in Communications, 22*(1), 121–133. doi:10.1109/JSAC.2003.818803

Tuong, C., Kuthethoor, S., & Hadynski, K., & Parker. (2010). Performance Analysis for SIP-based VoIP Services over Airborne Tactical Networks. *Aerospace Conference*, 1–8.

Tuong, L., Cook, S., Kuthethoor, G., Sesha, P., Hadynski, G., Kiwior, D., & Parker, D. (2010). Performance Analysis for SIP-based VoIP Services over Airborne Tactical Networks. *Aerospace Conference*, 1-8.

Turek, T., Zerawa, S.-A., & Anees, T. (2011). *Towards safety and security critical communication systems based on SOA paradigm*. Paper presented at the 2011 IEEE 20th International Symposium on Industrial Electronics (ISIE), Gdansk, Poland. doi:10.1109/ISIE.2011.5984337

Tychogiorgos, G., & Leung, K. (2014). Optimization-based resource allocation in communication networks. *Computer Networks, 66*, 32–45. doi:10.1016/j.comnet.2014.03.013

Tzvetkov, V. (2008). SIP Registration Optimization in Mobile Environments Using Extended Kalman Filter. *Communications and Networking in China, 2008. ChinaCom 2008. Third International Conference on*, 106-111. doi:10.1109/CHINACOM.2008.4684980

Uchoa, A. G. D., Demo Souza, R., & Pellenz, M. E. (2010). Type-I HARQ scheme using LDPC codes and partial retransmissions for AWGN and quasi static fading channels.*Proceedings of the 7th International Symposium on Wireless Communication Systems*, 571-575. doi:10.1109/ISWCS.2010.5624326

Uppal, H. A. M., Javed, M., & Arshad, M. J. (2014). An Overview of Intrusion Detection System (IDS) along with its Commonly Used Techniques and Classifications. *International Journal of Computer Science and Telecommunications, 5*(2), 20–24.

UUSee. (2015). Available at http://www.uusee.com/

Vaishampayan, V. A. (1993). Design of multiple description scalar quantizers. *IEEE Transactions on Information Theory, 39*(03), 821–834. doi:10.1109/18.256491

Vasu, K., Maheshwari, S., Mahapatra, S., & Kumar, C. S. (2012). QoS-aware fuzzy rule-based vertical handoff decision algorithm incorporating a new evaluation model for wireless heterogeneous networks. *EURASIP Journal on Wireless Communications and Networking, 2012*(322). doi: 10.1186/1687-1499-2012-322

Venkataraman, V., Yoshida, K., & Francis, P. (2006). Chunkyspread: Heterogeneous Unstructured Tree-Based Peer-to-Peer Multicast. *Proceedings of the 2006 IEEE International Conference on Network Protocols*, 2-11. doi:10.1109/ICNP.2006.320193

Vinay, P., Kumar, K., Tamilmani, K., Sambamurthy, V., & Mohr, A. E. (2005). Chainsaw: Eliminating Trees from Overlay Multicast. *4th International Workshop on Peer-to-Peer Systems,* 127-140.

Vivek, K. G. (2001). Multiple description coding: Compression meets the network. *IEEE Signal Processing Magazine, 18*(5), 74–93. doi:10.1109/79.952806

Voznak, M., & Rozhon, J. (2010). SIP back to back user benchmarking. *Wireless and Mobile Communications (ICWMC), 20106th International Conference on*, 92-96.

Voznak, M., & Rozhon, J. (2010). SIP Back to Back User Benchmarking. *Wireless and Mobile Communications (ICWMC),20106th International Conference on*, 92-96. doi:10.1109/ICWMC.2010.86

Voznak, M., & Rozhon, J. (2012). SIP Registration Stress Test. *6th International Conference on Communications and Information Technology*, 95-100.

Voznak, M., Kovac, A., & Halas, M. (2012). Effective Packet Loss Estimation on VoIP Jitter Buffer. In *Networking 2012 Workshops* (pp. 157–162). Springer Berlin Heidelberg. doi:10.1007/978-3-642-30039-4_21

Voznak, M., & Rozhon, J. (2010). Performance Testing and Benchmarking of B2BUA and SIP Proxy. *Conference Proceedings TSP*, 497-503.

Voznak, M., & Rozhon, J. (2012). SIP End To End Performance Metrics. *International Journal of Mathematics and Computers in Simulation*, *6*(3), 315–323.

Voznak, M., & Rozhon, J. (2013). Approach to Stress Tests in SIP Environment Based on Marginal Analysis. *Telecommunication Systems*, *52*(3), 1583–1593. doi:10.1007/s11235-011-9525-1

Vriendt, J. De, Vleeschauwer, D., & Robinson, D. C. (2014). QoE model for video delivered over an LTE network using HTTP adaptive streaming. *Bell Labs Technical Journal*, *18*(4), 45–62.

Waldman, M., Cranor, L., & Rubin, A. (2001). Publius. In *Peer-to-Peer: Harnessing the Power of Disruptive Technologies* (pp. 145–158). O'Reilly & Associates.

Wang, X. (2002). BuddyWeb: a P2P-based collaborative web caching system. *Proceedings of the Web Engineering and Peer-to-Peer Computing, Networking Workshops*, 247–251.

Wang, Z., Lu, L., & Bovik, A. C. (2004). Video quality assessment based on structural distortion measurement. *Signal Processing: Image Communication, 19*(2), 121–132.

Wang, S., Guo, W., Khirallah, C., Vukobratović, D., & Thompson, J. (2014). Interference allocation scheduler for green multimedia delivery. *IEEE Transactions on Vehicular Technology*, *63*(5), 2059–2070. doi:10.1109/TVT.2014.2312373

Wang, X. G., Min, G., Mellor, J. E., Al-Begain, K., & Guan, L. (2005). An adaptive QoS framework for integrated cellular and WLAN networks. *Computer Networks*, *47*(2), 167–183. doi:10.1016/j.comnet.2004.07.003

Wang, X., Dai, H., Zhang, H., & Li, F. (2012). Channel-aware adaptive resource allocation for multicast and unicast services in orthogonal frequency division multiplexing systems. *IET Communications*, *6*(17), 3006–3014. doi:10.1049/iet-com.2012.0111

Wang, Y. (2009). *Statistical Techniques for Network Security: Modern Statistically-Based Intrusion Detection and Protection*. Hershey, PA: IGI Global. doi:10.4018/978-1-59904-708-9

Wang, Y., Chen, F., & Chen, G. (2005). Adaptive subcarrier and bit allocation for multiuser OFDM system based on genetic algorithm. *Proc. IEEE International Conference on Communications, Circuits and Systems*, 1, 242–246.

Webster, D. (2015). *Cisco Visual Networking Index Predicts IP Traffic to Triple from 2014-2019; Growth Drivers Include Increasing Mobile Access, Demand for Video Services*. Retrieved from http://newsroom.cisco.com/press-release-content?articleId=1644203

Wei, S., Weihua, Z., & Yu, C. (2007). Load balancing for cellular/WLAN integrated networks. *IEEE Network*, *21*(1), 27–33. doi:10.1109/MNET.2007.314535

Wen-Mei, W. H. (2011). *GPU Computing Gems Emerald Edition*. Elsevier.

Wien, M., Schwarz, H., & Oelbaum, T. (2007). Performance analysis of SVC. *IEEE Transactions on Circuits and Systems for Video Technology*, *17*(9), 1194–1203. doi:10.1109/TCSVT.2007.905530

Winkler, S., van den Branden Lambrecht, C. J., & Kunt, M. (2001). *Vision and video: models and applications* (Vol. 10). Academic Press.

Wolter, K., & Moorsel, A. (2001). *The Relationship between Quality of Service and Business Metrics: Monitoring, Notification and Optimization*. Hewlett Packard (HP), HP Labs Technical Report HPL-2001-96.

Wong, C., Cheng, R., Lataief, K., & Murch, R. (1999). Multiuser OFDM with adaptive subcarrier, bit, and power allocation. *IEEE Journal on Selected Areas in Communications*, *17*(10), 1747–1758. doi:10.1109/49.793310

Wozencraft, J. M., & Horstein, M. (1960). Digitalised communication over two way channels. *The 4th London Symposium of Information Theory*.

Wozencraft, J. M., & Horstein, M. (1961). Coding for two-way channels. Tech. rep. Research Laboratory of Electronics, M.I.T.

Wu, F. (2010). *Synchronization and Resource Allocation in Downlink OFDM Systems* (Doctoral Thesis). University of Plymouth, Plymouth, UK.

Wu, N., Wen, M., Su, H., Ren, J., & Zhang, C. (2012). *A parallel H. 264 encoder with CUDA: mapping and evaluation*. Paper presented at the Parallel and Distributed Systems (ICPADS), 2012 IEEE 18th International Conference on. doi:10.1109/ICPADS.2012.46

Wu, J., Liu, S., Zhou, Z., & Zhan, M. (2012). Toward Intelligent Intrusion Prediction forWireless Sensor Networks Using Three-Layer Brain-Like Learning. *International Journal of Distributed Sensor Networks*, *2012*, 1–14. doi:10.1155/2012/243841

Xia, G., Gang, W., & Miki, T. (2004). End-to-end QoS provisioning in mobile heterogeneous networks. *Wireless Communications, IEEE*, *11*(3), 24–34. doi:10.1109/MWC.2004.1308940

Xiao, L. (2010). *Radio Resource Allocation in Relay Based OFDMA Cellular Networks* (Doctoral Thesis). Queen Mary, University of London, London, UK.

Xiao, X., Shi, Y., Gao, Y., & Zhang, Q. (2009). Layer P2P: A new data scheduling approach for layered streaming in heterogeneous networks. IEEE INFOCOM Proceedings, 603-611.

Xiaonan, W., & Shan, Z. (2012). Research on Mobility Handover for IPv6 Based MANET. *Transactions on Emerging Telecommunications Technologies*.

Xuebing, P., Tao, J., Daiming, Q., Guangxi, Z., & Liu, J. (2010). Radio-Resource Management and Access-Control Mechanism Based on a Novel Economic Model in Heterogeneous Wireless Networks. Vehicular Technology. *IEEE Transactions on*, *59*(6), 3047–3056. doi:10.1109/TVT.2010.2049039

Xuguang, L., Nanning, Z., Jianru, X., Xiaoguang, W., & Bin, G. (2007). A Peer-to-peer architecture for efficient live scalable media streaming on Internet. *Proceedings of ACM Multimedia Conference*, 783–786.

Xu, Y., Hu, R. Q., Wei, L., & Wu, G. (2014). QoE-aware mobile association and resource allocation over wireless heterogeneous networks. *Proc. IEEE Global Communications Conference*, 4695–4701. doi:10.1109/GLOCOM.2014.7037549

Yafeng, W., Lei, Z., & Dacheng, Y. (2003). Performance analysis of type III HARQ with turbo codes.*Proceedings of the 57th IEEE Semiannual Vehicular Technology Conference*, *4*,2740-2744.

Yahiaoui, S., Belhoul, Y., Nouali-Taboudjemat, N., & Kheddouci, H. (2012). AdSIP: Decentralized SIP for Mobile Ad hoc Networks. *Advanced Information Networking and Applications Workshops (WAINA),201226th International Conference on*, 490–495. doi:10.1109/WAINA.2012.151

Yang, B., B. & Garcia-Molina, H. (2003). Designing a super-peer network, *Proceedings.19th International Conference on Data Engineering*, 49-60.

Yang, F., Yang, Q., Fu, F., & Kwak, K. S. (2014). A QoE-based resource allocation scheme for multi-radio access in heterogeneous wireless network. *Proc. International Symposium on Communications and Information Technologies*, 264–268. doi:10.1109/ISCIT.2014.7011913

Yang, S., & Wang, X. (2010). An incentive mechanism for tree-based live media streaming service. *Journal of Networks*, *5*(1), 57–64. doi:10.4304/jnw.5.1.57-64

Yao, L. (2012). Measurement and Analysis of an Internet Streaming Service to Mobile Devices. *IEEE Transactions on Parallel and Distributed Systems*, *99*, 1–1.

Yasakethu, S. (2010). *Perceptual Quality Driven 3D Video Communications* (Doctoral Thesis). University of Surrey, Guildford, UK.

Yasakethu, S., Worrall, S., Silva, D., Fernando, W., & Kondoz, A. (2011). A compound depth and image quality metric for measuring the effects of packet loss on 3D video. *Proc. IEEE International Conference on Digital Signal Processing*,1–7. doi:10.1109/ICDSP.2011.6004998

Yasinovskyy, R., Wijesinha, A., Karne, R., & Khaksari, G. (2009). A Comparison of VoIP Performance on IPv6 and IPv4 Networks. *Computer Systems and Applications, 2009. AICCSA 2009. IEEE/ACS International Conference on*, 603-609. doi:10.1109/AICCSA.2009.5069388

Yi, C., & Klara, N. (2003). Layered peer-to-peer streaming. *Proceedings of ACM International Workshop on Network and Operating System Support for Digital Audio and Video*, 162–171.

Yi, C., Baochun, L., & Klara, N. (2004). On achieving optimized capacity utilization in application overlay networks with multiple competing sessions. *ACM Symposium on Parallelism in Algorithms and Architectures*.

Yi, C., Liang, D., & Yuan, X. (2007). Optimizing P2P streaming throughput under peer churning. *IEEE Global Telecommunications Conference*, 231-235.

Yi, C. T., Jianzhong, S., Mohamed, H., & Sunil, P. (2005). An analytical study of peer-to peer media streaming systems. *ACM Trans. Multimedia Computer, Communication, Application*, *1*(4), 354–376.

Yifeng, H., & Ling, G. (2009). Improving the streaming capacity in P2P VoD systems with helpers. *Proceedings of ICME*, 790–793.

Yu, X., Sun, K., & Yuan, D. (2010). Comparative analysis on HARQ with turbo codes in Rician fading channel with low Rician factor.*Proceedings of 12th IEEE International Conference on Communication Technology*, 342-345.

Zahran, A. H., Liang, B., & Saleh, A. (2006). Signal threshold adaptation for vertical handoff in heterogeneous wireless networks. *Mobile Networks and Applications*, *11*(4), 625–640. doi:10.1007/s11036-006-7326-7

Zander, J., & Kim, S. (2001). *Radio resource management for wireless networks*. Boston: Artech House.

Zezza, S., Magli, E., Olmo, G., & Grangetto, M. (2009). SEACAST: a protocol for peer-to-peer video streaming supporting multiple description coding. *Proceedings of the 2009 IEEE international conference on Multimedia and Expo*, 1586-1587. doi:10.1109/ICME.2009.5202819

Zhang, X., Liu, J., Li, B., & Yum, Y. S. P. (2005). DONet/CoolStreaming: a data-driven overlay network for live media streaming. Proc. of IEEE INFOCOM'05, 2102-2111.

Zhang, M. (2013). Major automatic repeat request protocols generalization and future develop direction.*Proceedings of the 6th International Conference on Information, Management, Innovation Management and Industrial Engineering*, 2,5-8. doi:10.1109/ICIII.2013.6703196

Zhang, M., Xiong, Y., Zhang, Q., Sun, L., & Yang, S. (2009). Optimizing the throughput of data driven peer-to-peer streaming. *IEEE Transactions on Parallel and Distributed Systems*, *20*(1), 97–110. doi:10.1109/TPDS.2008.69

Zhang, Q., Ji, Z., Zhu, W., & Zhang, Y.-Q. (2002). Power-minimized bit allocation for video communication over wireless channels. *IEEE Transactions on Circuits and Systems for Video Technology*, *12*(6), 398–410. doi:10.1109/TCSVT.2002.800322

Zhang, Y., & Letaief, K. (2004). Multiuser adaptive subcarrier-and-bit allocation with adaptive cell selection for OFDM systems. *IEEE Transactions on Wireless Communications*, *3*(5), 1566–1575. doi:10.1109/TWC.2004.833501

Zhao, B., Huang, L., Stribling, J., Rhea, S. C., Joseph, A. D., & Kubiatowicz, J. D. (2004). Tapestry: A Resilient Global-scale Overlay for Service Deployment. *IEEE Journal on Selected Areas in Communications*, *22*(1), 41–53. doi:10.1109/JSAC.2003.818784

Zhao, Q. L., Gan, Z., & Zhu, H. (2012). Adaptive resource allocation method over OFDMA system for H.264 SVC transmission. *IEEE International Conference on Signal Processing*,2, 1435–1438. doi:10.1109/ICoSP.2012.6491845

Zhao, Q., Mao, Y., Leng, S., & Jiang, Y. (2014). QoS-aware energy-efficient multicast for multi-view video in indoor small cell networks. *Proc. IEEE Global Communications Conference*, 4478–4483. doi:10.1109/GLOCOM.2014.7037513

Zhijie, S., Jun, L., Roger, Z., & Athanasios, V. V. (2011). Peer to Peer media streaming: Insights and new developments. *Proceedings of the IEEE*, *99*(12), 2089–2109.

Zhu, A. D., & Kocak, T. (2011). *Throughput and Coverage Performance for IEEE 802.11ad Millimeter-Wave WPANs*. Academic Press.

Zhuo, Z., & Ping, L. (2006). *A Highly Efficient Parallel Algorithm for H.264 Video Encoder.* Paper presented at the Acoustics, Speech and Signal Processing.

Zhu, W., Luo, C., Wang, J., & Li, S. (2011). Multimedia cloud computing. *Signal Processing Magazine, IEEE, 28*(3), 59–69. doi:10.1109/MSP.2011.940269

Zrida, H. K., Abid, M., Ammri, A. C., & Jemai, A. (2008). *A YAPI-KPN parallel model of a H264/AVC video encoder.* Paper presented at the Research in Microelectronics and Electronics.

About the Contributors

Khalid Al-Begain is a Professor of Mobile Computing and Networking at the University of South Wales, UK. He is also the Director of the Centre of Excellence in Mobile and Emerging Technologies (CEMET), a 6.45million R&D Centre partly funded by European Funding. He is currently seconded to lead as first President, the establishment of Kuwait College of Science and Technology, a leading private university in the Gulf He received his MSc and PhD in Communications engineering in 1986 and 1989, respectively, from Budapest University of Technology, Hungary. He also received Post Graduate Diploma in Management from University of Glamorgan in 2011 after finishing two years MBA course modules. He has been working in different universities and research centres in Jordan, Hungary, Germany and the UK. He has led and is leading several projects in mobile Computing, wireless networking, analytical and numerical modelling and performance evaluation He is the President of the European Council for Modelling and Simulation since 2006 and Past-President of the Federation of European Simulation Societies (EuroSim). He has over 200 publications including 2 authored and 18 edited books.

Ashraf Ali is a lecturer in the Electrical Engineering Department at The Hashemite University in Jordan, currently he is a researcher at University of South Wales working in a research project in developing mission critical communication systems over broadband 4G/5G communication Systems. along with his research interests, he is a lecturer of graduate modules such as Mobile Communication Technologies at University of South Wales and a lecturer for Electrical Circuits, Signals and Systems, Engineering Mathematics, Electronics, Data and Multimedia Communications at The Hashemite University. He worked as Research Assistant at Intracom Telecom and Athens Information Technology where he worked with the industry in developing multimedia and real time services over broadband communication technologies. He worked as research assistant in the computer engineering department at Jordan University for Science and Technology. He is an author of multiple research papers in the field of multimedia services modeling and broadband communications performance evaluation.

* * *

Anthony Adeyemi-Ejeye received the B.Sc. Applied Inofrmation Technology degree in 2006 from Sikkim Manipal University, India. The M.Sc. Computer Networking and Information in 2011, and Ph.D in Computing and Electronic Systems in 2015 from the University of Essex. He was a Senior Research Officer in Computer Science and Electronic Engineering School, University of Essex, and currently is a KTP associate (Post-Doctoral) in the Wireless and Multimedia Research Group, Kingston University. His research interests are; Ultra-High definition transmission, computer networking, telecommunication, quality of experience and quality of service in the field of multimedia streaming.

Mohammed Alreshoodi received the B.Sc. Computer Science degree in 2005 from Qassim University, Buraydah, Saudi Arabia. The M.Sc. (Hons.) Computer Networking and Information in 2011, from the University of Essex. In 2016, he achieved his Ph.D. degree in Computer Science and Electronic Engineering School, University of Essex, Colchester, UK. Currently, he is an assistance professor at Computer science college, Qassim University, Saudi Arabia. His research interests are; computer networking, telecommunication, quality of experience and quality of service in the field of multimedia streaming.

Mazin Alshamrani received the B.Sc in Computer Engineering in 2007 from Jordon University of Science and Technology, Jordon, then he received the M.Sc degree in Network Systems Engineering from the University of Plymouth, United Kingdom. In 2015, he achieved his Ph.D degree in Electrical and Electronic Engineering from the 5G Innovation Center (5GIC), University of Surrey, United Kingdom. He then extended his research experience and worked as a Research Fellow at the Center of Excellence for Mobile Applications and Services (CEMAS) at the University of South Wales, United Kingdom. His research interests is in Multimedia Applications over Hybrid Communication Systems, Mission Critical Communication Systems, individuals and vehicles tracking, monitoring, and guidance systems, Quality of Service, Mobile Applications Design and Implementations, Quality of Experience (QoE), Big Data, and Cloud Computing. He also has interest in Network Security Systems and DDoS. He works currently as a R&D Consultant and Projects Manager at the Ministry of Haj and Umra, Saudi Arabia.

Emad Danish received the B.Sc. Engineering degree in 1997, the M.Sc. (Hons.) Engineering degree in 2004, both in Electrical and Computer Engineering from King Abdulaziz University, Jeddah, Saudi Arabia. He received the Ph.D. degree in Electronic Engineering in 2016 from the Centre for Vision, Speech and Signal Processing, University of Surrey, Guildford, United Kingdom. Currently, he is a Senior Specialist Information Technology Infrastructure in Saudia Airlines IT, Jeddah, Saudi Arabia. His research interests include consumer's perception driven multimedia communications, with main focus on mobile video transmission, energy and bandwidth efficiency, Quality of Experience and Quality-of-Business.

Salima El Makhtari received her Engineer degree in Telecommunications from the National School of Applied Sciences of Tangier and Ph.D. degree in Electronics and Telecommunications in 2006 and 2016 respectively. Both from Abdelmalek Essaadi University, Morocco. Since 2006, she has worked for international companies, leaders in microelectronics and wireless communications. During the past years, she has participated in several projects establishing collaboration between the industry and Moroccan universities. Since 2007, she has supervised over 10 final-year projects of Engineer students. She has delivered lectures and courses at many Moroccan universities and engineering schools. She has authored a set of technical papers published in international journals and conferences. In 2014, she authored a paper titled "Improved turbo decoding using soft combining principle" and published in Wireless Personal Communications Journal, Springer US. In 2012, she authored a paper in ICMCS'12 studying the performance of AMC turbo coded OFDM for 3GPP Long Term Evolution downlink system. She has acted as a reviewer for several journals such as Telecommunication Systems of Springer and Wireless Personal Communications of Springer. Her research interests include: channel coding and decoding techniques, iterative channel decoding algorithms, digital modulation techniques, OFDM systems, HARQ protocols, physical layer simulations for 4G cellular standards, wireless communications, power optimization for ASICs and FPGA, VLSI circuits and embedded systems.

Ahmed El Oualkadi received Ph.D. degree in electronics from the University of Poitiers, France, in 2004. From 2000 to 2003, he was a research assistant in the Laboratoire d'Automatique et d'Informatique Industrielle - Ecole Supérieure d'Ingénieurs de Poitiers, Electronics & Electrostatics Research Unit, University of Poitiers, France. In 2004, he was an assistant professor at University Institute of Technology, Angoulême, France. During this period, he worked, in collaboration with EADS-TELECOM, on various European projects which concern the nonlinear analysis & RF circuit design of switched- capacitor filters for radio-communication systems. In 2005, he joined the Université Catholique de Louvain, Microelectronics Laboratory, Louvain-la-Neuve, Belgium, where he worked on the analog and mixed design of low power high temperature circuits and systems, in SOI technology, for wireless communication. During this period, he has managed and participated in several European and regional projects in the areas of wireless communication and sensor networking. Currently, he is an associate professor in the Abdelmalek Essaadi University, National school of applied sciences of Tangier, Morocco. His main research interest is the analog IC, mixed-signal and RFIC design for wireless communication, embedded system applications and information technology. He has authored or co-authored more than 60 papers in recognized journals and international conferences. He was the editor of the book "Filtres Agiles pour les Systèmes Radiomobiles Reconfigurables: cas des filtres à capacités commutées à très fort facteur de qualité", EUE 2011 and the author of 3 book chapters. He is a member of European Microwave Association (EuMA), IEEE Circuits & Systems Society, IEEE Electron Devices Society and IEEE Communications Society. He was also a member of the Organizing and the Scientific Committees of several international symposia and conferences. Furthermore, he is a referee of the IEEE Transactions on Circuits & Systems I (TCAS-I), IEEE Journal of Solid-State Circuits, Bentham Recent Patents on Electrical Engineering, Springer Multimedia Tools and Applications (MTAP) Journal, IET Micro & Nano Letters, Journal of ElectroMagnetic Waves and Applications (JEMWA), Progress In Electromagnetics Research (PIER, PIER B, C, M, PIER Letters), WSEAS Transactions on Systems, WSEAS Transactions on Circuits & Systems, WSEAS Transactions on Communications, International Journal of Electronics, WILEY International Journal Circuit Theory and Applications and International Journal of Science and Technology (Maejo). He is a member of the scientific and technical committee of WCECS and WASET, and a member of the editorial board of several journals: IETI Transactions on Computers (ISSN: 2414-1429), International Journal of Organizational Leadership (ISSN: 2383-1103), Wireless and Mobile Technologies (ISSN: 2374-2119) and International Journal of Latest Research in Engineering and Technology (ISSN: 2454-5031).

Farnaz Farid completed her PhD degree in Computing and Information technology from Western Sydney University. Prior to that she has worked in China as a web application developer and web business SME at IBM. Her research interests include wireless and cellular networking, web engineering, technology for development.

Geza Koczian is a Senior Research Officer at University of Essex. He received his MSD from the University of Essex. He has been involved in data communication network designs for many years. Specifically he is investigating high-speed Ethernet solutions for trains using both theoretical and advanced experimental techniques.

Karen Medhat received her Bachelor degree with honors in computer engineering in 2012 from the Faculty of engineering, Cairo University, Egypt. She is currently pursuing her Master degree in the Faculty of engineering, Cairo University, Egypt.She worked as a research assistant in the Electronics and Communications Engineering department at Faculty of Engineering, Cairo University. Karen joined IBM as a Software Development Engineer where she has been developing large-scale enterprise applications, related to Transportation, Electricity, automotive and insurance sectors, since 2013.

Mohamed Moussaoui received M.S. and Ph.D. degrees from the University of Valenciennes et du Hanaut Cambrésis (UVHC), France in 2001 and 2005 respectively with specialization in Wireless Communications. From 2005 to 2007, he worked as a research scientist at IEMN/DOAE laboratory, Group RDTS, Valenciennes, France. Since 2007, he joined the National School of Applied Sciences of Tangier (ENSAT) as an assistant professor, Abdelmalek Essaadi University, Morocco. He has authored and co-authored over thirty technical papers. He is author of a book « les techniques de détection multi-utilisateurs pour les systèmes CDMA » by Éditions universitaires européennes 2011. He has acted as a reviewer for several journals (IEEE communications Letters, IEEE Cloud Computing, Telecommunication Systems – Springer, Wireless Personal Communications- Springer...) and has been TCP member in several international conferences (WNGMN'09, IEEE IWCRN-2012, 8ème JFMMA-2013, ECCTD '12, ISTASC '12, CSCC '12, ISCGAV '12, EHAC '13, ISPRA '13, NANOTECHNOLOGY '13, CSCS '12, OEPT 2014, ICeND2014, 2014 WNC3, DigitalSec2014, ICACNI-2014, ISI'14, IEEE CloudNet'14, CloudNet'15, ICCSN 2014, CCA-2014, CHUSER 2014, SPIN-2014, INTECH 2014, ICCIDM-2014, ICIDRET 2014, ICECCS 2014). He actively involves the organizing committee of numerous leading international conferences, workshops and spring schools (EPRFTE'10-13, JFMMA'13, IEEE IWCNR'12, JFMMA'11, JDTIC'11, ISRETWMC'10, MMS'09, WNGMN'09, …). He is member of NEM, the Networked and Electronic Media Initiative and serves on Texas Instrument Expert Advisory Panel.. He has also served as expert consultant to the Francophonie University Agency, Middle Eastern Office.. His research interests include CDMA, Multi-user detection, MIMO/OFDM system, iterative decoding, HARQ, Cognitive Wireless Networks and chaotic systems.

Michael Parker received his BA (first-class) degree in electrical and information sciences from Cambridge University, UK in 1992, and his PhD from Cambridge in 1996. In 1997, he joined Fujitsu Telecommunications Europe, Ltd., Colchester, UK, and from 2000-2003 he was also with the Photonics Networking Laboratory, Fujitsu Network Communications, Richardson, Texas, USA. From 2003 to 2007, he was with Fujitsu Laboratories of Europe, Ltd. He was appointed Visiting Professor at the University of Essex, UK in 2004. Since 2008 he has been associated with Lexden Technologies Ltd. and the University of Essex. He has filed over 20 patents, and has authored more than 130 papers in international journals and conferences. He has been involved with the e-photon/ONe and BONE European FP7 Networks of Excellence, as well as the EPSRC NIRVANA, and EU FP7 OASE, STRONGEST, FIVER, SODALES, IDEALIST, research projects, and currently the H2020 iCIRRUS, and 5G-PPP CHARISMA (where he is the Technical Manager) projects.

Raad Raad graduated from the University of Wollongong, Australia in 1997 with a Bachelor of Engineering (Hon 1) in 1997. He went on to complete his PhD thesis entitled Neuro-Fuzzy Logic Admission Control in Cellular Mobile Networks in 2006. Dr. Raad has over 5 years of industrial research experience and another 5 years of experience in academic research. Dr. Raad is the author of five United

States patent filings of which three have been granted and over 50 refereed publications and technical reports. His expertise is in wireless communications with a focus on Medium Access Control (MAC) and bandwidth management protocols for wireless networks. Dr. Raad has led and collaborated on significant projects in the areas of sensor networks, IEEE 802.11, IEEE 802.15.3, Mesh LAN, RFIDs and cellular networks. The technical areas that he covered during the numerous projects include admission control, bandwidth management, low power MAC protocols and routing protocols.

Muhammad Salman Raheel received his Bachelors degree in Electrical (Telecommunication) Engineering from COMSATS University Lahore, Pakistan in 2010. He joined COMSATS university and worked under Optical and Microwave Communication research group as a Research Associate for an year. Afterwards, he went to finish his PhD degree in Peer to Peer Networks using Video Coding Techniques in 2016. Recently, he is working as an Associate Research Fellow at University of Wollongong. His research interest includes peer to peer systems, wireless or mobile computing and multimedia communications.

Rabie Ramadan is currently an assistant professor at Cairo University. He received his Ph.D. from Southern Methodist University, Dallas, TX, USA in 2007. His field of expertise is in mobile computing, networking, and computational intelligence. In 2005, he joined Banc of America Securities Company in Frisco, Dallas, TX as a senior developer. The company is part of the bank of America in which its responsibility is to implement all of the bank required software. Dr. Ramadan was involved in the implementation of many projects such as handling the real-time stock market information and pushing it efficiently to different customers. Dr. Ramadan was one of the founders of "Ibda" company in Egypt as Software Company from 1995 to 2003 where he was serving as the head of the web programming department. He has also great experience dealing with networks where he was working as a network administrator at International telecommunication Union (ITU) regional office. He is a co-founder of the IEEE Computational Intelligence Egypt chapter. Currently, he is serving as the chapter chair.

Chun Ruan received her PhD degree in Computer Science in 2003 from the University of Western Sydney. Currently she is a lecturer in the School of Computing, Engineering and Mathematics at Western Sydney University. Prior to that, she worked as an associate professor at the Department of Computer Science, Wuhan University, China.

Hassan Samadi received the Ph.D. degree in Applied Mathematics from the University of Franche-Comté, Besançon, France, in 1996. Since 2001, he joined the National School of Applied Sciences of Tangier (ENSAT), Abdelmalek Essaadi University, Morocco, as a Professor-Researcher, where he is also a member of the Research Committee and a Permanent member of Technologies of Information and Communication research laboratory. Ex-Head of Systems of Information and Communication Department at ENSAT. He has taught many courses related to Applied Mathematics, Operational Research, and Tools facilitating decision making. He has supervised many thesis of Ph.D. and Master of Engineering students. He is the author and co-author of many technical papers published in international journals like: R.Acad.Sci.Paris, C.R.Math.Rep.Acad.Sci.Canada, International Journal of Engineering Science, Mathematical Methods in the Applied Sciences, Mathematik und PhysiK, ZAMP, Wireless Pers Commun, IJCA. His research interests include: Operational Research and Tools facilitating decision making, elecommunication Systems Modelling,Partial Differential Equations, Problems for linear and non-linear boundary, Homogenization and Reinforcement problems.

Seyed Shahrestani completed his PhD degree in Electrical and Information Engineering at the University of Sydney. He joined Western Sydney University in 1999, where he is currently a Senior Lecturer. He is also the head of the Networking, Security and Cloud Research (NSCR) group at Western Sydney University.

Ihab Talkhan, IEEE member since 1984, is currently a Professor and Chairman of Computer Engineering Department, Faculty of Engineering, Cairo University, as well as a Professor at Electronics Engineering Department, American University in Cairo. He got his Ph.D. in 1993 jointly between Oakland University (Rochester, Michigan, USA) & Cairo University. Fields of interest: Computer Engineering, Microelectronics, Digital System Design and Testing, VLSI Device modeling. Dr. Talkhan got a certificate of Merit for outstanding Achievement from Ministry of Health in 2007, and Certificate of Merit for outstanding Academic Achievement from the Mathematical Sciences Department, Oakland University in 1989. Dr. Talkhan is also the Deputy Director of the Computation Scientific Center, Cairo University Also; he was the IT Senior advisor for both Minister of Health and Minister of Communication and Information Technology. He has been an IT consultant for different entities UNDP, Egyptian Electric Utility & Consumer Protection Regulatory Agency, Ministry of Electricity & Energy, Ministry of Foreign Trade, Ministry of Foreign Affairs, Ministry of Interior and Others.

Stuart D. Walker was born in Dover, U.K. in 1952. He received the B.Sc. (Hons) degree in physics from Manchester University, Manchester, U.K., in 1973, the M.Sc. degree in telecommunications systems and Ph.D. degree in electronics from Essex University, Colchester, U.K., in 1975 and 1981 respectively. From 1981-82, he was a post-doctoral research assistant at Essex University. From 1982-87, he was a research scientist at BT Research Lab, and from 1987-88 he was promoted to Group Leader of Submarine Systems Design. He joined Essex University in 1988 as a Senior Lecturer, and was promoted to Reader in 2002 and to Full Professor in 2004. At Essex University, he manages a laboratory concerned with all aspects of Access Networks.

Index

N

O

P

Q

S

T

U

V

Printed in the United States
By Bookmasters